THE
COLONIAL HISTORY
SERIES
No. 85

General Editor
D. H. Simpson
Librarian of the Royal Commonwealth Society

MISSION TO VITIAN OR FIJIAN ISLANDS

COLONIAL HISTORY SERIES

This series of reprints aims at presenting a wide variety of books; their link is that they all deal with some aspect of the relations between European powers and other parts of the world—including such topics as exploration, trade, settlement and administration. Historical studies, and books which furnish the raw material of history, will find a place, and publications will not be restricted to works in English. Many titles reprinted will have new introductions by authorities on the subject.

ALSO AVAILABLE IN THIS SERIES

S. Bannister. Humane Policy or Justice for the Aborigines. (Series No. 38)

J. E. Erskine. Journal of a Cruise Among the Islands of the Western Pacific. (Series No. 6)

H. Vere Evatt, The Hon. Mr. Justice. Rum Rebellion. (Series No. 19)

R. L. Stevenson. A Footnote to History: Eight Years of Trouble in Samoa. (Series No. 7)

VITI:

AN
ACCOUNT OF A GOVERNMENT MISSION
TO THE
VITIAN OR FIJIAN ISLANDS
1860–1861

BY

BERTHOLD SEEMANN, Ph.D., F.L.S., F.R.G.S.

WITH A NEW INTRODUCTION
BY
PHILIP A. SNOW, M.A., F.R.A.I., J.P.

and
DAWSONS OF PALL MALL
Folkestone, England
1973

LIBRARY
WAYNE STATE COLLEGE
WAYNE, NEBRASKA

Reprinted in 1973
DAWSONS OF PALL MALL
Cannon House
Folkestone, Kent, England

Distributed in the U.S.A. 1973 by:
Harper & Row Publishers, Inc.
Barnes & Noble Import Division

ISBN: 06–496160–5

*Printed in Great Britain
by Photolithography
Unwin Brothers Limited
Old Woking, Surrey*

INTRODUCTION

by

PHILIP A. SNOW, M.A., F.R.A.I., J.P.

As I write, it is 100 years bar one since Viti, or Fiji as it came to be called, was ceded to Queen Victoria as a Colony. This work of Seemann's is of special importance in that it is an examination of, a commentary on, the first offer by the Fijians of cession which, though refused, was to lead fifteen years afterwards and ninety-nine years ago—in 1874—to a further offer that was not declined, to the final Cession of Fiji to Britain and to the beginning of a remarkable (one has some justification for saying unique) relationship between the two countries.

Viti ought to be read more or less at the same time as, or in conjunction with, two other records: (a) the authorised examination of the Cession offer by Colonel (later General) W. J. Smythe, the Commissioner specifically appointed from Britain for the purpose; an important aspect of Seemann's *Viti* is that it includes Smythe's Report as an Appendix, and (b) the less official impressions of Colonel Smythe's wife, Mrs. S. M. Smythe, in her *Ten Months in the Fiji Islands* (please see notes on further reading at the end of this Introduction), where there are a number of accounts and views contributed by Smythe himself.

A pivotal aspect of Seemann's *Viti* is that it is all for Cession (that ultimately, if belatedly, was to occur), whereas Smythe, the Commissioner, was against it. This being the position, the British Liberal Government under Lord John Russell, Prime Minister, had to accept the recommendations of its Commissioner, establishing "the impolity of appropriating the islands" because of the expenditure involved, the possibility of war with Fijian tribes and international repercussions. As events turned out, however, it was judgement on the lines of Seemann's that was to prevail in 1874 when Britain accepted a further offer of association with Fiji which was to endure and flourish for close on the last hundred years, composing in itself an example of as near a model Colonial relationship as could ever be likely to exist. With hindsight, Seemann proved to have conspicuous imaginative judgement because the story of Fiji since the Cession in 1874 to the present has been singularly successful. And it is of interest that Smythe, years after his Report, was to change his mind and agree that he should have recommended that Britain accept the earlier offer of Cession in 1860.

In 1970 Fiji passed harmoniously from being a Colony to a Dominion, to independent status with the maximum goodwill and encouragement of the British Colonial Government.

It must be related, as an aftermath to the publication of *Viti*, that when the first offer of Cession was turned down in 1861 Fiji attempted to go its own way. An autonomous Kingdom set up under the leading chief, Cakobau, ran into troubles of all kinds, principally financial, and was in constant jeopardy of being overwhelmed by Ma'afu, the Tongan prince who had conquered much of Fiji with an ease that made it seem likely that he could take the whole of Fiji. It was these circumstances which, thirteen years after the first attempt to give Queen Victoria the Group, led to the desperate repetition of an offer of Cession by Cakobau and his advisers: and this time, a century ago, it suited Britain to accept.

Apart from what I have already stated, or in further emphasis of the points made in support of this major piece of Colonial history, there are multiple reasons for the reprinting of this work, not least the fact that, nearly 110 years old, it is very scarce indeed. Most importantly, it is the first account of the Fiji Islands of a generally historical and scientific nature by other than a missionary, trader or explorer. It has consequently less bias and more depth than the very few accounts that preceded it. A decade earlier, a general book on Fiji, *Life in Feejee*, by A Lady, had been published. Of historical value, it inevitably lacks the authoritativeness and breadth of *Viti*, which distributes its interest through all aspects of earliest recording of Fiji. Reading *Viti*, one never ceases to be astonished that Seemann's stay in the Islands was as short as six months. Assigned through the offices of his patron, Sir William Hooker, Director of the Royal Herbarium at Kew, to accompany Colonel Smythe, the Commissioner sent out by the Duke of Newcastle, Secretary of State for the Colonies, to investigate the offer of the Fijians to cede their islands, Dr. Seemann's is a very clearly written account of the background (and the foreground) of the mission. He was able to tour more extensively than the Commissioner who was almost wholly absorbed in political negotiations: Seemann did so with his eyes very wide open, making invaluable contributions to ethnological records of customs at the critical stage before they were to come under the impact of a heavy measure of European culture in certain areas quite shortly after. His observations on cannibalism, one of the first major ways of Fijian life to disappear, are markedly vivid. Specially graphic too is Seemann's description of the leading contenders for power, Cakobau and Ma'afu, while his tracing of the early relations between Tonga and Fiji is of real historical importance, as are his records of primitive religion and legend passing into the beginnings of Fijian modern history.

Since Seemann was one of Britain's most distinguished botanists—his *Flora Vitiensis* published after his visit to Fiji remains a classic—and as such was recognised by four columns in the *Dictionary of National Biography*, it is not surprising that there is a good deal of natural history in *Viti*. Indeed, scientific enquiry and assessment were his particular terms of reference on the Commission. His observations on plants, trees and the food of the Fijians were to be of real significance in Fiji's future, both commercially and socially, and in their economic and political life. That there is in his records a good deal that has become extinct gives his work increased value. In addition to flora, fauna is described with a sharp eye and, on top of all this, his work includes the attractive illustrations that we expect of the pre-camera period but which in fact are particularly rare where Fiji is concerned.

His general attitude is best summed up in his own concluding words:

"I have simply written an unvarnished account of all I heard and saw, and refrained from discussing the rejection of so fine a country from a political point of view. I have no doubt as to the future of Fiji. The importance of the Group once recognised, nothing will stop our race from taking possession of it and replacing barbarism and strife by civilization and peaceful industry".

One might have wished for more on the European personalities whom he was working with. While there is much penetrating study into the characters of Cakobau, Ma'afu and the other indigenous leaders, the civilities of the time precluded any remarks or inferences on his co-travellers, Colonel and Mrs. Smythe. His narrative just conveys a soupçon of a tilt at possible pompous predilictions by Smythe himself who felt it important to adorn himself with the full dress and plumes of his Artillery uniform in the cannibal bush. And who is to cast doubt on what Smythe felt? Subsequent Governors and Commissioners held that thick drill uniforms with tight trousers, high neckbands, ceremonial swords and topees in the sweltering atmosphere up to the last Colonial days of but yesterday subscribed to the image. Seemann's narrative does not contain much of mirth but there is this:

"I had just been speculating on the cause of the Fijian, in common with other insular floras, being poor in gay-coloured, and rich in green, white and yellow, flowers when lo! a look in the valley revealed bushes covered with a perfect mantle of scarlet and blue, thrown up to great advantage by the bright rays of the sun. I saw my travelling companions had made a halt near the very spot where nature had condescended to refute a deeply-rooted generalisation. I clambered down the hill as fast as the condition of the ground would admit and for a while lost sight of the gay display by intervening objects. A few more steps and I stood before a startling sight—Colonel Smythe's artillery uniform hung up to dry in the sun!"

There is nothing elaborate about the style of the author of *Viti*, but his mastery of English was early and complete.

Berthold Carl Seemann, a Hanoverian (born in the capital of that State in 1825), wrote his first botanical paper when he was 17. He left Germany for Kew at the age of 19 and,

only 21, was recommended by Sir William Hooker, the head
of Kew, as Naturalist to H.M.S. *Herald* on a hydrographical
survey of the Pacific. He explored the Panama Isthmus,
discovering a great deal, not least some curious hieroglyphics.
The *Herald's* cruise lasted from 1845 until 1851, giving
Seemann botanical experience of the West coasts of North
and South America and Mexico where he narrowly escaped
the Comanche and Apache Indians, who in all probability
were those who successfully liquidated his friend of later
Fiji days, W. T. Pritchard. Then, in direct contrast, he
collected many plants and anthropological specimens of the
Esquimaux (and more Pacific material in the Sandwich
Islands) when the *Herald* was ordered to the Arctic to
search for Sir John Franklin. *The Botany of the Voyage of
H.M.S. Herald* was written by him between 1852 and 1857.
He published his *Narrative of the Voyage of H.M.S. Herald*
in two volumes in English in 1853: it was translated later
into German. Göttingen, where he had graduated, made
him a Ph.D. With his brother, W. E. G. Seemann, who
died in 1868, he started a German journal of botany,
Bonplandia, at Hanover that lasted for a decade from
1853, after which he immediately began *The Journal of
Botany, British and Foreign*. Then followed in 1860 his being
invited to accompany Smythe in Fiji. Seemann's visit to
Panama caused him virtually to foresee a canal there which
undoubtedly affected his prognostications on the prospects
of Fiji, lying as it did right on the line of the Pacific route
from Central America to Australasia. (Commodore Erskine
had earlier thought deeply of the effect of a Panama Canal
on Pacific harbours.) *Viti* appeared in 1862 and in the
same year also his entertaining, light-hearted account, *Fiji
and its Inhabitants*, in Sir Francis Galton's *Vacation Tours
and Notes of Travel in 1861*.

Throughout this time he had been preparing, on Sir
William Hooker's advice, his magnum opus. With its title
page "*Flora Vitiensis: a description of the plants of the Viti
or Fiji Islands with an account of their history, uses and
properties*, by Berthold Seemann, Ph.D., F.L.S., F.R.G.S.",
it came out in ten parts over several years (1865–1873), the
Colonial Office giving him no financial assistance, and
Seemann having to make three arduous voyages to Central
America for the purpose of finding sufficient funds for

publication. It is a superb production, mostly in Latin but with much discerning analysis in English on Fijian customs, ancestry, proverbs and a miscellany of detailed information that added significantly to European knowledge of the Islanders. Apart from the comments under the Latin descriptions of the plants, bushes and trees, additionally his Preface, Historical Notice and Introduction in *Flora Vitiensis* are all valuable reading to supplement a study of *Viti*, expanding on many of the subjects mentioned in this earlier work. Nine parts of *Flora Vitiensis* were published before his death. Its hand-coloured plates, a hundred in all, are a distinguished feature of the work which also contains an illuminating picture of Seemann but his career is in itself evidence enough of his forcefulness, determination and enterprise. Both the *Botany of the Voyage of H.M.S. Herald* and *Flora Vitiensis* have been reprinted this year.

Venezuela was another country he was to know well when in 1864, diverted from botany to geology, he was sent there by Dutch and French capitalists to examine its resources: he discovered a valuable bed of anthracite before going on to Nicaragua. *Dottings on the Roadside in Panama, Nicaragua and Mosquito* was then published. Still detached from the botany that he hoped to return to (he particularly wanted to revisit Fiji), he was appointed managing director of a gold mine in Nicaragua and manager of a sugar estate in Panama. Seemann died of fever and cardiac complications on October 10th, 1871, exactly three years to the day before Fiji was ceded to Britain. He was buried close to his house at the Nicaraguan gold mine "in the little patch of industry and civilization his energy had called into existence in the primeval forest, and surrounded by the tropical vegetation he knew so well".

A Vice President of the Anthropological Society and Fellow of the Linnean Society, Seemann married an Englishwoman who died during one of his absences in Central America. He left an only daughter.

It is understating history to record that Seemann had been prolific in his publications. Articles on general literature and politics are estimated to have amounted to several thousands in English, German and other languages. Many were the scientific treatises to which he contributed, and in 1866 he wrote a notable Preface to W. T. Pritchard's important *Polynesian Reminiscences*. Pritchard had been British Consul in Fiji when the first offer of Cession was made. Smythe disliked Pritchard, Seemann liked him. One of the by-products of Smythe's Report was to have Pritchard removed as Consul. Seemann sought to defend Pritchard but without success. Pritchard, never reinstated, left the South Pacific for Mexico and is believed to have been killed by American Indians while travelling to California. The opposite sides taken by Smythe and Seemann over Pritchard and Cession indicate that all was not harmonious on the Mission and, although neither were uncivil enough to say so at the time publicly, one senses in both *Viti* and Smythe's contribution to his wife's book that Seemann's presence was not always welcome. Smythe on one occasion expresses surprise that Seemann, thought to have left earlier, turned up at one of the meeting places when Smythe was still negotiating. The reference strikes one as chilly.

Versatility, conscientiousness and superb scientific application were manifest qualities in Seemann: the first of these was further exemplified by his composing music and writing three plays which enjoyed some popularity in Hanover.

Before Seemann visited Fiji in 1860 and after his attachment to H.M.S. *Herald* off America up to 1851, the ship was to make a scientifically (and artistically) important call at the Fiji Group in the course of a commissioning to survey some of the little known group of islands in the South Polynesian Ocean. Seemann had thought it expedient not to have all his research connected with the *Herald's* movements and had left the ship: it is curious that he was to follow the *Herald* to a remote part of the world which was to be the inspiration for the production of his greatest work.

On restudying Seemann's *Viti* (published by Macmillan when they were principally based in Cambridge), a number of points are noted which are worth, I believe, individual comment, quite apart from the ones made earlier on the general purpose and content of *Viti*.

When the Commission to investigate the first Cession offer was set up the Duke of Newcastle decided, on Hooker's advice, that it was important that an exhaustive survey of the botanical side of Fiji should be added to what was intended to be at first a strictly political inquiry. Hence the selection of Seemann who complains in his Preface to *Flora Vitiensis* that, apart from his free passage to the Islands, the remuneration was inadequate for outfitting and travel and subsistence in the Group: he accepted the attachment to the Mission, hoping that the pecuniary loss would be covered by sale of duplicate specimens. In the event this sale partly met the collecting expenses: the original specimens were deposited at Kew. Seemann in turn chose as his assistant Jacob Paul Storck (page 2), paid for by him. Storck had been born in Hesse in 1838 and went to Sydney Botanical Gardens in 1853, joining Seemann in 1860. In his Historical Notice in *Flora Vitiensis*, Seemann laments that Storck was incapacitated throughout almost their time together by various kinds of illnesses but praises his zeal and ability. When Seemann left Fiji in November, 1860, to reach England four months later and "very much shattered from a violent attack of dysentery on the homeward voyage" Storck decided to stay permanently in Fiji, acquiring with Frederic and William Hennings, also Germans, an island in the Rewa River Delta on which to grow cotton. He introduced tea, vanilla and coffee as well. Later he set up a more elaborate plantation on the Upper Rewa, introducing the China bamboo which has been consistently successful. When cotton collapsed on America ceasing to be internecine he turned to sugar but never had any personal luck, despite his enterprise. He married a German and his descendants in Fiji have all contributed to the Island's needs in varying capacities, as undertakers, gaolers and, in most recent times, as stenographers.

Seemann raises (at page 15) an original question about tapa, the product of the paper-mulberry tree (which is still such a useful and skilful art in the South Pacific), conjecturing whether it can have been derived by Fijians from China where it had long been made. I know of no one making this suggestion before or since Seemann.

At page 21 the drinking of coconut milk is expressed as the most likely cause of elephantiasis. It was to take almost a further half century for this theory to be exploded and the cause tracked down to the bite of the culex mosquito (as compared with the anopheles, malaria-carrying mosquito) which can also carry yellow fever. Nevertheless, although the cause was wrongly guessed in Seemann's time, the only cure known then, removal to a colder climate—not easy for the Pacific indigenous races—remains valid until another antidote is found.

Seemann refers at page 23 to the good behaviour in his time of mosquitoes at Somosomo on Taveuni (which remains, by the way, the only place where land snakes are ritually important). They never showed any politeness to me after the passage of much more time since Seemann's day. On the Rewa Delta (page 91) Seemann was driven to having dinner on the river rather than on land so as to escape them: this is still a judicious move in the Rewa River area. In fact, mosquitoes have only been controlled in two or three small areas of modern Fiji (notably the capital, Suva, and at the International Airport at Nadi).

While Seemann found (at page 25) that Fijians were fond of books, this predilection has not been developed or even been maintained, not necessarily a reflection on Fijians, more perhaps an omission in the trend of educational policy. Fijians were fascinated by the upgraded beachcomber, Harry the Jew (John Humphrey Danford), reading to them in their own language *Arabian Nights* and *Aladdin* (page 195). These and similar fantasies still fascinate Fijians: one can only deplore that any growth of literary interests has been discouraged by a shortage of translated works. (*Aladdin* was not translated until a century after its introduction to them by Harry the Jew).

The study in *Viti* of Harry the Jew is notable: he was clearly a man of parts, the first European to live in the interior (infinitely more dangerous than on the coast), and no other observer has given him his due. Baron von Hügel, the young Austrian traveller of singular intrepidity and initiative, was to come across him 15 years later and be intrigued by him.

It is satisfactory to record that virtually all the species of trees so frequently referred to by Seemann still dominate the Fijian scene—the casuarina or ironwood (nokonoko), the screw-pine or pandanus (balawa), the Tahitian chestnut (ivi), the weeping fig or banyan (baka), the tree-fern (balabala; is there any item of tropical flora more graceful?), the greenheart of India (vesi), the acacia (vaivai), the kauri-pine (dakua and spelt by him kouwrie), the mangrove (dogo), the barringtonia (vutu); all continuing to beautify the view and at the same time to make material contributions to the economy and life of the islanders. His suggestion (pages 44 and 86) that the fan palm (which he named after Pritchard and is conspicuously attractive but not all that common) might be the national emblem has not quite been realised: the profuse coconut palm tree is frequently the main feature of badges of representative cricket, football and hockey teams touring outside Fiji.

Much of Seemann's interest is directed to the potentialities of cotton-growing in Fiji as the Civil War had put America, the world's principal supplier, out of the market. It is because America was able to recapture it when the War ended that Fiji's chances of making this a main crop evaporated after only a brief span of importance.

Quite new to me was his advocacy of a combination of arrowroot and port for curing dysentery (page 67): this never received the backing of the medicos when they came to establish themselves on the setting-up of the Government Health Service after Cession.

A typical piece of Seemannesque percipience and soundness of judgement was his backing of Suva (page 70) as the best site for the capital of the Group. It was to become clear about 20 years later that Levuka, principal settlement at the time of his visit, with its mountainous backdrop would be too constricting.

More doubtful is his assertion (page 74) that Cakobau had a "perfect right" to call himself King of Fiji. It was always a tenuous title, in dispute so long as Ma'afu continued absorbing large and important sections of the Fiji Archipelago. There are arguments both ways as to the validity of the claim. Smythe in a letter to the Duke of Newcastle went probably too far in declaring:

"In several important points I have found the information furnished by Mr. Pritchard to Her Majesty's Government and embodied in my instructions incorrect. Thakombau (the Vunivalu of Mbau) has no claim to the title of King of Fiji".

The fact was that it suited some, if not most, Europeans, particularly, of course, the anti-Ma'afuites, to elevate Cakobau and perhaps the wish was father to the thought that with Cakobau king Fiji would settle more rapidly into a state of peace and order when many were tired of the occasional anarchical atmosphere which suited the less reputable Europeans.

Seemann's mastery of English, German and Latin, together with the fact that this very early work of his on Fiji is remarkably free of spelling mistakes in Fijian (Lomolomo at pages 243 and 245 should be Lomaloma), must all indicate that he was a linguist of natural ease and brilliance. (Mrs. Smythe's book is much less accurate.) His tribute on page 80 to Hazlewood's *Dictionary*, from which he learnt his accurate control of the language, is worth noting: that work, published only ten years before Seemann's visit, has not been surpassed to date. A new linguistic work is about to come out: a century later, Hazlewood is at last likely to be improved upon, if that seems possible.

At page 85 Seemann notices the native Wesleyan teachers' use of illustrations such as "Now when you eat a human hand you are . . . ." This was for congregations just coming out of cannibalism. Illustrations of this nature would not have gone on many years more, perhaps not later than the 1880s, before shame at reminders of the practice led to absolute reticence over the custom which did not survive to the twentieth century.

On page 98 Seemann refers to the kava root, from which the characteristic South Pacific drink is made, being "masticated by young men and tasting like soapsuds, jalap and magnesia". This combination of taste would not have occurred to me, but I confess to not knowing what jalap is. Kava is like a mild, muddy-looking toothpaste perhaps— but stimulating and splendid in countering tropical heat and encouraging amiability.

Fijians have never taken to eggs: Seemann noted this indifference (page 37) at the outset of European contacts.

So much of what Seemann noticed in the 1860s still applies. Here are but four examples: (i) No more effective method of getting rid of the giant flying cockroaches from boats (page 121), despite modern fumigants and insecticides, has been devised than that of sinking boats in the sea. (ii) That the copies of *Illustrated London News* which Harry the Jew produced as a source of endless delight to Fijians (page 159) have until recent years maintained their popularity is no mean record for a journal. (iii) It still remains the greatest insult for a Fijian to tell another "I will eat you" (page 181). (iv) At page 221 it is recorded that Storck was affected with rheumatism and treated with dilo oil, a practice still much adhered to.

For once (at page 158) a prognostication made by Seemann has not materialised. He thought that, when Fiji became a Colony, Namosi would, by reason of the coldness of its higher altitude, become a resort and an escape from the day-long, night-long humidity of so much of Fiji throughout the year. Namosi has in fact only ever been visited by the very occasional traveller. Seemann did not know of Nadarivatu, much further inland and higher in the mountains which aspired to be something of a hill station from the 1880s on, but even this, to my own surprise, never became a mini-Simla or Darjeeling. The interiors of the two largest islands in the Fiji Archipelago are a significant blank in Seemann's map.

Seemann's prescience about sugar (page 280) was more accurate than his assessment of the future of cotton or tamarinds, coffee or tobacco. Sugar became the real basis of Fijian economy from the 1890s to the present (and tobacco has picked up a little recent success).

An observation (on page 165) that struck me particularly was that the Fijians at Namosi attending the official meetings with Smythe did so with their faces painted half black, half red, in various stripes. "None of the influence which civilization and missionary teaching have had on the Fijians were here perceptible" is a revealing phrase of Seemann's who sided with Pritchard in reserving judgement as to the totally beneficial effect of missionaries, a judgement not withheld by Smythe. Even so, the remarkable dignity of Fijians, which they still retain, was distinctly recorded.

An important ethnological observation is made by Seemann at page 329 in his contemplation of the origin of Polynesians. As toddy was not known in Polynesia, he concludes that they could not have come from Malaya and Asia where the fermented coconut juice was a historical traditional drink.

Finally, I cannot withhold complete admiration for Seemann's extent of accurate observation of, and imaginative comment on, a huge variety of every kind of subject described at pages 274–416, a large section of the book composing a remarkably concise and important summary of the essence of Fiji in all its aspects. As regards Colonel Smythe's Report (Appendix), it is difficult to understand why Smythe, having considered three reasons for accepting sovereignty, namely, the possibility of Fiji lying on an important route from Australia to Panama, the cotton potentialities as they then were, and the power and security that possession would bring, should have judged it inexpedient not to accept the unanimous request of the Fijians whom it was possible to consult. Only the second reason was of doubtful strength. It is an anticlimax that he was to confine his recommendations virtually to "improving relations" and to Suva being considered as the capital (and to having Pritchard deprived of his Consulship). Time was to prove him wrong in turning down the first offer of Fiji to Britain and, while he was later to admit the error of judgement, one cannot but appreciate all the more Seemann's sympathetic and imaginative assessment of the situation from a junior position and to regret that his view was not listened to more seriously by the Government of the time.

1973. PHILIP A. SNOW.

NOTES ON FIJIAN SPELLING AND PRONUNCIATION

There are two forms of Fijian spelling, one for local (Fiji) use and the other for overseas use. The difference between the first and the second is:

Examples.

Local	Overseas	Local	Overseas
C	Th	Cakobau	Thakombau
B	Mb	Cakobau	Thakombau
		Bau	Mbau
		Bureta	Mbureta
G	Ng	Gau	Ngau
Q	Ngg	Qamea	Nggamea

P.A.S.

SOME FURTHER RECOMMENDED READING

1861. Account of the Isles of Berthold Seemann, quoted from the Athenaeum of 1861, and of the Annexation of the Group.
Nautical Magazine, Pp. 257, 282, 470.

Allen, Percy S.
1907. *Cyclopaedia of Fiji.*
Sydney, McCarron and Stewart. Pp. 334.

Belcher, Sir Edward
1843. *Narrative of a Voyage Round the World, performed in Her Majesty's Ship "Sulphur", during the years 1836–1842, including details of the Naval Operations in China from Dec., 1840, to Nov., 1841; published under the authority of the Lords Commissioners of the Admiralty.*
London, Hy. Colburn. 2 Vols.
Reprinted Dawsons of Pall Mall 1970
(Colonial History Series Nos. 74 and 75)

Bonplandia.
Zeitschfrift für die gesamte Botanik. Officieles Organ der K. L.-C Akademie der Naturforscher.
(Vol. I, 1853–Vol. 10, 1862).

Brenchley, Julius Louis
1873. *Jottings during the Cruise of H.M.S. Curaçoa among the South Sea Islands in 1865.*
London, Longmans, Green. Pp. xxvii, 487.

Cooper, H. Stonehewer
1880. *Coral Lands.*
London, Richard Bentley & Son. 2 Vols.

Derrick, Ronald Albert
1946. *A History of Fiji.*
Suva, Printing and Stationery Department. Pp. vii, 250, xxviii.

Diapea, William
1928. *Cannibal Jack: the true autobiography of a white man in the South Seas.*
London, Faber and Gwyer. Pp. xxiv, 242.

Erskine, John Elphinstone
1853. *Journal of a Cruise among the Islands of the Western Pacific, including the Feejees and others inhabited by the Polynesian negro races, in Her Majesty's Ship "Havannah".*
London, John Murray. Pp. vii, 488. Reprinted Dawsons of Pall Mall 1967 (Colonial History Series No. 6.)

Farmer, Sarah Stock
1855. *Tonga and the Friendly Islands: with a sketch of their mission history written for young people.*
London, Hamilton Adams. Pp. vii, 427.

Fiji Islands, The. General Remarks on the Aspect, Climate,
1861. Soil and Flora. From Dr. Seemann's Report.
The Cotton Supply Reporter, Vol. II, Aug. 15, pp. 601–602.

Forbes, Litton
1875. *Two Years in Fiji (with a visit to Rotumah, and remarks on the Polynesian labour traffic.)*
London, Longmans & Green. Pp. xii, 355.

Fraser, A. A.
1863. Review of Viti by B. Seemann.
Anthropological Journal, Vol. I, p. 355.

Gray, Asa
1861. Notes upon a portion of Dr. Seemann's recent collection of dried plants gathered in the Feejee Islands.
Proceedings of the American Academy of Arts and Science, Vol. 5, pp. 314–352.

1862. Plantae Vitiensis Seemannianae. Remarks on the plants collected in the Vitian or Fijian Islands by Dr. Berthold Seemann.
Bonplandia, 10, 34–37.

Hazlewood, David

1850. *A Feejeean & English and an English & Feejeean Dictionary: with examples of common and peculiar modes of expression, and uses of words. Also, containing brief hints on Native Customs, Proverbs, the Native Names of Natural Productions and Notices of the Islands of Feejee, and a list of the foreign words introduced.*
Feejee, Vewa, Wesleyan Mission Press. Pp. 350.

im Thurn, Sir Everard Ferdinand

1925. *The Journal of William Lockerby, Sandalwood Trader, in the Fijian Islands during the years 1808–09. (Includes also (a) Samuel Patterson's Narrative of the Wreck of the "Eliza" in 1808, (b) Journal of the Missionaries put ashore from the "Hibernia" on an islet in the Fiji Group in 1809, (c) Capt. Richard Siddon's Experience in Fiji 1809–1815).*
Cambridge University Press, Hakluyt Society Publ. LII. Pp. 250.

Lady, A. (Wallis, Mary Davis Cook)

1851. *Life in Feejee: or, Five Years Among the Cannibals.*
Boston, W. Heath. Pp. 422.

Legge, Christopher Conlagh

1966. William Diaper: A Biographical Sketch.
Journal of Pacific History, Vol. I, pp. 79–90.

Macdonald, Sir John Denis

1857. Proceedings of the Expedition for the Exploration of the Rewa River and its Tributaries in Na Viti Levu, Fiji Islands.
Geographical Journal, Vol. XXVII, 232–268.

Martin, John
1817. *Account of the Natives of the Tonga Islands in the South Pacific Ocean with the original grammar and vocabulary of their language compiled and arranged with extensive communications of Mr. William Mariner, several years resident in those islands.* London, James Murray. 2 Vols.

Meade, Hon. Herbert George Philip
1870. *A ride through the disturbed districts of New Zealand; together with some account of the South Sea Islands.* Edited by R. H. Meade. London, John Murray. Pp. xii, 375.

Pritchard, William Thomas
1866. *Polynesian Reminiscences, or Life in the South Pacific Islands.* Introduction by B. C. Seemann. London, Chapman & Hall. Pp. xii, 428. Reprinted Dawson of Pall Mall 1968 (Colonial History Series No. 32.)

Rejection of Fiji by Great Britain, The. Review of Viti by B. Seemann. *Quarterly Review*, Vol. XX, p. 35.

Ricci, James Herman de
1874. *How about Fiji? Or Annexation versus non-annexation.* With an account of the various proposals for . . . and a short . . . of the natural aspects of the Group. London, Edward Stanford. Pp. 81.

1875. *Fiji, our new Province in the South Seas.* London, Edward Stanford. Pp. viii, 332.

Rochas, Victor Henry de
1861. Les Iles Viti ou Fidji. Paris, *Nouvelles Annales des Voyages*, avril, pp. 5–19.

Rowe, George Stringer (Editor)
1861. *Fiji and the Fijians.* Vol. I. The Islands and their Inhabitants by T. Williams. Vol. II. Mission History by J. Calvert. London, Alexander Heylin. Pp. 266.

Seemann, Berthold Carl
1858. *Acht Jahre in Asien und Afrika . . . Nebst einem Vorworte von B. Seemann.*
1862. Antiaris Bennettii.
Bonplandia, Vol. 10, pp. 3–4.
1862. On Antiaris Bennettii, a new species of upas-tree from Polynesia.
Annual Magazine of Natural History, III, 9, pp. 405–407.
1852–57. *The Botany of the Voyage of H.M.S. Herald under the command of Capt. H. Kellett during the years 1845–51.*
London.
1860. *The British Ferns at one view.*
London.
1861. Cyrtandra Pritchardii.
Bonplandia, Vol. 9, pp. 364–365.
1867. *Die gelben Rosen. Dramatischer Zufall in I Aufzuge.* Frei nach A. Ebeling. Hannover.
1861. Die giftigen Pflanzen der Viti-oder Fiji-Inseln.
Hamburg, *Gart. Blumenzeit*, 17, pp. 437–442.
1852. *Die in Europa eingeführten Acacien, mit Berücksichtigung der gärtnerischen Namen.* Mit zwei färbigen Kupfertafeln.
Hannover.
1851. *Die Volksnamen der amerikanischen Pflanzen. . . .* The popular nomenclature of the American flora.
Hannover.
1869. *Dottings on the Roadside to Panama, Nicaragua and Mosquito.*
1844. *Endlicher's Paradisus Vindobonensis,* Abbildungen seltener und schönblühender Pflanzen. . . . Erläutert von Dr. B. Seemann.
1861. The Feejee Islanders: their Religion, Laws, manners and customs.
Lado, Nov. 13, 1860. From the Athenaeum.
Nautical Magazine, May, pp. 257–263.

SOME FURTHER READING

1862. Fiji and its Inhabitants in *Vacation Tours and Notes of Travel in 1861*, edited by Sir Francis Galton, pp. 249–292.

1862. Fiji Islands. American Journal of Science, Series 2, 34, pp. 366–367.

1865–73. *Flora Vitiensis: a description of the Viti or Fiji Islands, with an account of their history, uses and properties. . . .* With one hundred plates by W. Fitch. London, Lovell Reeve. Pp. 453.

1862. *Hannoverische Sitten und Gebräuche in ihrer Beziehung zur Pflanzenwelt, ein Beitrag zur Culturgeschichte Deutschlands.* Populäre Vorträge. Leipzig.

1862. Lindenia vitiensis. Bonplandia, Vol. 10, pp. 33–34.

1861. More about the Feejees. From the Athenaeum. Nautical Magazine, Sept., pp. 478–487.

1853. *Narrative of the Voyage of H.M.S. Herald during the years 1845–51, under the command of Capt. H. Kellett—being a circumnavigation of the globe, and three cruises to the Arctic Regions in search of Sir J. Franklin.* London. 2 Vols.

1861. Notes made during a government expedition to the Viti or Fiji Islands. Gardeners' Chronicle, pp. 599–600, 622–625, 649.

1862. Noticen uber Südsee-Pflanzen. Bonplandia, Vol. 10, pp. 153–155.

1862. Pimia Rhmnoides und Disemma caerulescens, zwei neue Südsee-Pflanzen. Bonplandia, Vol. 10, pp. 366.

1861–62. Plantae Vitiensis. Bonplandia, Vol. 9, pp. 253–262, 1861; Vol. 10, pp. 295–297, 1862.

1862. Podocarpus dulcamara Seem. Bonplandia, Vol. 10, pp. 365–366.

1863. Podocarpus vitiensis, a new coniferous tree from the Viti Islands. London, *Journal of Botany*, Vol. I, pp. 33–36.

1861. Poisonous plants of the Viti or Fiji Islands. *Gardeners' Chronicle*, p. 697.

1856. *Popular History of the Palms and their allies.* London.

1862. Pritchardia pacifica. *Bonplandia,* Vol. 10, pp. 309–310.

1858. *Reise um die Welt und 3 Farten der Königlich Britischen Fregatte Herald nach dem nördlichen Polarmeer zur Aufsuchung Sir J. Franklin's in den Jahren 1845–51.* Hannover, Zweite Auflage.

1862. Remarks on a Government Mission to the Fiji Islands. *Geographical Journal*, Vol. 32, pp. 5–62.

1868. Revision of the Natural Order Hederaceae, being a reprint, with numerous additions and corrections, of a series of papers published in the *Journal of Botany*. London.

1862. Smythea pacifica. *Bonplandia*, Vol. 10, pp. 68–70.

1862. Solanum anthropophagorum. *Bonplandia*, Vol. 10, p. 274.

1861. Storckiella Vitiensis. *Bonplandia, Vol.* 9, pp. 363–364.

1861. Suva, die kunftige Hauptstadt der Viti Inseln. *Petermann's Mitt.*, pp. 359–360.

1862. *Synopsis plantarum Vitiensium. Systematic list of all the Fijian plants at present known.*

1861. *Twentyfour views of the vegetation of the coasts and islands of the Pacific, with explanatory descriptions.*

SOME FURTHER READING

1862. Vegetable productions and resources of the Vitian or Fijian Islands. *Appendix to Parliamentary Paper, Correspondence relative to Fiji Islands.*

1867. *Wahl macht Qual. Schwank in I Aufzuge.* Hannover.

Smythe, Sarah Maria
1864. *Ten Months in the Fiji Islands*, with an Introduction and Appendix by Col. W. J. Smythe, Royal Artillery, late H.M. Commissioner to Fiji. Oxford & London, John Henry and James Parker. Pp. xviii, 282.

Snow, Philip Albert
1969. *A Bibliography of Fiji, Tonga and Rotuma.* Canberra, Australian National University Press, and Miami University Press. Pp. xliii, 418.

Turner, George
1861. *Nineteen Years in Polynesia: missionary life, travel and research in the Islands of the Pacific.* London, John Snow. Pp xii, 548.

Twyning, John Poyen
1850. *Shipwreck and Adventures of John P. Twyning among the South Sea Islanders: giving an account of their feasts, massacres, &c., with the certificates of Wesleyan missionaries who lived in the islands.* London, Printed for the author and sold by Dean & Son. Pp. 174.

Waterhouse, Joseph
1866. *The King and People of Fiji, containing a life of Cakobau, with notices of the Fijians, their manners, customs and superstitions, previous to the great religious reformation in 1854.* London, Wesleyan Conference Office. Pp. xii, 435.

Waterhouse, Joseph
1857. *Vah-Ta-Ah, the Fijian Princess, with occasional allusions to Feejeean customs and life.* London, Hamilton Adams. Pp. 164.

Wilkes, Charles
1844. *Narrative of the United States Exploring Expedition during the years 1838, 1839, 1840, 1841, 1842.* Philadelphia, C. Sherman. 5 vols.

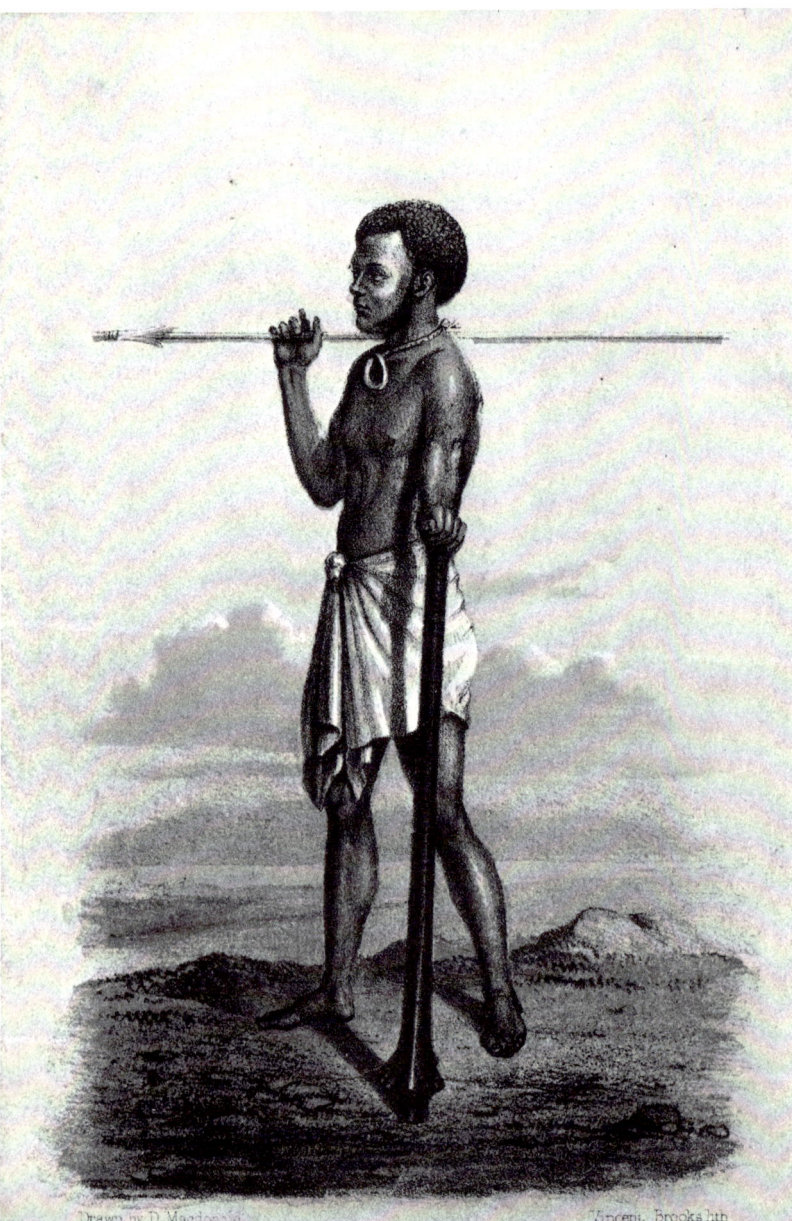

RATU VAKARURU, A CHIEF OF VITI LEVU.

VITI:

AN

ACCOUNT OF A GOVERNMENT MISSION

TO THE

VITIAN OR FIJIAN ISLANDS

IN THE YEARS 1860–61.

BY

BERTHOLD SEEMANN, Ph.D., F.L.S., F.R.G.S.,

AUTHOR OF THE NARRATIVE AND THE BOTANY OF H.M.S. HERALD,
'POPULAR HISTORY OF PALMS,' ETC. ETC.

With Illustrations and a Map.

Cambridge:
MACMILLAN & CO.,
AND 23, HENRIETTA STREET, COVENT GARDEN, LONDON.
1862.

The right of translation is reserved by the Author.

JOHN EDWARD TAYLOR, PRINTER,
LITTLE QUEEN STREET, LINCOLN'S INN FIELDS.

TO

SIR RODERICK IMPEY MURCHISON,

D.C.L., LL.D., F.R.S.,

DIRECTOR OF THE GEOLOGICAL SURVEY,

ETC., ETC., ETC.,

WHOSE PRE-EMINENCE IN SCIENCE HAS ALWAYS
BEEN COUPLED WITH A GENEROUS ENCOURAGEMENT OF THE
LABOURS OF OTHERS,

This Work is Dedicated

WITH FEELINGS OF HIGH REGARD AND ESTEEM

BY

THE AUTHOR.

PREFACE.

In 1859 Mr. W. T. Pritchard, H.B.M. Consul in Fiji, son of the Rev. George Pritchard, formerly of Tahiti, arrived in England with a document purporting to be the cession of Fiji—or rather Viti—to the Queen of Great Britain. The cession had been made by Cakobau (= Thakombau), the principal chief of Bau and king of the whole group, and with the consent of the leading chiefs. The importance of accepting the proffered sovereignty was insisted upon by parties capable of taking a comprehensive view of the question. The Legislative Assembly of New South Wales, on the motion of Mr. M'Arthur, voted an address to the Queen in support of this proposal. Captain Towns, a patriotic citizen of Sydney, fully impressed, like many of his countrymen, with the importance of acquiring the islands, generously offered a cheque for the whole Fijian debt, in order to remove at least one of the possible obstacles in the way of the cession. Nor is it any secret that the occupation of the islands has been recommended by Captains Fremantle, Denham, Erskine, and Loring, and Admirals Washington* and Sir Edward Belcher; in

* See Appendix.

fact, by all naval men who knew anything about the subject. Men high in office were equally favourably inclined towards the cession. However, before coming to any definite decision, the Government determined to obtain more ample information than was at hand, and early in 1860 I was asked to join a "Mission to Viti" dispatched for that purpose.

Whilst in Fiji, I was induced to write a series of letters on the country, its people, and productions, to the 'Athenæum,' which that journal did me the honour to publish, and which, whole or in part, found their way into several other home and colonial papers, were translated into German and French, and altogether obtained a circulation for which their original place of publication alone can account. On my return to London I was urged to make additions to this series, and I acceded to this wish by bringing the subject before the Royal Geographical Society, and writing papers for the 'Gardeners' Chronicle' and Galton's well-known 'Vacation Tourists and Notes of Travel.' But a good deal of matter remained still unpublished, which, together with the pith of all I have previously made known, will be found in the following pages.

In order that the public may have the means of forming a correct judgment on the Fijian question, I have reprinted in the Appendix Colonel Smythe's Official Report, at variance as it is with all that has been written on the islands. My impression of Fiji and its inhabitants was most favourable, and I am convinced that, under judicious management, the country would become a flourishing colony,—an opinion shared by almost all who have visited the group, as was again proved at a crowded meeting at the Geographical Society when the subject was discussed.

Desirous of collecting as many productions of the country as possible, I neglected to investigate several subjects which fell not within my assigned province. It was only after the publication of Colonel Smythe's 'Report,' that I became aware of the full importance of my neglect. For instance, it would have been very important to know how many thousand acres of land had passed out of the hands of the natives. As a great many islands and vast tracts of country have already been purchased by British subjects, statistics on these points would probably have materially influenced the decision of Her Majesty's Government with respect to the acceptance of the cession.

Amongst other things I brought home a comprehensive collection of plants, which, together with those already in this country, chiefly accumulated by Government expeditions, furnish ample materials for a Flora of Fiji, a Flora Vitiensis. I expended a good deal of my own money in order to make these collections as complete as possible, and was in hopes that the Government would see fit to assist me in publishing such a work, especially as my report on the resources and vegetable productions of the islands had been presented to both Houses of Parliament by command of Her Majesty, and the nature and possible value of the projected publication .must have become evident. His

Grace the Duke of Newcastle, ever ready to advance science, fully sharing these hopes, made an application to the Treasury to that effect, but was "very sorry to inform me that his application had been unsuccessful." Thinking what had been collected with so much expense, under great difficulties, and in a country only partially reclaimed from cannibalism, was also worth making known, I resolved to incur the risk of publishing the work at my own cost. It will consist of 400 pages of letter-press (quarto), and 100 coloured plates, all representing objects hitherto unknown to science, and drawn by the skilful pencil of Mr. Fitch. The work will take about three years to bring out, and its publication will commence immediately.

All the native names are spelt according to the system of orthography laid down in Hazelwood's 'Fijian Dictionary' (London: Trübner and Co.), and wherever any deviation should be discovered, it may be regarded as a mistake of mine, unless particularly noticed. Nothing but endless confusion will be the result if every nation is allowed to write Fijian names according to its own orthography. For the illustrations of my present work I am indebted to Mrs. Smythe, Dr. Macdonald, and Captain Denham, to whom I beg to tender my best thanks, as well as to those friends who, since my departure from Fiji, have kept me supplied with the latest intelligence from that group.

BERTHOLD SEEMANN.

London, September 30, 1862.

CONTENTS.

CHAPTER I.

Departure from England.—Arrival at Sydney.—Voyage to Fiji.—The 'John Wesley.'—The Pitcairners at Norfolk Island.—First Glimpse of Fiji.—Lakeba.—The Tonguese.—Visit to a Mission Station.—First Botanical Excursion.—Hints to Collectors.—Native Church.—Bark-cloth Manufacture.—Tomb of a Chief.—Missionary Life.—Departure from Lakeba 1

CHAPTER II.

Island of Taviuni.—The King of Cakaudrove.—Elephantiasis.—Kind Offer of Mr. Waterhouse and Captain Wilson.—Somosomo, its Advantages and Disadvantages.—Queen Eleanor.—Ascent of Summit of Taviuni.—A Royal Escort.—Sylvan Scene.—Arrival at the Top.—Singular Swamp of Vegetable Turtle Fat.—Dinner.—Timidity of the Natives.—Chief Golea's Return from a Military Expedition.—Polygamy.—The Rotuma-Men.—Wairiki.—Arrival of the 'Paul Jones'. 19

CHAPTER III.

Fiji as a Cotton-growing Country.—Cotton not Indigenous but Naturalized.—Native Names.—Number of Species.—Average Produce of the Wild Cotton.—Excellence of Fijian Cotton acknowledged at Manchester.—Efforts of British Consul and Missionaries to extend its Cultivation.—The First Thousand Pounds of Cotton sent Home.—Establishment of a Plantation at Somosomo, Wakaya, and Nukumoto.—Prospects of Cotton-growing in Fiji . . . 48

CONTENTS.

CHAPTER IV.

Departure from Somosomo.—Island of Wakaya.—The Balolo.—Arrival at Levuka.—H.B.M. Consul.—The Late Mr. Williams.—Lado and its Origin.—Site for the New Capital.—The King of Fiji.—Bau.—Causes of its Supremacy.—Viwa 58

CHAPTER V.

The Wai Levu or Great River.—Canal Dug by Natives.—Mataisuva.—Institution for Training Native Teachers.—Sacred Groves, Trees, and Stones.—Mosquitoes.—Island of Naigani.—Mr. Eggerström's Kindness.—Feuds at Nadroga.—Nukubalawu.—Taguru.—Navua River 82

CHAPTER VI.

Stay at Navua.—Chief Kuruduadua's Household.—"Harry the Jew."—A Prince as he was Born.—Massacre Prevented.—Kuruduadua's Character.—Statement of Mr. Heekes Respecting the Namuka Outrage.—Town and Bures of Navua.—Tatooing.—Return to Lado. 97

CHAPTER VII.

Arrival of Colonel Smythe from New Zealand.—The 'Pegasus' and 'Paul Jones.'—Visit to Bau.—Quarrelsome Disposition of the Chief of the Fishermen.—Cession of Fiji to England.—First Official Interview with the King. 120

CHAPTER VIII.

Excursions to Koroivau and Namara.—Departure from Bau.—Passage through the Great River of Viti Levu.—Buretu.—Apostate Christians.—Rewa.—Arrival at Tavuki, Kadavu.—Whale Ships.—Attempt to ascend Buke Levu.—The Isthmus of Kadavu.—Ga Loa or Black Duck Bay.—Departure for Navua 133

CONTENTS. xiii

CHAPTER IX.

Page

Departure from Kadavu.—Arrival at Navua.—A Court of Justice.—Starting for the Interior.—The Navua River.—Its Fine Scenery.—Rapids.—A Canoe upset.—Town of Nagadi.—Hospitable Reception.— Soromato. — Kidnapping. — Family Prayers. — Heathen Temple.—A Large Snake to be Cooked.—March across the Country.—Vuniwaivutuku.—A Difficult Road.—A Purse Lost.—No Thieves.—Arrival at Namosi.—Danford's Establishment.—His Usefulness as a Pioneer 146

CHAPTER X.

Popular Ideas Respecting the Interior of Viti Levu.—Malachite and Antimony.—Ascent of Voma Peak.—Visit to a Heathen Temple. —" Spirit Fowls."—Official Meeting with Kuruduadua and his Subjects.—A Rebellion to be Suppressed.—Presentation of Food. —" The Oldest Inhabitants."—A Court-Fool and his Tricks.—Mr. Waterhouse Preaching.—Departure of Colonel Smythe and Messrs. Pritchard and Waterhouse, for Nadroga 160

CHAPTER XI.

Fijian Cannibalism.—The Great Cauldron.—Naulumatua and his Appetite for Human Flesh.—Bokola.—Vegetables Eaten with Cannibal Food.—The Ominous Taro.—Approximate Number of Bodies eaten at Namosi.—Ovens for Baking Dead Men.—Suspension of the Bones.—Not all Fijians Cannibals.—Efforts of the Liberal Party to Suppress Anthropophagism.—Aided by Europeans.—Real Significance of Eating Man only Partly Understood.—Concessions to Humanity.—Abolition of Cannibalism throughout Kuruduadua's Dominions 173

CHAPTER XII.

Stay at Namosi Prolonged.—The Governor's Attention.—" Crown Jewels."—The Clerk of the Weather.—Sorcerers.—Fijian Family Life.—Story-Tellers Popular.—A Fijian Tale 186

CHAPTER XIII.

Departure from Namosi.—Vuniwaivutuku.—The "Veli."—Mode of Tatooing the Mouth.—Passing down the Navua River.—Nagadi cleared out by its Vasu.—Our Canoe Capsized.—Return to the 'Paul Jones.'—Kuruduadua's Character.—Leaving Navua.—Bega. —Mr. Storck's Illness.—Return to Kadavu.—Ascent of Buke Levu.—Rewa.—Immigrants from New Zealand.—Mr. Moore's Powerful Sermon.—Arrival at Lado.—Office Drudgery . . . 202

CHAPTER XIV.

Voyage around Vanua Levu.—Departure from Lado.—East Coast of Viti Levu.—Nananu Island.—The Fijian Mount Olympus.—Bua.— Naicobocobo.—Nukubati.—Naduri.—Interview with the Chief.— Discontent of his Subjects.—Bêche-de-mer Trade.—Mua i Udu and its Superstitions.—Na Ceva Bay.—Arrival at Waikava.—Visit to my Cotton Plantation.—Meeting at Waikava.—Departure . . 222

CHAPTER XV.

History of the Tongamen in Fiji.—Their Physical Superiority over the Fijians.—Their Arrogance.—Captain Croker's Defeat.—Early Intercourse between Tonga and Fiji.—Increase of Tonguese Immigration.—Chief Maafu.—King George of Tonga visits Fiji.—Conquest of Kaba and Rabe.—Arrival of British Consul.—Cession of Fiji.—Maafu's Attempted Conquest.—Ritova and Bete.—Maafu's Ambition Curbed.—Peace Restored.—Ritova Installed in his Estates. —Tonguese Intrigues Renewed.—Bete's Death.—Commodore Seymour's Visit.—Termination of the Wars between Fijians and Tongans 236

CHAPTER XVI.

General Remarks on the Aspect, Climate, Soil, and Vegetation of Fiji. —Colonial Produce.—Staple Food.—Edible Roots.—Kitchen Vegetables.—Edible Fruits.—National Beverages.—Kava . . . 274

CHAPTER XVII.

Vegetable Poisons.—Medicinal Plants.—Scents and Perfumes.—Materials for Clothing.—Mats and Baskets.—Fibres used for Cordage.—Timber.—Palms.—Ornamental Plants.—Miscellaneous . . 332

CHAPTER XVIII.

Remarks on the Fauna of Fiji.—Mammals.—Birds.—Fishes.—Reptiles.—Mollusks.—Crustacea.—Insects.—Lower Animals . . 381

CHAPTER XIX.

Fijian Religion.—Degei, the Supreme God.—Inferior Deities.—Worship of Ancestors.—Idolized Objects.—Temples.—Creation and Ultimate Destruction of the World.—A Great Flood.—Immortality of the Soul.—Conception of Future Abode.—Props of Superstition 389

CHAPTER XX.

Historical Remarks on Fiji.—Discovery of the Islands.—Sandal-wood Traders.—Early White Settlers.—Missionaries.—Foreigners at present Residing in the Group.—Departure from Fiji in the 'Staghound.'—Terrific Storm off Lord Howe's Island.—Arrival in Sydney.—Return to England.—Conclusion 404

APPENDIX.

I.—Report of Admiral Washington, R.N. 419
II.—Report of Colonel Smythe, R.A., to Colonial Office . . . 421
III.—Systematic List of all the Fijian Plants at present known . 431

VITI:

AN

ACCOUNT OF A GOVERNMENT MISSION TO THE VITIAN OR FIJIAN ISLANDS.

CHAPTER I.

DEPARTURE FROM ENGLAND.—ARRIVAL AT SYDNEY.—VOYAGE TO FIJI.—THE 'JOHN WESLEY.'—THE PITCAIRNERS AT NORFOLK ISLAND.—FIRST GLIMPSE OF FIJI.—LAKEBA.—THE TONGUESE.—VISIT TO A MISSION STATION.—FIRST BOTANICAL EXCURSION.—HINTS TO COLLECTORS.—NATIVE CHURCH.—BARK-CLOTH MANUFACTURE.—TOMB OF A CHIEF.—MISSIONARY LIFE.—DEPARTURE FROM LAKEBA.

HAVING left Southampton on the 12th of February, 1860, by the overland mail, and having touched at Mauritius, King George's Sound, and Melbourne, I arrived at Sydney on the 16th of April, where I was to join Colonel Smythe, R.A.,—who had gone out by the previous mail,—and proceed with him in her Majesty's ship 'Cordelia,' it was supposed, to Fiji. The first news heard was, that a war had broken out in New Zealand, in consequence of which all available naval force had been dispatched to the scene of action. This altered our plans considerably. Colonel Smythe, thinking that the outbreak of native discontent would be only of short duration, and that after its termination he should still be able to obtain a Government vessel for Fiji, resolved to proceed by the mail steamer to New Zealand. He came on board the 'Benares' to communicate this resolution to me, but I, having made an attempt to find him on shore, was absent, and as his steamer left soon after the English mail had been transferred, I did not meet with him until three months afterwards.

Sir William Denison, to whom I had letters from the home Government, advised me either to go to New Zealand and wait there for an opportunity, or else direct to Fiji, in the missionary vessel 'John Wesley,' about to sail that day. Wishing to economize my time as much as possible, I preferred the latter. In communicating with the Rev. John Eggleston, General Secretary of the Wesleyan Mission, that gentleman kindly postponed the departure of their vessel a few days, in order to afford me time for making the necessary preparations for future explorations. He supplied me besides with letters of introduction to residents in the Fijian islands, books, and a list of articles used as barter, all of which proved highly acceptable. In reply to Sir William Denison's asking for a passage for me and my assistant, Mr. Jacob Storck, Mr. Eggleston cheerfully granted a free passage to both of us, at the same time reminding the Governor-General that the Wesleyans as a body felt under obligations to the Government for so frequently allowing their vessels to assist their missionaries in the Pacific Ocean, rendering them timely aid, and supplying them with medicines, and bringing them home when ill. With the assistance of Mr. Charles

Moore, Director of the Botanic Gardens at Sydney, I was enabled to complete all my arrangements without loss of time. When embarking, I had accumulated a whole cart-load of luggage, containing none save the most necessary things, and surveyed by me with a heavy heart when thinking of the difficulty of transporting them from island to island. None save those who have experienced it, can have any conception of travelling in countries where no money is current, and all is paid for in kind. How easy is moving about when one can carry a whole year's travelling expenses in the waistcoat pocket! But think of people never doing a thing for you unless you have counted out, or measured off, the requisite number or amount of your stock in trade.

All being ready and the wind fair, I left Sydney Harbour on Friday, April 20, 1860, on board the 'John Wesley,' Captain Birkenshaw. There were, in all, six passengers,—Captain Wilson, from Sydney, about to look after his cocoa-nut oil establishment at Somosomo; Mr. and Mrs. Harrison, a missionary and his wife, for Fiji; Mr. Storck and myself, and a Fijian native teacher, who had come to Sydney with the view of proceeding to England, but who, after reaching New South Wales, had become so home-sick, that he was obliged to return to his native country. Though having been only a few thousand miles, he would be regarded as a mighty traveller on his return, and doubtless looked upon himself as such. For, as the Italian would wish " to see Naples, and die," or the Spaniard declares that—

" El que no ha vista Sevilla
No ha vista maravilla "—

so the South-Sea Islanders would say, "Let me behold Sydney, and go home again."

No one should speak ill of the bridge that carries him over, or look a gift-horse in the mouth; but I have been so frequently asked about the 'John Wesley,' that I may be exculpated when saying a few words about the vessel as she appeared to me. The 'John Wesley' was launched in 1846, having been built by Messrs. White and Sons, of Cowes, and being paid for by charitable contributions. I have read high eulogiums on her, but anybody who has sailed in her will not be inclined to endorse them. It has never been my misfortune to be on board a vessel behaving worse than she did. She is about thirty feet too short, and never easy, let the wind be ever so favourable and the sea as smooth as a pond. In a slight gale the pitching is awful, and the rolling terrific. We were often watching and wondering what would be her next move after all these had been going on for awhile, when perhaps she would shake her rudder so violently that one almost feared it must come out. In consequence of her constant uneasiness, the wear and tear in ropes and spars is considerable, and the annual expenditure must be much greater than might be expected from a vessel of her size. Nearly every morning there was something gone, and we used to chaff the captain about the superior behaviour of his craft; but he, like a true sailor, would defend her through thick and thin. In rough weather she had, besides, the bad quality of leaking; and, as some of the cocoa-nut oil carried in her on a former occasion had oozed out of the tanks and casks and become rancid, the stench was quite overpowering. It requires a peculiar constitution not to become sea-sick on board, and this is perhaps the most serious inconvenience that the missionaries and their families suffer when going backwards and forwards in her to the Colonies, or from island to island. When we left Sydney Harbour, I observed several of our men in unfurling sails, sea-sick, a sight I never before beheld; and Mr. and Mrs. Harrison were ill during nearly the whole passage. Nor is she, with all these drawbacks, a fast or a good sailer. We were twenty-three days from Sydney to Fiji, a distance of 1,735 miles, and I believe that may be considered a fair average passage. The crew was an extremely mongrel set. There were men of all colours, countries, and religions: black Africans, copper-coloured Chilians, and white Englishmen; Heathens, Mahometans, Roman Catholics, and Protestants. I expressed my surprise that in a vessel belonging to a religious society there should be so mixed a ship's company; but the Captain thought it rather an advantage than otherwise, offering, as it did, a field for missionary labours during the voyage. Indeed, when not suffering from sea-sickness, Mr. Harrison made some attempts in that direction.

We endeavoured to make Norfolk Island, but could not fetch it within about one hundred miles. I should have liked to look at that charming spot, which, no longer a convict station, as in days of yore, has lately been given by the Government to the Pitcairners,—those much-petted descendants of 'Bounty' mutineers and Tahitian women,—because their own little island began to be too small for the growing community. The Pitcairners landed on the 8th of June, 1856, from the 'Morayshire,' a vessel belonging to Mr. Dunbar, of London, commanded by Mr. Joseph Mathers, and under the agency of Acting-Lieutenant G. W. Gregorie, of her Majesty's ship 'Juno.' They numbered in all 194 souls, one of whom died soon after landing; the rest comprising 40 men, 47 women, 54 boys, and 52 girls. Almost an entire week was employed in disembarking all the seventy years' gathering of chattels, including almost every moveable article, even to the " gun " and " anvil " of the ' Bounty.' On landing they were greeted individually by the commissariat officer and Captain Denham, of her Majesty's ship ' Herald,' who happened to be there, and then conducted to their comfortably-prepared quarters, until they should be able to make their own selection from the commodious dwellings erected for them. Dr. Macdonald instructed the islanders essentially in the resources of the ample dispensary at their use, whilst the artificers of the ' Herald ' imparted to them the uses of a variety of tools and implements, comprising the wind and water mills; indeed, everything was done to make them comfortable. The first provident step for future provision was taken by planting their favourite sweet-potato, and, pending harvest time, which they gave six months to come about, the ' Herald ' left the newly-transferred community provided with 45,000 lbs. of biscuit, flour, maize, and rice, with groceries in proportion, and abundance of milk at their hands; whilst their live stock consisted of 1300 sheep, 430 cattle, 22 horses, 10 swine in sties, 16 domestic fowls, and a quantity of wild pigs and fowls. Even 16,000 lbs. of hay and 5000 of straw were left them; and, lest their first crop should be late or fall short, a list of additional supplies was sent to the Governor-General.* According to all accounts the Pitcairners do not display themselves to advantage in their new home, and most visitors are anything but pleased with them. As might have been expected, the numerous presents given and sent to them have had a bad effect, making them accomplished beggars, who state their case in such a way as will most readily induce the hearer to give them some present or influence others to do so. They are besides said to be an indolent set, who, rather than fetch fuel from the woods, will burn the floors, doors, and window-frames of the fine buildings erected by the convicts, and generously placed by Government at their disposal. If report be true, Sir William Denison, on his visit to the island, gave them a severe and well-deserved lecture on this head. Several of them are said to have already returned to Pitcairn Island, where they seem to have felt more comfortable, though cramped for space, and a few are said to have embarked in whaling operations. Let us hope that the whole community, about which so much truth and fiction has been written, may gradually be led to habits of industry, and learn to rely more upon its own resources than the charitable contributions of others.

On the 10th of May we got the trade wind, and on Saturday the 12th, about eight o'clock in the morning, caught the first glimpse of Fiji. We had left Sydney

* See Captain Denham in 'Hydrographic Notice,' n. 5.

on the 20th of April, and had thus been twenty-three days on the passage, four of which we had strong gales and were compelled to heave to. We bantered the Captain a good deal about the long passage, and ascribed it all to his having left on a Friday, at the same time accumulating instances where departures on that unlucky day had been followed by as disastrous consequences as when thirteen sit down to table. But he thought it high time that such vestiges of superstition should be rooted up, and said there was no more in them than in the Flying Dutchman. On the following day we were off Lakeba (Lakemba). It being Sunday, Captain Birkenshaw would not give offence by sending a boat on shore on the Sabbath. I suggested that we might all go to church as soon as landed, but he maintained that it was as much as his place was worth to entertain such an idea; so we had the mortification of stopping another day on board, and sail backwards and forwards between the islands of Lakeba and Olorua. I enjoyed much the fine sight that thus was offered. The sky was clear and bright, and a number of little islands and islets were rising from the blue sea, the waves breaking on their rocky shores, or forming curly crests on the long reefs that encircle many of them. They were all more or less elevated, and covered with vegetation, here with patches of grass or brake and other hard-leaved ferns, there with brushwood or larger trees; the presence of countless screw-pines and ironwood (*Casuarina*) trees imparting to them their peculiar Polynesian character. Well may it be said, that the graceful waving iron-wood bears on its very face the proof of its being at home in a country and in situations continually agitated by the trade winds. Any other tree would become stunted and unsightly under such circumstances, whilst the iron-wood is rendered only more graceful by them.

The next morning we endeavoured to effect a landing, no easy task, as the sea was running rather high, and we had to search amidst a heavy surf for a channel through the reef encircling Lakeba, and on which Colonel Smythe's vessel, the 'Pegasus,' struck, when paying a visit a few months afterwards. I have often admired the grandeur of the South Sea reef, when the water breaks with all its force on that mighty fabric of coral and volcanic rock; and wondered why such a grand sight has not as yet been immortalized by some great painter in search of a fitting subject for his brush. It is certainly overpowering to sit down before Niagara, and watch the mighty masses of water steadily pouring into a gigantic basin. Impossible, one thinks, that such tuns and tuns can be discharged without the supply becoming exhausted. Nevertheless there is no abatement. As the sun rises it shines upon the foaming mass, and its last rays kiss the same spectacle. Like eternity, it is endless; and our thoughts, taken captive as we gaze and gaze on the massive volumes, are wandering towards those realms whence no traveller has returned. The sight of a great South Sea reef is something equally grand, but produces a rather different effect. Besides being influenced by wind and tides, the surf assumes almost every moment a different aspect. Now it is little more than a long line of silent ripples, now it is lashed into wild spray to great height, speaking in hollow roars, and showing a variety of tints which the pen must ever despair of depicting. So far from becoming absorbed in thought at such a sight, as at the monotonous grandeur of Niagara, one longs to stir, to push on, to become active like the never-resting element.

Though we got a good wetting, and might have been swamped had it not been for the skilful steering of our mate, we landed in safety. As soon as the boat was near shore fifty or sixty natives plunged into the water to carry us on their backs to the beach, when we shook hands with Mr. Fletcher, one of the Wesleyan missionaries stationed here. The natives were nearly all fine strapping fellows, some of them quite six feet high, and all Fijian, with the exception of a couple of Tonguese or Tonga men, inhabitants of a neighbouring group of islands. One of the latter was Charles, the son of the Tonguese chief, Maafu, a mighty man in Polynesian annals, and the source of much trouble, both in Tonga and Fiji. When most people read of "natives" they imagine them to be types of unsightliness, if not downright ugliness; of many races, not Caucasian, that may in some measure be true, but whoever goes to the South Seas will have reason to change his opinion entirely. Some of these islanders are really very handsome, both in figure and face; and all entitled to pronounce an opinion on the subject have agreed that there are few spots in the world where one sees so many handsome people together as in Tonga. I have never been in Circassia, and can therefore not speak from personal experience; but, if what one reads be correct, Tonga may fairly be classed with the Tyrol and Circassia, for its male population. I do not include the females, because, according to our taste, the women of Tonga, like those of the Tyrol, are too masculine and robust to please our conceptions of feminine beauty. When I looked at these Tonguese, with their fine athletic body, symmetrical, handsome faces, and rich dark hair, I could not refrain from thinking what caricatures civilization has made us. The gait of such a man is something to wonder at, and sculptors would find him a fine subject for study. Here they might obtain models almost approaching their notions of ideal perfection, instead of copying, as they now too often are compelled, the body of a life-guardsman, the head of a footman, and the hands and feet of some of higher-bred types.

Charles Maafu, I was informed, had been sent to Lakeba by his father, as a punishment for several larks the young rascal had been up to. I don't wonder there should have been a great deal of temptation in his way, for, besides being the son of a powerful chief, a lineal descendant of one of the royal houses of Tonga (Finau), he was about eighteen years of age and extremely handsome. He wore only a few yards of cotton cloth around his loins, and an ornament made of mother of pearl. King George, of Tonga, had proposed to have his own son and Charles educated at Sydney. The offer was unfortunately declined by Maafu, and the young man had thus learnt nothing except what he had been able to pick up in the missionary schools of the islands.

Through a fine grove of cocoa-nut palms and breadfruit trees, Mr. Fletcher kindly conducted us to his house, a commodious building, thatched with leaves, surrounded by a fence and a broad boarded verandah, the front of the house looking into a nice little flower-garden, the back into the courtyard. The ladies gave us a hearty welcome, no doubt being glad to look once more upon white faces and hear accounts from home. We had brought, besides provisions and stores for the next year, batches of letters and newspapers; and those who have been in out-of-the-way places, and obtained after long intervals news from home, will be able to enter into the joy that prevailed. After being cramped on board a vessel for so many weeks, and tossed and rocked about night and day, it was a rare pleasure to us to sit down once more in a comfortable house on shore; and comfortable the house certainly was. Though the thermometer ranged more than 80° Fahrenheit, the thick thatch kept off the scorching rays, and there was a fresh current of trade-wind blowing through the rooms. It was a pleasing sight to see everything so scrupulously neat and clean, the beds and curtains as white as snow, and everywhere the greatest order prevailing. There were all the elements of future civilization, models ready for imitation. The yard was well stocked with ducks and fowls, pigs and goats, the garden replete with flowers, roses in full bloom, but alas! with little scent, cotton shrubs twelve feet high, and bearing leaves, flowers, and fruit, in all stages of development. These missionary stations are fulfilling all the objects of convents in their best days. When all around was barbarism, strife, and ignorance, they afforded a safe refuge to the weary tra-veller,—as they still do in the East,—and cultivated science and religion at a time when scarcely any one thought of them. When you have reached a convent in the East, or a mission-station in the South Sea, you seem to be nearer home. You feel that you are amongst people whose sympathies incline into the same direction as your own, the mode of living also beginning to tell upon your animal spirits, and you fly to the library, limited though it may be, to have an hour with the great minds of civilization.

Our stay at Lakeba being restricted to a few hours, I made all possible haste to collect specimens of the vegetation. Quite a troop of boys followed, carrying baskets which they made in an incredibly short space of time, out of the leaves of the cocoa-nut palm. Determined to collect everything we could lay hands on, we accumulated about fifty different species, forming quite a load for our young attendants. The true secret of making comprehensive collections, whether of objects of any kind or details of information, is to secure them if possible the first time on coming in contact with them. One has it always in his power to reject what is worthless. To go on the principle that you may come to a place where you can get them better, is an unsound one to adopt, and one that often leads to mortification. Not only do the eye and ear get accustomed to the objects or facts of search, and the hand neglects to secure them, because they no longer strike us as new, but it often happens that they are extremely local, and are never met with again. When I take up my abode in a district, for the purpose of exploring it botanically for instance, I begin by gathering the plants that grow around my abode, instead of rushing at once to distant parts, where no doubt fine treasures may be expected. The first day I shall probably not get any plants save the most common weeds, and most likely not venture out of sight of head-quarters. But after I have collected the objects with which under any circumstances I must become familiar, and would most likely fancy I had in my collection, because they were so common, I am able on the second and third day to venture a good deal further, and when at last I make more distant excursions, I am at least certain that in bringing home anything, I am not carrying coals to Newcastle or owls to Athens.

The boys were quite indefatigable in assisting me to collect, and telling me the different local names of the plants. A great number of these names I was already acquainted with, having learnt them from the Fijian dictionary, and it did not take many weeks before I was familiar with all the vernacular nomenclature of the most generally diffused organized beings. This feat the natives could never comprehend. They thought it strange that at a time when my whole knowledge of Fijian amounted to little more than yes or no, and a few sentences absolutely forced upon me, I should be able to pronounce the names of almost anything they held up to my admiring gaze. The Lakeban boys also took us to a ravine, where some years ago Dr. Harvey, of Trinity College, Dublin, had collected a fine fern (*Dipteris Horsfieldii*, J. Smith), which has magnificent fan-shaped leaves, when growing in favourable situations, from eight to ten feet high, and four feet across. The plant is found in all parts of Fiji, New Caledonia, and various other islands, and has never been introduced into our gardens, where it would be a great ornament, nor did any of my specimens survive being taken out of their native soil.

Mr. Fletcher showed us over the town, famous as the first spot in Fiji where Christianity was triumphant and a printing-press established. The church, constructed in native fashion, is a fine substantial building, capable of holding about two hundred and fifty people. On the open place before it was spread out one of the largest pieces of native bark-cloth I have ever seen, being about one hundred feet long and twenty feet wide. This was the only cloth worn before the recent introduction of cotton fabrics. Considering that it was manufactured without the aid of any machinery, simply by peeling the bark of the paper-mulberry, when the tree is scarcely thicker than a little finger, and then soaking and beating the different pieces in such a way that they expand and all join together in one large mass, the piece was well deserving to be examined. But perhaps the most curious fact is that not only did the Fijians, as indeed most Polynesians, know how to make such cloth, but they also printed it in many different colours and patterns, probably exercising the art of printing ages before Guttenberg, Coster, or whoever else may lay claim to its invention in Europe, were dreamt of. Was it of endemic growth, or did the Fijians derive it in some way from China, where it seems to have been practised from time immemorial?

Not far from the church was the tomb of a departed chief, a series of slabs placed perpendicularly and forming a square filled up by mould, over which a kind of shed was erected. A dense grove of iron-wood trees, so much reminding us, by their sombre aspect, of our pines, form an appropriate accompaniment to the place. The wind playing in the branches, caused a wailing melancholy sound, fully impressing me with the idea that even the savages who planted these trees must have had some sparks of poetry in their composition. It is a strange ethnological fact, that most nations surround the tombs of those dear to them with trees belonging to the pine tribe, or at least trees partaking, as the iron-wood does, of their physiognomy. The Greeks and Turks think the cypress a befitting expression of their grief; the Chinese, the beautiful *Cupressus funebris;* and the Germans and English, the arbor-vitæ and yew. All attempts to convince people that a graveyard ought to have as cheerful a look as such a drear lonely spot can ever be expected to assume have in the long-run proved a failure. Ivy-clad church walls, mossy tombstones, and sombre-looking yews, are in better keeping with it than gay flower-beds or bright tinsel.

The mission-station at Lakeba is close to a great swamp, and cannot be very healthy. Many more salubrious spots might doubtless be found, but the missionary, in order to do the greatest amount of good, should live amongst his flock, and avoid every kind of isolation. He should mix with them as freely as he possibly can, and, on the principle that example is better than precept, exhibit as much of his daily family life as is compatible with necessary privacy. From that point of view, the place has been well chosen; but it is certainly a great deal to expect from an ill-paid missionary, to expatriate himself, and take up his abode in such localities as these. I felt the greatness of the sacrifice expected, on seeing here the widow of a poor fellow who had died only a short time before our arrival. Though the climate of Fiji cannot be termed unhealthy, the Wesleyans have lost a good number of their labourers in this field. In some measure this calamity may be accounted for by their having selected men physically unfit to embark in such an enterprise. Excessive zeal should not be the only qualification. To expect from the Great Giver and Preserver of life, that it would please Him to grant a body constitutionally unqualified for the trying climate of the tropics perfect health and long life, would be a miracle, outside religious circles regarded as little short of impiety. Nor from an economical point of view would it seem wise to go to the expense of sending out men, whose lives, on their being transferred to the tropics, would in all human probability not be worth five years' purchase.

On departing, our kind friends loaded us with fresh vegetables, yams, taro, and plantains, branches of Chinese bananas, heaps of cocoa-nuts, lemons, eggs, and bottles full of milk,—highly acceptable presents after nearly a month at sea. Mrs. Harrison, who had been sea-sick almost the whole voyage, seemed quite to recover at the very sight of them, and the pleasure they caused on board much reminded me of the foraging parties we used to have amongst the Eskimos, Kamtchadales, and American Indians, in days gone by, when, sick and tired of salt beef and pork, we would willingly part with any article of barter we happened to have about us, in order to obtain fresh provisions.

It was a fortunate forethought on the part of our Lakeban friends to provide us in this way, for our voyage to the next station, Wairiki, situated on the north-western shores of Taviuni, was to be rather a long one, a misfortune which we did not fail to attribute to our starting on a Friday, though the captain again protested. We soon made Vuna Point, the southern extremity of Taviuni, but there were so baffled by variable winds and dead calms, that it was deemed prudent to stand off and on, to keep clear of the reefs, which render the navigation of this, as well as most parts of the Fijian group a matter of some caution. It was not until Tuesday, the 22nd of May, more than a week after our departure from Lakeba, that we entered the Strait of Somosomo, and cast anchor off Wairiki, native town and mission-station. In a general map of the world the Viti group looks an insignificant speck, and one might fancy that a boat would quickly pass from island to island. But how one is deceived! The narrow channels widen into broad seas, in which the largest vessels, under proper guidance, have ample sea-room; the little islands expand into small continents, inhabited by untold thousands of human beings, covered with mountains often four thousand feet high, and traversed by rivers that may be followed for days without reaching their source.

CHAPTER II.

ISLAND OF TAVIUNI.—THE KING OF CAKAUDROVE.—ELEPHANTIASIS.— KIND OFFER OF MR. WATERHOUSE AND CAPTAIN WILSON.—SOMOSOMO, ITS ADVANTAGES AND DISADVANTAGES.—QUEEN ELEANOR.—ASCENT OF SUMMIT OF TAVIUNI.—A ROYAL ESCORT.—SYLVAN SCENE.—ARRIVAL AT THE TOP.—SINGULAR SWAMP OF VEGETABLE TURTLE FAT.—DINNER.— TIMIDITY OF THE NATIVES.—CHIEF GOLEA'S RETURN FROM A MILITARY EXPEDITION.—POLYGAMY.—THE ROTUMA-MEN.—WAIRIKI.—ARRIVAL OF THE 'PAUL JONES.'

THE island off which we were now anchored is properly called Taviuni, erroneously Vuna by Wilkes and the latest Admiralty charts. It is the third island in size of the Vitian group, being about twenty-four miles long and nine broad, running from south-west to north-east, and being traversed by a chain of mountains about two thousand feet high, the tops of which are nearly always enveloped in clouds. Stately cocoa-nut palms gird the beach, whilst the mountain-sides are covered by dense forests full of fine timber, and abounding in wild pigeons and the Kula, a species of paroquet (*Coriphilus solitarius*, Latham), valued on account of its scarlet feathers, by the Tonguese, and still more by the Samoans, for ornamenting mats. Numerous streams and mountain-torrents, fed principally by a lake at the summit, descend in every direction and greatly add to the beauty of the scenery. The northern shores especially, forming in conjunction with the opposite island of Vanua Levu the Straits of Somosomo, teem with vegetation, and present a picture of extreme fertility. The trees and bushes are very thick, and everywhere overgrown by white, blue, and pink convolvulus and other creepers, often entwined in graceful festoons. Here and there the eye descries cleared patches of cultivation, or low brushwood, overtopped by the feathery crowns of magnificent tree-ferns; villages nestling among them. The air is laden with moisture, and there is scarcely a day without a shower of rain. The north-western side of the island being moreover, from its geographical position, deprived of the direct action of the trade wind, the temperature feels warm when in other parts of the group it is comparatively cool. In consequence of this, few whites have taken up their residence in Taviuni, and the missionaries were about removing to Waikava, on Vanua Levu, nearly opposite Wairiki, where their houses would have the benefit of the trade wind and the sea breezes. Not mere fancy made them leave Wairiki. Their health was giving way, and their poor children suffered severely from a disease of the eyes. Besides, Taviuni is now thinly inhabited in comparison to formerly. The towns of Vuna, Somosomo, Weilangi, Wainikeli, and Bouma have only a small population. From Wilkès's description, for instance, I expected to find Somosomo, in 1840, the capital of the island as well as the kingdom of Cakaudrove, a large place, instead of a mere collection of ten houses, with neither heathen temple, Christian church, nor respectable strangers' house. The King of Cakaudrove, whose official title is Tui Cakau, had removed his court from Somosomo to Wairiki, and left the government of Somosomo to his younger brother, Golea. Tui Cakau is a miserable-looking man, without any chief-like attributes. He is below the middle height, —in the eyes of Fijians, who entertain a great contempt for little men, a serious blemish; suffering, besides, from elephantiasis and cutaneous diseases, his whole appearance is not prepossessing. Elephantiasis, incidentally mentioned, is one of the diseases to which Fijians are subject, and a fearful sight it certainly is, when the feet assume dimensions and shapes that make them more like those of elephants than human beings. The disease, however, is generally speaking, very local, and seems to be particularly prevalent in low, damp valleys. I remember going up a small river opposite the island of Naigani, where almost every inhabitant was afflicted by this calamity. Again, I have seen large bodies of natives, without noticing a single case. I have not heard of any white settlers having suffered from elephantiasis in Fiji, though it is well known that the whites in Samoa, Tahiti, or other Polynesian groups, are not free from this visitation. No one knowing the cause of the disease, there are of course many hypotheses respecting it. Every white man has his own, and one pretty generally diffused is, that it is brought on by drinking cocoa-nut milk. Yet there was a European who, acting on this belief, and scrupulously avoiding the tempting beverage, nevertheless became a victim, and had instantly to leave for colder climes, the only known remedy for checking its progress.

Mr. Joseph Waterhouse, the chairman of the Fijian district of Wesleyan Mission, kindly asked me to take up my residence at his house during my stay in Taviuni; but, as both himself and Mr. Carey, his coadjutor, were about to proceed to the annual meeting of their brethren in Bau, I declined the offer, and accepted instead that of Captain Wilson, my fellow-voyager from Australia. Mr. William Coxon, the captain's nephew, and manager of the cocoa-nut oil establishment which Captain Wilson and M. Jaubert, of Sydney, had a few years ago planted at Somosomo, came in his boat to fetch us, bringing with him several Rotuma natives, who had been employed in the establishment, and were willing to work their passage in the 'John Wesley' to Sydney, thence to watch for a vessel to their island home.

The distance from Wairiki to Somosomo is only six miles, and a fine breeze soon brought us there. The water off the latter place is shallow, leaving a large flat of rocks at ebb-tide. Captain Wilson warned me not to expect any but the roughest accommodation, as no proper dwelling-house had as yet been erected. I was quite contented with what I found; two sheds, one containing a hydraulic press for making oil, a large house for drying the cocoa-nuts, which also served for drying my plants, and a small dwelling-house, all built in native fashion, and thatched with the leaves of the sugar-cane. A grove of stately cocoa-nut palms diffused an agreeable shade over the place, and trees laden with bread-fruit, lemons, and oranges were dotted about. Almost immediately behind the house rose a small hill of rich vegetable mould, covered with beautiful tree-ferns, over which different kinds of convolvulus—blue, white and purple—were hanging in natural garlands. Following the gravelly beach for about a hundred yards on either side of the premises, one would come to a mountain stream, splashing, foaming, and murmuring in its rocky bed, and offering capital accommodation for bathing.* The ground, for some miles distant gently rising, passes abruptly into steeper mountains. There was little cleared land, though the soil is fertile, and there being few paths the woods were difficult to penetrate.

Fortunately a person need not be on the look-out for wild beasts,—there are none to molest him. Snakes, about four feet long, and of a light-brown colour, frequenting trees, especially cocoa-nut palms, to feed upon the insects attracted by the flowers, are the only animals that now and then startle him. Perhaps another source of annoyance in this earthly paradise, are the myriads of flies that follow one in the woods, and keep him constantly employed; but as a set-off against this must be put the good behaviour of the mosquitoes, which are neither very numerous nor keep late hours, but leave at dusk, and do not appear again till after breakfast. Somosomo has, besides, the reputation of producing dysentery, which the natives, in the belief that it was un-

---

* Here a spiny fresh-water shell I discovered abounds, called, in honour of Mr. Consul Pritchard, *Neritina Pritchardii*, Dohr., by one of our rising conchologists.

known before the visits of white men, term "the white man's disease." However, none of us were attacked by it during our stay, though we were constantly exposed to sun and rain, and ultimately out of biscuit, which served us for bread. The natives also believe dysentery catching, and hence will carefully avoid contact with a person suffering from that infliction. They will never sit down on a seat or lie down on a mat one of these invalids has occupied, and moreover often compel the poor sufferers to retire into the depths of the forests until they shall have recovered. Curiously enough, those Polynesian islands free from dysentery, as, for instance, the Samoan group, are visited by fever, and those free from fever, as Fiji and others, are liable to dysentery.*

Chief Golea was absent on a fighting expedition to Vanua Levu, but his wife Eleanor was at home, and paid us a visit on our arrival, accompanied by two young women, also wives of Golea. Eleanor is the niece of Cakobau (= Thakombau), King of Fiji and Chief of Bau. She is much higher in rank than her husband, who is only a younger son of a king under the suzerainty of her uncle. Bau has always understood how to

---

* The early stages of dysentery are easily checked by eating basinfuls of the native arrowroot (*Tacca pinnatifida* and *sativa*) so plentiful about Fiji, especially on the sandy beaches, and by avoiding bananas and plantains, which I quite agree with Rumphius and Forster in considering as helping to bring on this disease. The arrowroot should be made so thick that a spoon will stand upright in it, and taken with a little nutmeg, and if possible white sugar. I found no arrowroot to be so effective as that of the South Sea, and when, after my return from Fiji, I had a serious attack of dysentery in London, and was unable to get my favourite remedy, no shop having it genuine, I had an illness of several months, which nearly proved fatal.

guard against the centrifugal tendency of Fiji and preserve its political superiority; and giving Bauan women of rank to petty chiefs has been one of the means employed. A queen thus married would still hold the same position she did before marriage, and her sons would, as " *vasus*," have great privileges at Bau, and be identified with her prosperity. Eleanor was a tall, fine-looking woman, of much lighter colour than the generality of her countrywomen, a cheerful countenance, and possessed of dignity and self-possession. Considering the scantiness of her dress, this is saying very much in her praise. Though her husband and most of his other wives were still heathens, she was a Christian, and I believe a sincere one, judging from the almost frantic manner in which she endeavoured to obtain a Fijian Bible seen in my possession. She exhausted every argument to get it, and her joy was indescribable when her wishes were acceded to. It was much increased by the volume being the Viwa edition, which is preferred to the London, not only because it is a larger book and printed in the islands, but also because in the recent London edition some changes have been introduced of which the natives do not approve. The Fijians are fond of books, especially large ones, even if written in languages not understood by them. Some of the whites maintain that this is simply because they use them as cartridge paper, but I do not believe this to be generally the case. I had several good offers for Endlicher's 'Genera Plantarum,' and other large well-bound volumes, though never any for the bales of botanical drying-paper I carried about with me. Eleanor, notwithstanding her high rank, did not seem to exempt herself from any of the duties devolving upon Fijian women. I often saw her go fishing on the reef, and being up to her waist in water. One night, when all was silent, and we were sitting in the house reading and writing, we heard her call loudly for help, and on rushing down to the beach, we found that she and two other women had caught a large turtle in their net, and were holding on to the splashing animal with all their might, until assistance could be obtained.

On the 30th of May, we ascended for the first time the summit of Somosomo; Captain Wilson, Mr. Coxon, and several men kindly sent from the mission at Wairiki, accompanied us, carrying baskets, for making collections. The Queen of Somosomo, hearing of our intention, joined the expedition with her whole court. At daybreak we found her train waiting for us, on the banks of a river, all fully equipped for the occasion. A few strokes of the pen will describe their dress. The Queen wore two yards of white calico around her loins, fern-leaves around her head, the purple blossom of the Chinese rose in a hole pierced through one of her ears, and a bracelet made of a shell. No other garment graced her stately person, and yet she looked truly majestic. Her attendants dispensed with the calico altogether, and were simply attired in portions of banana and cocoa-nut leaves fresh from the bush, which was so far convenient to them as they were ordered to push ahead, make a road, and shake the dew and rain from the branches obstructing the way. In our European clothes, we stood no chance in keeping up with them.

They were always a long distance ahead, waiting for our coming up, and enjoying themselves in opening cocoa-nuts, and smoking cigarettes, made with dry banana leaves instead of paper.

The ascent was rather steep, and Mr. Storck had the misfortune to hurt himself rather seriously from falling down a considerable precipice, just when in the act of gathering some botanical specimens. The road was very bad, the forest being so thick that no glimpse of the sun could fall upon a soil saturated with excessive moisture. Large trees and abundant underwood of small palms and tree-ferns produced a solemn gloom, and made us long for a look at the sky. Wild pigeons of a brown colour, and in very good condition for eating, there abounded, and a number were brought down by our guns. As we were pushing on, collecting all that came in our way, and now jumping over rivulets, now climbing over rocks, we suddenly arrived at an open space, exhibiting a beautiful view of the whole Straits of Somosomo. The eye passing over a dense belt of forest, espied the islands of Rabi, Kioa, and Vanua Levu, the reefs showing very plainly by the surf breaking upon them, whitish fleeting clouds occasionally passing between us and this fine panorama.

The women had kindled a fire, and thought it a good place to take refreshment. The Queen was seated on the top of a rock, the maids of honour grouped around her. It was a pretty sight. The dark beauties, the really artistic effect of their ornamental leaves and flowers, the easy grace of their movements, made them look like so many nymphs that one reads of in classic story, but never seems to meet with nowadays. As we were taking our luncheon, the Queen asked numerous questions about our system of monogamy. For her part, she could never bring herself really to esteem a man contented with one wife, and she was glad her husband was a polygamist. Of course we tried to convince her of our way of looking upon the subject, but, having fairly refuted our assumption that women do not like to see their husband's affection distributed over a whole harem, she almost got the best of the argument.

After another hour's scramble we reached the summit, and found it to all appearance a large extinct crater filled with water, and on the north-eastern part covered with a vegetable mass, so much resembling in colour and appearance the green fat of the turtle, as to have given rise to the popular belief that the fat of all the turtles eaten in Fiji is transported hither by supernatural agency, which is the reason why on the morning after a turtle-feast the natives always feel very hungry. This jelly-like mass is several feet thick, and entirely composed of some microscopic cryptogams, which, from specimens I submitted to the Rev. M. J. Berkeley, a weighty authority in these matters, proved to be *Hoomospora transversalis* of Brebisson, and the representative of quite a new genus, named *Hoomonema fluitans*, Berkl. A tall species of sedge was growing among them, and gave some degree of consistency to the singular body. We were not aware until it was too late that these strange productions were only floating on the top of the lake and forming a kind of crust, or else we should not have ventured upon it. On the contrary, we took it to be part of a swamp, that might safely be crossed, though not without difficulty, for we were always up to our knees, often to our hips, in this jelly. All this caused a great deal of merriment. A little hunchback, who carried a basket swinging on a stick, looked most ludicrous in his endeavours to keep pace with us. Now and then, when one or the other was trying to save himself from sinking into inextricable positions, he had to crawl like a reptile, and the others were not slow to laugh at his expense. The first symptoms of danger were several large fissures which occurred in the crust we were wading through. The water in them was perfectly clear, and a line of many yards let down reached no bottom. These fissures became more and more numerous as we advanced, until the vegetable mass abruptly terminated in a lake of limpid water full of eels. The border was rather more solid than the mass left behind, and all sat down to rest, from the great exertion it had required to drag ourselves for more than a mile and a half through one of the worst swamps I ever crossed. As it was getting quite a fashionable hour for dinner, and our appetite was becoming more keen every minute, we determined not to postpone it any longer; cold yams, taros, and fowls, washed down with a bottle of Australian wine mixed with water from the lake, constituted our meal.

The sides of the lake were covered with scarlet myrtles and a fine feathery palm (*Kentia exorrhiza*, Herm. Wendl.) closely allied to those of New Zealand and Norfolk Island, but different. There were, besides, many other plants, too numerous to be enumerated here, that yielded a rich harvest. I should have liked to tarry much longer than I did, but the natives became desirous of returning, and as the sun was gradually declining, there was no retaining them. Our company dwindled down to a few faithful attendants, and even these were speedily reduced to one, Ambrose, a native teacher, and a man deservedly valued by the missionaries. Having to be in the forest late in the evening is to the Fijians something terrible. They see ghosts and evil-intentioned spirits start up in every direction, and to escape falling victims to their anger, they yell and shout at the top of their voice, like children when left in the dark at night. We regained Somosomo, dreadfully tired and covered all over with mud, but well satisfied with our day's excursion, and it was not long before we were in bed, under two blankets, which in June and July are never found too warm in Fiji.

On the 31st of May, Golea, the chief of Somosomo, returned from his fighting expedition. It was a fine scene; six war-canoes with their large triangular sails skimming before the wind, the warriors on board, dancing, shouting, singing, and sounding the conch-shell. Eleanor, accompanied by the whole seraglio of the chief, hastened to the beach, in order to welcome their lord and master by clapping of hands, dancing, and singing. There being no men at home, the little hunchback of Golea's establishment came breathless to our place, begging Mr. Coxon to pull the trigger of a pop-gun which was to be fired the moment his highness stepped on shore, but which no one had the courage to touch. Golea, soon after landing, paid us a visit. He was a fine man, about twenty years of age, and more than six feet high, with intelligent features, and as melodious a voice as I ever heard. Like most of his fighting-men, his face was blacked with charcoal obtained from the Qumu-tree (*Acacia Richei*, A. Gray). Over his luxuriant head of hair he wore the *sala*, made of a very fine piece of white native cloth, and looking somewhat like a turban. Around his loins he wore a narrow strip of bark-cloth, done up in the T-bandage fashion. Arms and legs were decorated with bands made of the bleached leaves of the Voivoi, a species of screw-pine; whilst a boar's tooth, nearly circular, was suspended around his neck. Golea, flushed with victory, gave us a rather circumstantial account of his recent exploits, the first I believe he had ever been engaged in on his own account, and, being a young man, he made the most of them. His object had been to punish some district of Vanua Levu for having, three years ago, killed his brother. He had taken nine towns, which he assured us had been a great achievement. Soon afterwards we heard another version of the affair, according to which the inhabitants, not appreciating the idea of being clubbed, had adopted the maxim of running away in order to live to fight another day. This fully accounted for only two killed, one an old woman, the other a child; and malice, as venomous in Fiji as elsewhere, added that even these two had only been knocked down and would probably recover. We may rejoice that no more serious calamities attended Golea's expeditions, which may be said to have closed a long line of murders. Golea's father, Tui Kilakila, in February 1854, was murdered, by the hands of, or, as some assert, at the instigation of, his own son, who then succeeded him to the throne of Cakaudrove. A second brother, to avenge his father's foul murder, committed fratricide, and was in his turn assassinated by the people whom Golea had just returned from punishing.

Golea, on my asking him when he would follow his eldest brother in embracing Christianity, replied that his religion was fighting, and that he did not as yet think of becoming a disciple of the new faith. One of his great objections seemed to be its allowing him only one wife, whilst now he had an extensive harem, to which he continually made new additions. The Wesleyans have invariably refused to admit as members of their society, any professed native Christians who would not give up polygamy. Of course, among Protestants, any sect is at perfect liberty to adhere to whatever rules and regulations it may think fit to impose upon itself, and no words should be lost upon the discussion of it by laymen. But when taking a common-sense view of the case, whether polygamists on becoming Christians should put all save one wife away, it assumes a different aspect, which the Bishop of Natal has done good service in ventilating. To say that discarded wives of a polygamist may find husbands argues nothing; so may fallen women of our own country. According to the *lex loci*, the wives enjoy a legitimate existence before the general adoption of Christianity. By declaring them illegitimate, a serious wrong is inflicted upon them. And why do evil that good may come? These women, suddenly deprived of the consciousness that they are legitimate and respectable, and, without their fault, becoming illegitimate and outcasts, are driven from a home to which they are bound by many ties. Had less objection been offered to polygamy, far greater progress might have been made in christianizing Polynesia and many other parts of the world, where a man is estimated in a great measure by the number of his wives, and it becomes a serious thing to ask him to lower himself in public estimation by putting away all his wives save one. Had or were the broad principle admitted, that a man might remain a polygamist on becoming Christian, but not add to his number, many would have been induced to join the Christian community who, under present circumstances, hung back as long as they possibly could. The whole question has often presented itself; and, in the earlier stages of Christianity, the Church distinctly proclaimed the necessity of admitting polygamists. Of course, as all males born of the newly-converted would at once become Christians, and only be allowed to have one wife, polygamy would die out altogether in one generation. I am persuaded that this is the right view to take of the subject, whatever some theologians may argue to the contrary. When at Bau, the subject of succession to the throne was discussed, and the missionaries were for seeing it descend upon Cakobau's youngest son, because he was the son of his Christian wife, a boy of very tender age; and to fix the stigma of bastardy upon his eldest son, the child of the highest woman of his household, and to whom the king was not married by Christian ritual, yet legitimately united according to

Fijian customs. Were the case tried before any competent tribunal, no doubt it would be given in favour of the eldest son,—a fine manly fellow, who would well deserve the honour he was to be deprived of.

Golea asked for grog,—which the natives term "*Yaqona ni papalagi*," or foreign Kava,—but was told that there was none in the house. He then begged to be supplied with a cup of tea, which was cheerfully given. Some of the Fijians are gradually acquiring a taste for intoxicating drinks, as most other Polynesians have done, and there is not a more painful task than to be obliged to refuse supplying them. However, I do not think that the dark-coloured races of Polynesia, including amongst others the Fijians and New Caledonians, have that intense longing for spirits characteristic of the Hawaiians, Samoans, Tonguese, and other light-coloured races, who are great slaves to it, notwithstanding all that is done to check a habit which helps so materially to decimate them. Yet, whether this difference is merely owing to the fact that the former have not had such unrestricted intercourse with the whites as the latter, or whether sobriety is to them a virtue as easy to exercise as it is to the Spaniards and Italians in comparison to the Teutonic nations, the future alone will show. The lower class of whites are setting them a bad example, and one has often reason to blush for his own race. Whilst I was in the islands the first grog-shops were opened at Levuka, and several others have since been established in Bau, and other parts of the group. What has always surprised me is, that considering the Fijian to be a tropical climate, most of these great drunkards enjoy such a long life. They boast—whether it be true I had no means of testing—that they are often intoxicated two months at a time. One of the oldest white settlers always bought a large cask of spirits whenever he had the chance, and, as he did not know when he should have another, he took the daily precaution to fill up the cask with as much water as he had drunk spirits.

On the 1st of June, one of the Rotuma men, working in the establishment, died. His countrymen seemed to feel his loss very much, as he had been a petty chief among them, and they proceeded to bury him in their own fashion. The body was wrapped up in cloth, and a mound raised about two feet above the ground, large stones being placed all around, and the inside filled up with gravel from the beach. Rotuma is a small island three hundred miles north of this group, and belonging to the Fijian Consulate. Some years ago, the Wesleyans endeavoured to establish a permanent mission there, but, although succeeding in making a few converts, they were forced to abandon the field. The ruling chief, described as a fine young fellow, having made a voyage to Sydney, where he was well received,—even, if report be true, at Government House,—had been persuaded by some whites and a New Zealander, who gained influence over him, that if he wished to preserve the independence of his country he must not admit missionaries, as they proved invariably the harbingers of national annihilation. The Wesleyans therefore received intimation to withdraw their Tongan teachers, and the few native converts returned to their former religion, the principal features of which seem to be a belief in a Supreme Being, and the worship of ancestors. The French have been more successful in the neighbouring island of Fotuna, where the Roman Catholic priests established a flourishing mission.

The Rotuma men can nearly all speak a little English; they are a good-looking people, with as light a skin as the Tonguese, rich black, often curly, hair, worn very long, and regular, frequently Jewish, features. The latter peculiarity has been remarked by all who have visited Rotuma, and amongst the men working on the Somosomo estate there was one who bore the nickname of "Moses," in consequence of his undeniable resemblance to an unadulterated Hebrew. They circumcise, tattoo around the loins, and perforate the left ear, into which they put a gay flower, or the rolled up leaf of the *Dracæna terminalis*. The Rotuma men are a hardworking set, and, if Fiji should become a European colony, their island will be likely to supply a good number of useful hands. I have seen them pull an oar all day long under a broiling tropical sun, or work away at the mill and oil-presses, without ever losing their good temper or complaining. True, in Somosomo they were well fed, and had as much as they liked to eat of yam, pork, or fish. Hardly a day elapsed without a pig being clubbed for their especial benefit. One of them invariably attended to the cooking, not only for the men but also for us. He gloried in the name of Koytoo, and was the youngest and best-looking of the lot, with rich curly hair, and a figure as symmetrically formed as a sculptor could desire to copy. Two yards of blue striped calico was his simple garb. When I first took up my abode under Captain Wilson's hospitable roof, Koytoo could not even be termed a plain cook. He excelled in boiling and roasting yam, and in frying pork in the European fashion, but beyond that his acquirements did not extend. It was I who gave him the benefit of the culinary experience gained during my long travels, by initiating him into the mysteries of making coffee, tea, pancakes (without eggs), fritters, chicken and turtle soup. For a yard of calico the Queen would sell us six fowls in the bush; but here we found how true was the old proverb, "A bird in the hand is worth two in the bush." As will be explained in another place, the Fijian fowls are far from being domesticated; they are to all intents and purposes wild. Now and then they show themselves near the dwellings, to pick up the offal, but as soon as any one makes an attempt to catch them they are off, and the only expedient to get them is by shooting. In the tropics, to eat day after day pork and yam, the usual food of Fiji, is not very tempting, and we therefore endeavoured to introduce some diversity into our mode of living, by obtaining as many fowls as we could. Often and often did Messrs. Storck and Coxon leave their, I cannot say soft, couch at dawn to have a crack at them; but the birds were so cunning that no sooner did they creep near the place whence the crowing proceeded, than they were silent or had decamped. Eggs were but seldom seen. The Fijians consider it babyish to eat them, and cannot be induced to look for them. The turtle-flesh was always sent to us as a present, either from the chief or his head wife, and after I had instructed Koytoo into the mysteries of concocting it into soup, with which neither he nor the Fijians were previously acquainted, the chief would never fail to appear at the very moment the soup was put on our table. In fact there were always boys of his loitering about the kitchen, eagerly watching the moment that it was ready, and then running as fast as they could to inform their chief of the important event.

Koytoo was an expert climber, and thought nothing of ascending a tree to collect some specimens of flower or fruit for me. We often made excursions together, and I have frequently admired the way in which he would walk up the smooth trunk of a tall cocoa-nut palm, in order to knock down a few fruits for refreshing ourselves. Without closely embracing the tree, as we are wont to do in climbing, he actually walked up, his feet and hands just touching the trunk, and his body being far off. He was scarcely seated on the leaves forming the feathery crown of the palm, when down came a number of nuts, all of which he had carefully tapped with his fingers to ascertain by the sound whether they had arrived at that stage of maturity which I preferred for drinking; for there is a great difference in the taste of the cocoa-nut as it advances towards maturity, and for every one of these stages the natives have a distinct term. What is yet still more remarkable, they at once know the stage by merely tapping at the nut with their fingers. As the transition from one stage to another, from insipid to sweet, and very slightly acid, is brought about in a day or so, it requires a well-trained ear to detect the difference, and, though trying very hard, I never could master it. No sooner were the nuts down than Koytoo stood again on *terra firma*, cutting a stick about three feet long and one inch thick, which he placed obliquely in the ground, and used for shelling the nuts. Thus divested of their thick outer fibrous covering, the hard shell of one nut was used as a hammer for knocking a hole in the other, and so nicely was this done, that the hole was hardly larger than a shilling, and scarcely a drop of the milk was spilt. We used to empty a great number of nuts in this state without ever experiencing any bad effects. We who wear clothes ought to have a steady hand, for should any of the milk be spilt, it will, on running over the few remaining fibres of the husk, become astringent, and produce an indelible stain in linen and cotton, having exactly the appearance of iron-mould.

On the 4th of June, I paid a visit to Korovono, on Vanua Levu, Mrs. Waterhouse obligingly lending me the mission boat and crew to take me across the Straits of Somosomo. My object was to examine the Kowrie pines and wild nutmegs of that place. We left Somosomo early in the morning, and reached our destination at three o'clock in the afternoon. Jetro, an old Manila man, who had come to Fiji years ago, and spoke Spanish with some difficulty, met us on the beach, and conducted us to a fine grove of Kowrie pines (*Dammara Vitiensis*, Seem.) shortly to fall a prey to the axe. European sawyers had already cut down a number of the best trees, yet some good specimens were still standing, and

I took exact measurements of them. They were from eighty to a hundred feet high, and, four feet above the base; the largest was eighteen feet in circumference! The Fijian Kowrie, or Dakua, as the natives term it, does not form entire forests by itself, like some of our pines, but grows intermingled with other trees, in Korovono with myrtles and wild nutmegs. These nutmegs are also stately trees, with fine oblong leaves; and their produce, though it will never be able to enter into competition with the cultivated nutmeg of the East Indies, is sufficiently aromatic to be employed for home consumption. One of the men climbed up the highest Kowrie pines by means of a creeper, that hung like a rope from the uppermost branches, and he threw down a good supply of fruit, and also a snake five feet long, which had taken up its abode there.

On returning to the beach we kindled a fire to make a cup of tea, and the natives brought us plenty of cocoa-nuts and bananas. Our camp was pitched under a couple of magnificent Dilo trees (*Calophyllum inophyllum*, Linn.) the thick, glossy, green foliage of which was set off to advantage by the numerous white blossoms with which the tree was crowded. The branches, densely covered with ferns and orchids, were quite overhanging the water; indeed all the beaches of the Strait of Somosomo are characterized by this peculiarity. The vegetation, instead of receding from the sea, as in most parts of the group, is quite bent over the briny fluid. We had intended to stop for the night at Korovono, but at dusk the mosquitoes began to be very troublesome, and, as we had omitted to bring curtains for our protection, sleep would have been out of the question. A council of war being held, it was thought preferable, notwithstanding the wind being dead against us, to beat out of the bay and pull the rest of the way. Leaving without further delay, we passed, about midnight, Kioa, or Owen Island, as it is sometimes called, from having become the property of Mr. Owen, an enterprising Australian gentleman, who endeavoured to form a settlement on it. Mr. Owen was for some time a member of the Victorian Legislature, at Melbourne, where he was often alluded to as "Member for Fiji." Though taking advantage of every slight breeze, we had to be at sea all night and did not reach Somosomo until six o'clock the next morning, and were heartily glad when Koytoo, the Rotuma cook, brought the breakfast, as usual consisting of yams, pork, and coffee.

On the 5th of June, a small island schooner came in belonging to a half-caste, and manned by a crew of the same mixed origin. They brought all the news of the group, and complained bitterly of the missionaries injuring their trade by inducing the natives to contribute cocoa-nut oil towards the support of the Wesleyan Socicty, an article which formerly passed direct into the hands of the small traders. When a native became Christian, he was made to give every three months eight gallons of oil, or thirty-two a year, equal to £4 sterling. Notice was given a few days before the oil was due; and when a trader visited a place he found none but empty casks,—the church had swallowed it all up. This statement, like many others heard in the islands,

I found only partially true; indeed, I have never been in a country where it is more difficult to arrive at real facts than Fiji. To say nothing about those who make it a point to diffuse absolute untruths, nearly everybody seems to rejoice in overstating a case or giving a most partial version of it; and it requires no slight discrimination to keep on good terms with those with whom one wishes to stand well, so fearfully rampant is the gossip. The most outrageous stories were unblushingly circulated about the different consuls and missionaries; and sometimes I felt hot and cold, while having to be an unwilling listener to scandal of this description. People in civilized countries do not know how much they owe to the laws that protect them, at least against the grossest libels. Talk of village scandal, it is nothing to it. Of course, in a society of whites so limited, this state of affairs might be expected, but a new feature in the history of gossip is that all the tittle-tattle of the other groups of the Pacific was dealt out as so many delicious morsels in Fiji. The doings of known personages in Tahiti, Samoa, and Tonga were discussed with avidity. Fancy, we in Europe troubling ourselves with the small talk of places more than a thousand miles distant.

Before the arrival of the British consul, several of these small island schooners carried on a profitable traffic in human beings. They used to go to the large islands, and purchase young women, for whom from five to ten dollars in barter were usually given. These women were sold again to whites in other parts of the group, often for fifty dollars each. Several women were pointed out to me as having been bought in this way to become housekeepers of European settlers, and, as their new lords and masters clothed, fed, and treated them better than their Fijian, they had cheerfully stayed with them. Mr. Pritchard's presence has in a great measure put a stop to these and to several other iniquities, or at all events prevented their being carried on in open daylight; but until the home government shall think fit to lighten the consul's duties, by placing a fast-sailing schooner at his disposal, and allow him some abler assistance than he has hitherto obtained from his clerks, similar shortcomings must be expected.

On the 12th of June I went for a few days to Wairiki. The premises occupied by the mission of that place are very commodious; there are two large dwelling-houses, built about two hundred yards apart, one occupied by Mr. Waterhouse, the other by Mr. Carey. On the second day of my stay there, those two gentlemen returned from Bau, bringing a message from Mr. Pritchard, the British consul, to the effect that Colonel Smythe had as yet not arrived, and that a little schooner should be sent for me, in case I did not reach Ovalau by the 12th instant. Mr. Carey showed me his collection of native curiosities, including a fine set of clubs, spears, bows, and arrows. I also saw here for the first time a fan made of the leaf of a beautiful palm, a tree which had proved quite new to science, and which in honour of Mr. Pritchard, and as a grateful acknowledgment of the invaluable assistance he rendered to me, the name of *Pritchardia pacifica* has been given by Mr. Wendland and myself,—the specific name being justified by its growing in various groups of the Pacific, and Mr. Pritchard's untiring efforts to preserve the peace of that region. Fans made of this palm are used exclusively by the chiefs, and forbidden to be carried by the common people. Should Fiji ever choose a national emblem, the claims of this palm to be regarded as such, should not be overlooked.

Mrs. Waterhouse made me a present of an Orange Cowry, or Bulikula as the natives term it (*Cyprœa aurantium*, Martyn), the first I had seen there. This shell has hitherto been found exclusively in Fiji, where it is confined to the islands and shores of North-west Viti Levu; it is worn as an ornament around the neck by natives of rank. Not many years ago, a couple of these cowries would fetch as much as £50 in Europe, but at present a pair without the least flaw, and of the deepest tint the shell is known to assume, may be bought in London for £6. Hugh Cuming, Esq., the possessor of the largest conchological collection ever brought together, is my authority. This statement will doubtless be received with surprise by the Fijian traders, who ask a much higher price on the spot, and still fancy great profits might be realized, in the European markets. It should however be remembered, that though the Orange Cowry is extremely local in its geographical range, and will consequently always be a rare shell, specimens have found their way to every public museum and every private cabinet of importance long ere this, and the principal demand having thus been met, the price has necessarily declined.

The road from Wairiki to Somosomo leads for several miles along a fine sandy beach, underneath a bower of stately trees, and then branches off inland. I passed magnificent groves of Tahitian chestnuts (*Inocarpus edulis*, Forst.), growing on the banks of rivulets and diffusing a delightful shade and coolness, whilst their grooved trunk and knobby root, always rising above the ground, are conspicuous objects. Although it was now the dry season, nevertheless I was completely drenched by several showers. Indeed there were few fine days during the whole time I was staying in Taviuni, and I may as well add that 1860 was as unusually wet in Fiji as that year proved in Europe and other countries. The land between Wairiki and Somosomo does not appear to be very rich, the soil being rather stony; the extreme luxuriance of the vegetation must therefore principally be ascribed to the great quantity of rain that falls almost throughout the year.

One day, Messrs. Storck and Coxon made a large kite, to the great amusement and entertainment of the Fijians, who, chief and all, turned out to see it. They called it a "*manumanu*" (bird), and had never beheld such a thing before; our Rotuma men, however, said they knew it, and in their island often made it of Ivi (*Inocarpus*) leaves. Great was the joy when the "*postilions*" reached their destination, and, as there was a fine breeze, the trick was always successful. So much were they gratified that they came for several days in succession to beg that the kite might be brought out, till at last the toy got such a bore that the makers were obliged to destroy it.

In accordance with my request, Mr. Consul Pritchard

sent, on the 19th of June, the 'Paul Jones,' a schooner of nine tons,—built in the islands by Mr. Jones, an Englishman formerly residing at Levuka,—and entirely of native woods, Dilo (*Calophyllum inophyllum*, Linn.) and Vaivai (*Serianthes Vitiensis*, A. Gray), with masts of Fijian Kowrie-pine. The crew were all half-castes, mostly sons of Englishmen who had taken up their residence in Fiji. They could speak English more or less fluently, having had some instruction at the different missionary schools. The late Mr. Hunt, one of the most distinguished champions of Christianity in these parts, seemed to have taken considerable interest in their education, and they always spoke in the highest terms of him. It was amusing to hear some of their English. In Fijian, B, N, and G, are combinations of two distinct consonants, sounding like Mb, Nd, and Ng. Joe, our cook, a very good-natured fellow, had the greatest difficulty in steering clear of these letters. In spite of all our pains, he would insist in telling us that the "yams were quite *ndone*," and that "*mbreakfast* was ready."

The captain of the 'Paul Jones' brought a letter from the consul informing me that Colonel Smythe had not yet arrived, and advising me to hasten my departure from Somosomo if I wished to take advantage of an excursion he had arranged to the dominions of Kuruduadua, a powerful heathen chief, hitherto inaccessible to all missionary influence, and residing on the large island of Viti Levu. My mind was at once made up. In a few hours, all my baggage was packed, and embarked.

During my stay at Somosomo, many of my things had

been left in an open shed, and in boxes that could not be locked every time they had to be opened; yet I did not lose a single article, though the hatchets, knives, and cotton prints must have been invaluable in the eyes of the natives. On the whole, the Fijians confirm Captain Cook's opinion, according to which the light-coloured Polynesians have thievish propensities, the dark-coloured not. The Tannese, a dark-coloured race, he must either have looked upon as an exception to his rule, or else they must not have been in those days the set of expert thieves they are at present.

The extreme fertility of the soil about Somosomo induced me to establish there an experimental cotton plantation; and before fairly embarking on board the 'Paul Jones' for Ovalau, I must insert a short chapter on cotton, which those who think it a subject no amount of literary skill can make attractive, may skip without losing the thread of the general narrative.

CHAPTER III.

FIJI AS A COTTON-GROWING COUNTRY.—COTTON NOT INDIGENOUS BUT NATURALIZED.—NATIVE NAMES.—NUMBER OF SPECIES.—AVERAGE PRODUCE OF THE WILD COTTON.—EXCELLENCE OF FIJIAN COTTON ACKNOWLEDGED AT MANCHESTER.—EFFORTS OF BRITISH CONSUL AND MISSIONARIES TO EXTEND ITS CULTIVATION.—THE FIRST THOUSAND POUNDS OF COTTON SENT HOME.—ESTABLISHMENT OF A PLANTATION AT SOMOSOMO, WAKAYA AND NUKUMOTO.—PROSPECTS OF COTTON-GROWING IN FIJI.

COTTON was one of the subjects to which attention was principally directed by my instructions; and I have endeavoured to collect every information which might prove useful in forming a correct estimate of the Fijis as a cotton-growing country. If I understand the nature and requirements of cotton aright, the Fijis seem to be as if made for it. In the whole group there is scarcely a rod of ground that might not be cultivated, or has not at one time or other produced a crop of some kind, the soil being of an average amount of fertility, and in some parts rich in the extreme. Cotton requires a gently undulated surface, slopes of hills rather than flat land. The whole country, the deltas of the great rivers excepted, is a succession of hills and dales, covered on the weather-side with a luxuriant herbage or dense forest; on the lee-side with grass and isolated screw-pines, more immediately available for planting.

Cotton wants sea-air. What country would answer this requirement better than a group of more than two hundred islands surrounded by the ocean as a convenient highway to even small boats and canoes, since the unchecked force of the winds and waves is broken by the natural breakwater presented by the reefs which nearly encircle the whole? Cotton requires, further, to be fanned by gentle breezes when growing, and a comparatively low temperature; there is scarcely ever a calm, either the north-east or the south-east trade-wind blowing over the islands keeps up a constant current, and the thermometer for months vacillates between 62° and 80° Fahrenheit, and never rises to the height attained in some parts of tropical Asia, Africa, or America. In fine, every condition required to favour the growth of this important production seems to be provided, and it is hardly possible to add anything more in order to impress those best qualified to judge with a better idea of Fiji as a first-rate cotton-growing country.

Cotton is not indigenous in any part of the group. Independent of its introduction being alluded to in various works as having taken place in the early part of this century, there is no proper vernacular name for it. In all such cases, the Fijian language borrows that of an indigenous plant resembling the introduced one as closely as possible; thus the Cassava root received the name of "Yabia ni papalagi" (*i.e.* foreign arrowroot), the bird's-eye pepper that of "Boro ni papalagi" (*i.e.* foreign nightshade), and the pine-apple that of "Balawa ni papalagi" (*i.e.* foreign screw-pine). By the same rule, cotton became known as "Vauvau ni papalagi" (*i. e.* foreign Vauvau), from its close resemblance to the Bele, or Vauvau (*Hibiscus* [*Abelmoschus*] *Manihot*, Linn.), a cultivated species, the leaves of which are eaten as a potherb. It is true that when foreigners speak about "Vauvau" the natives of the coast know cotton is meant, but in districts where cotton has not yet penetrated, as for instance at Namosi, Viti Levu, one is sure to get the edible *Hibiscus*, if Vauvau, without adding "ni papalagi" (foreign), be asked for.*

Yet, notwithstanding cotton being undoubtedly an introduced plant, and although until lately no attention whatever was paid to its cultivation, it has spread over all the littoral parts of Fiji, and become in some localities perfectly naturalized. Six different kinds have come to my knowledge, all of which are shrubby, and produce flower and fruit throughout the whole year, though the greater number of pods arrive at maturity during the dry season, from June to September. There are two kinds of kidney-cotton, one (*Gossypium Peruvianum*, Cav.) having naked, the other (*Gossypium sp. nov.?*) mossy seeds. A third kind (*Gossypium Barbadense*, Linn.) has disconnected naked seeds; a fourth (*Gossypium arboreum*, Linn.) has disconnected seeds covered with a greenish moss and long staple; a fifth is probably an inferior variety of the preceding one, and only differs from it in the length of the staple; and a sixth (*Gossypium religiosum*, Linn.), being the Nankin cotton, valuable only in certain foreign markets. The four first-men-

---

* In Tahiti *Gossypium Barbadense* is known as "Vavau," a name evidently identical with the Fijian "Vauvau." Nankin cotton (*G. religiosum*) was found wild in Tahiti by Forster.

tioned kinds, especially *Gossypium Peruvianum* and *Gossypium arboreum*, are the most frequent in the group; the fifth seems confined to Laselase, some miles from Namosi; and the sixth (Nankin) has been met with on Kadavu by Mr. Pritchard, and on the Rakiraki coast by Colonel Smythe.

There is scarcely any difference in the look of the four first-mentioned kinds which a person not botanically trained could readily detect. Left to themselves, and never subjected to the pruning knife, these cotton shrubs become as high as a tall man can reach, and each shrub spreads over a surface of about fourteen feet square. I have had no opportunity of counting the number of pods produced throughout the year by a single specimen, but that found in July was on the average seven hundred per plant. Twenty pods of cleaned cotton weighed 1 oz.; thus each plant would yield 2 lbs. 3 oz. Allowing fourteen feet square for each plant, an acre would hold 222 plants, yielding at the rate of 2 lbs. 3 oz. per individual plant, 485 lbs. 10 oz. Even fixing the price of sorts, worth more than 1s. at Manchester, as low as 6d. per pound on the spot, an acre would realize £12. 2s. 9¾d. When it is borne in mind that Fijian cotton brings forth ripe fruit without intermission throughout the year, but that this calculation is based *solely* upon the number of pods *found at one time only*, and that the pods were gathered from plants upon which *no attention whatever had been bestowed*, the result will be still more striking; double, even treble the above quantity may safely be calculated upon as their annual crop. When it is further remembered that Fijian cotton is not an annual, as it is in the United States, and all other countries, when killed by frost or too low a temperature, and that the *plants will continue to yield for several years* without requiring any other attention than keeping them free from weedy creepers and pruning them periodically, the encouragement held out to cultivators will be pronounced very great.

Until the excellence of Fijian cotton had been acknowledged at Manchester, and the mercantile value of the different sorts been ascertained to be $7d.$ to $7\frac{1}{2}d.$, $8d.$, $9d.$, $11d.$, and even $12d.$ to $12\frac{1}{2}d.$ per pound respectively, no attempt had been made to cultivate the plant. It was almost entirely left to itself, and perhaps only here and there disseminated by the natives, in order to furnish materials for wicks. But when in November, 1859, Mr. Pritchard returned from England to Fiji, with the valuation printed in the Manchester 'Cotton Supply Reporter,' for March, 1859, he induced the most influential chiefs to give orders for planting it; and the Wesleyan missionaries, without any exception, zealously aided in these endeavours by recommending the cultivation, both personally and through the agency of their native teachers. Thus, cotton has been thickly spread over all the Christianized districts, and imparts to them a characteristic feature, occasionally very striking in places having a mixed religious population. In Navua, for instance, that part of the town inhabited by Christians is full of cotton, whilst that inhabited by the heathens destitute of it.

To guard against misconceptions, it must be stated that

cotton has as yet been cultivated by the natives in their peculiar style. Those who would look in the islands for broad square acres covered with any given produce will be seriously disappointed. The Fijian cultivator has such an abundance of good land at his command, and holds such stringent notions about the fallows to be observed, that he selects patches here and there only, which after an annual or biennial occupation, are deserted for others cleared for the purpose. When cotton was recommended to him, he followed his old cherished system, and the isolated patches now beheld are the result. These patches are of various sizes, but I have not seen any containing more than fifty plants. In Namara, and other districts subject to Bau, isolated specimens, often as many as twenty, are met with on the margins of every taro, banana, and yam plantation. On the island occupied by Bau, the Fijian capital, Mr. Storck, my assistant, counted four hundred shrubs, growing in the streets and squares. The number of plants thus dispersed all over Fiji must be considerable, though nobody could venture to give any approximate estimate of them; and their aggregate produce, if attentively collected, would doubtless amount to a quantity scarcely expected from such sources. Mr. Pritchard, in order to open the trade, pledged himself, before leaving England, to his Manchester friends, to forward 1000 lbs. of cleaned cotton within twelve months' time, and he experienced no difficulty in obtaining from Kadavu, Nadroga, and Bau an amount exceeding that promised before the time fixed for its dispatch,—the first ever sent home. Now that a demand has been established, there will be a marked increase in the crops, when the numerous young plants added to the old stock at Mr. Pritchard's investigation begin to produce their harvest.

On leaving England in February, 1860, the Manchester Cotton Supply Association, through their able secretary, Mr. Haywood, furnished me with a large quantity of New Orleans and Sea Island cotton-seeds, together with printed instructions for their cultivation. Distributing a fair share of the seeds and papers amongst white settlers, who, I felt persuaded, would make use of them, I myself was enabled to establish a small cotton plantation on the Somosomo estate of Captain Wilson, and M. Joubert, of Sydney, in the island of Taviuni. None of the seeds of the Sea Island sort possessed any germinating power; but those of the New Orleans cotton were very good, and readily grew. Sown on the 9th of June, they began to yield ripe pods within three months, and I was thus enabled to take home a crop from the very seed I brought out, though my absence from England only amounted to thirteen months altogether. This may truly be termed growing cotton by steam. When I paid a second visit to Somosomo, on the 18th of October, my plants were from four to seven feet high, full of ripe pods and flowers, which in the morning were of a pale yellow, but towards evening turned pink. Koytoo, the Rotuma native, whom I had desired to look after the plantation, said that the field only required weeding once; after that the cotton-plants grew so rapidly that they kept down the weeds, and he had no further trouble.

Simultaneously, Dr. Brower, United States Vice-Consul, had succeeded in raising New Orleans cotton on his estate, in the island of Wakaya, twelve pods of which weighed an ounce; whilst the seeds distributed by me amongst various people had evidently not fallen on barren soil. Of course, my plantation could only be a small one, but nevertheless it proved so far beneficial that it convinced those white settlers who had lately repaired to the group what quick returns cotton would yield, and some of them resolutely set about establishing plantations. The mail brought the news that some of them had as many as fifteen acres planted. Mr. Storck, my assistant, who went from Sydney with me to Fijis, made up his mind to remain behind when I came away, in order to devote his energies to cotton-growing. Mr. Pritchard supplying him with land, he commenced a plantation at Nukumoto, on the island of Viti Levu; and if the experiment should prove remunerative, more land will speedily be brought under cultivation.

The fact that cotton will grow, and will grow well, being established, the success of this and similar attempts will chiefly depend upon the supply of manual labour. Those best acquainted with the condition of the group, and the character of its people, confidently look forward to a steady supply of it. In Rewa, Ovalau, and other districts longest frequented by whites, the natives go round asking for employment. This is quite an innovation, and shows that the Fijian is becoming gradually accustomed to labour for fixed wages; and, when the chiefs shall have either voluntarily relinquished or been compelled to give up their claim to all the property accumulated by the lower classes, a favourable result will be the immediate consequence, and a fresh impulse be imparted to all branches of industry. Let the common people once be assured that nobody can legally take their fair earnings away from them, and that the little comforts with which they have managed to surround themselves may be openly displayed without the danger of being coveted by the chiefs and their favourites, and they will doubtless be eager to engage in any work that does not require any great mechanical skill or violent exertion, and at the same time will yield them reasonable returns.*

* Whilst these sheets were passing through the press, the Fijian contribution to the Great Exhibition of 1862 has arrived, which Mr. Consul Pritchard, in a letter to me, dated Levuka, Fiji, March 12th, 1862, accompanies with explanations, of which the following have an important bearing upon the cotton question:—" The box No. 1 contains eight samples of cotton. Of these samples, No. 1 is New Orleans cotton, from the plantation you established at Somosomo, which since your departure has been sadly neglected; the trees are half withered and overgrown with bush, and I fear the quality has much deteriorated. No. 2 is kidney cotton, grown by Mr. Storck on his plantation at Nukumoto (Rewa River). It was planted in July and gathered in December last. No. 3 is kidney cotton, native-grown at Rewa. No. 4 is native-grown, from Burebasaga (Rewa River). No. 5 is Sea Island cotton, grown on Nukulau, the little island in the Rewa roads, and planted by an Englishman, Mr. Smytherman, in January, and collected in August, 1861." I should here add, that Mr. M'Clintock, nephew of Sir Leopold M'Clintock, sowed some Sea Island cotton at Rewa ; in twenty-four hours it was up, with the first two leaves quite open; in two months and twelve days it was in full blossom, and is now almost ready to gather, not having been planted three months! " No. 7 is from Mr. Eggerström's plantation at Nagara, and was gathered four months after planting. No. 8 is native-grown."

Sea Island cotton delights in sandy soil impregnated with saline particles, and localities wafted by sea-breezes, such as Rewa and Nukulau are. With the high prices now commanded by this kind, and the prospect of continuance of civil wars in the United States, speculators would find it highly remunerative to hire or purchase land about Rewa, or localities similarly situated, for the cultivation of Sea Island cotton.

It is well known, both from public journals and the 'Correspondence relating to the Fiji Islands,' presented by command of her Majesty to both Houses of Parliament, May, 1862, that from samples submitted by Mr. Pritchard, the Executive Committee of the Manchester Cotton Supply Association resolved, " That these samples are of qualities most desirable for British manufacture; that such a range of excellent cotton is scarcely now received from any cotton-growing country; and that the supply obtained from the United States does not realize nearly so high an average value as this Fijian cotton." It must be borne in mind, that these and similar opinions were arrived at in 1859, long before my visit to the islands and the publication of the favourable report I made.* Doubtless the same Committee would now be prepared to pronounce a still higher opinion, if that were possible. The Fijian samples sent to the Great Exhibition of 1862 would furnish capital material for renewed examination, and amongst them would be found some of Sea Island cotton, the sort which, having the largest staple and fetching the highest price, was hitherto exclusively grown in perfection on the coast of South Carolina, Georgia, and a small part of Florida. Fiji has now supplied every sort of cotton, from the cheapest to the very best, and capitalists would do well in directing their attention to it.

* My report was sent by the Colonial Office to Manchester, and first published in No. 71 of the 'Cotton Supply Reporter,' of August 1st, 1861.

CHAPTER IV.

DEPARTURE FROM SOMOSOMO.—ISLAND OF WAKAYA.—THE BALOLO.—ARRIVAL AT LEVUKA.—H.B.M. CONSUL.—THE LATE MR. WILLIAMS.—LADO AND ITS ORIGIN.—SITE FOR THE NEW CAPITAL.—THE KING OF FIJI.—BAU.—CAUSES OF ITS SUPREMACY.—VIWA.

THE 'Paul Jones' had been seven days on her voyage from Port Kinnaird to Somosomo, having had to beat up, but in going back she had a fair though not a very strong wind. We left Somosomo in the afternoon of the 20th of June, and called at Wairiki to wish good-bye to the missionaries, and return them several articles they had kindly lent us. The first night we anchored in a small bay on the southern coast of Vanua Levu, and went on shore the next morning to botanize. The town, built near a great swamp, consists of about forty houses. We had scarcely shown our white faces in the first house when all the little children set up a perfect scream, and nothing their parents said or did could pacify them. If they had seen the "old gentleman" himself *in propriâ personâ*, they could not have been more frightened. The piercing screams brought children of all the other houses out, till the whole formed one great yelling chorus, so terribly grating on our ears that we made all possible haste to escape into the woods. Our excursion produced several plants not previously noticed, and also resulted in the discovery of an entirely new genus of *Rhamnaceæ*, which I have called, in honour of Colonel Smythe, R.A., *Smythea pacifica*.*

Steering in a south-westerly direction, we sighted the island of Koro, or Goro as some charts erroneously term it, where an immense number of yams are grown, and the souls of all the pigs killed in the group are supposed to go. A little further on we passed Wakaya, a small island belonging to Dr. Brower, and the site of a settlement chiefly composed of half-castes, who, besides attending to the sheep and cattle, look after the plantations of sugar, coffee, and cotton the enterprising Doctor has established. The most remarkable fact connected with Wakaya is its being one of the places where the *Balolo*, a curious annelidan, makes its periodical appearance. Of the very existence of this singular animal naturalists knew nothing, until a few years ago Dr. Gray, of the British Museum, described it under the name of *Palolo viridis*, adopting its Samoan and Tonguese vernacular name for the genus; and Dr. Macdonald wrote on its anatomy. The time when the Balolo comes in may be termed the Fijian whitebait season. It is watched for with the greatest anxiety, and predicted with unerring certainty from the phases of the moon. The first of these worm-like creatures floating on the surface of the ocean are seen in October,

* A coloured plate and a full description of this singular genus, closely allied to *Ventilago*, with which it agrees in habit to a remarkable degree, but differing by having a veritable dehiscent capsule, instead of a drupe, has been published in 'Bonplandia,' vol. x. p. 69, tab. 9. Additional particulars will be found in my 'Flora Vitiensis.'

hence termed Vula i Balolo lailai, *i. e.* the little Balolo month. Myriads appear about the latter end of November, generally on the 25th, which from that fact is known as the Vula i Balolo levu, or great Balolo month; and the natives of the coast are particularly busy in catching and forwarding the delicacy of the season to friends residing in places deprived of it,—presents all the more appreciated as a whole year must elapse before the same boon can again be conferred.

In a letter dated Levuka, Fiji, December 6th, 1861, and addressed to her friends, an English lady gives the following account:—" In November we all went for a few days to Wakaya, about ten miles east-north-east from Ovalau, in order to see the Balolos, which rise out of the reefs just before daylight, first in small numbers, but about sunrise in such masses that the sea looks more solid than liquid. As they were to appear on the morning of the 25th, we retired to rest at an early hour the night before, and rose with the moon, about one o'clock in the morning. An hour's pull in the whale boat brought us to the very spot they were to come. We found several natives already collected there in boats and canoes, all anxiously looking out who should get the first. This they discovered by sitting with their hands in the water as the canoe was gently paddled about. Presently there was great shouting,—nets were put out, the excitement commenced. At first our nets did very well, but soon the Balolos became too numerous for them to be of any use, and they were caught by the hand and thrown into the baskets with which the boats were filled. We placed a white handkerchief about four inches below the surface of the water, but the little creatures were so thick above it that it was quite invisible. At first I could not make up my mind to touch them, but seeing every one else doing so, I summoned up all my courage, plunged in my hands, and grasped a goodish number, of which, however, I got rid as quickly as possible. The little slimy things twist round the hand in half a second. They are, of course, perfectly harmless, swim very fast, and the longer ones have sometimes five or six coils in the body. When at the thickest they are all entangled one in another, which gives a very curious appearance, as they are of various colours, green, red, brown, and sometimes white. As the sun gains power they dissolve, and about eight or nine o'clock you scarcely find one. It is always in November they come in such masses, just after the last quartering of the moon, and they rise with the tide. As soon as the natives have gathered all they can, they make fires and ovens to cook them. Small quantities of Balolos are tied up in bread-fruit leaves, and have to lie in the oven from twelve to eighteen hours. When all is cooked, the natives expect a heavy shower of rain, as they say to put out the fires of their ovens. Should there be no rain, a bad yam season is predicted."

Several of the white residents eat Balolo, and a strong-minded English lady assured me it was quite a relish; however, everybody knows the old proverb, "De gustibus," etc., and if in the Samoan, Tongan, Fijian, or New Hebrides group—in all of which the Balolo is found—a dish of this description should be served up, strangers must exercise their own discretion whether

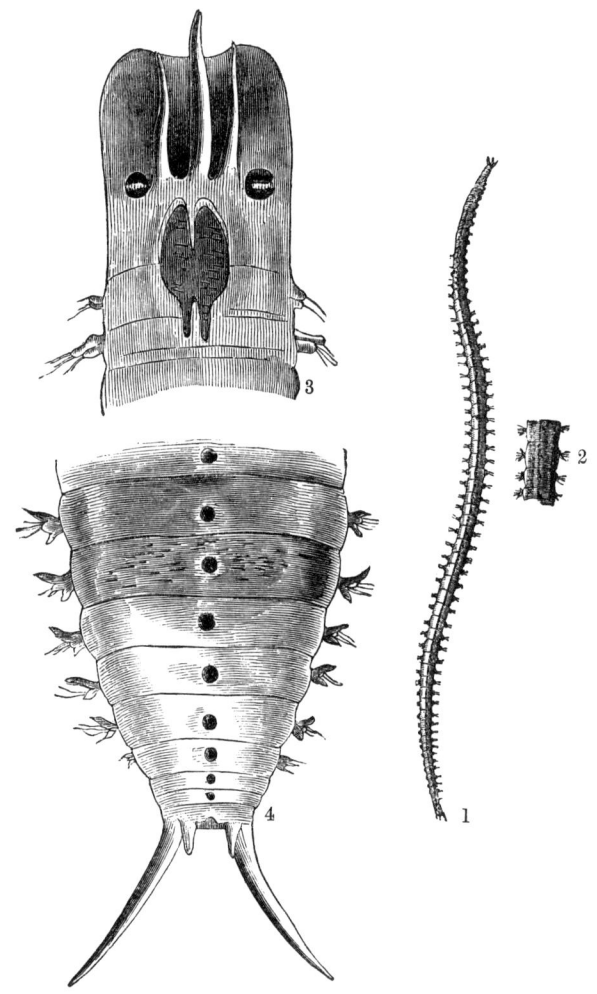

The Balolo (*Palolo viridis*, E. Gray).—Fig. 1. The entire animal, natural size; 2. Portion of the body slightly magnified; 3. Magnified figure of the head, with its three frontal tentacula and eyes; the position of the retracted jaws is shown in the central dark space behind the tentacula; 4. Posterior extremity of the Balolo, dorsal aspect; figures 3 and 4 copied from Macdonald's paper in 'Linnean Transactions,' xxii.

these little, creeping, crawling things, with their cylindrical, jointed body, are a delicacy to be recommended or a nuisance to be avoided.

The most singular portion of the natural history of the Balolo is the regularity of its periodical appearance. About Hanover I have often observed devout Roman Catholics going on the morning of St. John's day to neighbouring sandhills, gathering on the roots of herbs a certain insect (*Coccus Polonica*) looking like drops of blood, and thought by them to be created on purpose to keep alive the remembrance of the foul murder of St. John the Baptist, and only to be met with on the morning of the day set apart for him by the Church. I believe the life of this insect is very ephemeral, but by no means restricted to the 24th of June. But there is an Australian bird (*Psittacus undulatus*) which is known to lay its eggs always on the 17th and 19th of December, and forms another instance of certain actions in the life of an animal being performed, with unerring certainty, on particular days.

On the 22nd, at four P.M., we entered the harbour of Levuka, the principal port of the island of Ovalau. Captain Wilson, who had left Somosomo a few days before me, was standing at the beach, and conducted me to the office of the British Consulate, where I found Mr. William Pritchard, by whom the cession of Fiji to England has been brought about, and to whom I delivered a letter from Earl Russell. Mr. Pritchard is the son of the Rev. George Pritchard, formerly British Consul at Tahiti, at the time when the French, against the wish and will of the natives, assumed the protectorate of that group, treated Queen Pomare with unusual harshness, and the British representative in a manner that nearly brought about a war between France and England. Born in Tahiti, and thoroughly acquainted with the Samoan and most other Polynesian groups, Mr. Pritchard enjoys the peculiar advantage of being perfectly familiar with all native modes of thought. During my stay in Fiji I had frequent opportunities to see how successfully he was able to deal with these islanders, whenever any difficulty arose.

We called together on Mr. Binner, who has for years filled the office of training-master to the Wesleyan mission at Levuka, and also manages the commercial affairs of this religious society in Fiji. We thence went to Dr. Brower, the American Vice-Consul, who received me with great kindness, and whenever I visited Levuka I always took up my quarters under his hospitable roof. Mr. Williams, the American Consul, had died a few days before my arrival. I should have liked to have seen him, in order to form an independent estimate of a man about whom so many contradictory statements were afloat. He did not live on good terms with the missionaries, and controversies were carried on between them in the Australian and American newspapers, which, as is usual in such cases, proved advantageous to neither party. Mr. Williams bought considerable tracts of land, and it was maintained that the purchase was not in all instances a fair one, and that the natives had only from fear of American men-of-war given their assent to these transactions. It is impossible to say whether in all cases the sellers were satisfied with the bargain; yet I remember, quite in the interior of Viti Levu, Chief Kuruduadua publicly declaring at an official meeting that his brother had sold land to Mr. Williams, and that he, regarding the purchase as valid, had no wish to dispute it. This was a great deal from a man like Kuruduadua, who had a violent dislike to Americans, as some of them had burnt Navua, his seaside residence, a few years previously. Towards the natives Mr. Williams appears to have been very kind, and would not refuse them anything. I heard of a bet which a chief made, that he would obtain a waterproof coat just sent out to Mr. Williams, merely by asking for it, and which was won by him who trusted in Mr. Williams's generosity. The whole of the land on which the mission-station at Mataisuva is built, an extensive piece of ground, was presented by Mr. Williams to the Wesleyan body at the very time when some of their members were engaged in the hottest polemical struggle with him.

Dispatching my collections made in the eastern parts of the group by a vessel about to sail for Sydney, I started with Mr. Pritchard, in the consular gig, for Lado alewa, a little rocky islet on the western side of the island of Ovalau, which we reached about sunset, after a sail of about an hour and a half, and which Mr. Pritchard kindly invited me to look upon as my home during my stay in the islands.

Let me tell the history of this rock:—Once upon a time, a god and goddess, who rejoiced in the name of Lado (= Lando) were directed to block up the Moturiki passage leading into Port Kinnaird and the Bau waters, in order to stop the rolling surf from disturbing the nightly repose of the great Fijian deities. They resolutely set about it; but having, in common with other spiritual beings, a decided objection to daylight, they threw the two enormous rocks collected for that purpose in the middle of Port Kinnaird as soon as they began to " smell the morn ;" or, according to another version, their noble selves became changed into rocks, as were the villagers in the Bohemian legend of Hans Heiling,—now bearing the names of Lado alewa, the female Lado; and Lado tagane, the male Lado. The latter version seems to be the most rational,—if reason has anything to do with such things,—for once transformed into stone the two spirits were unable to stir again, whilst, if they had merely thrown down their burden, they might have been made to resume their labours, like Sisyphus of old. However, be that as it may, the fact is, that we were now on the rock identified with the name of the goddess—the larger of the two; and I trust that whatever intentions the Fijian Olympus may formerly have entertained respecting the two Lados in general, and the one we had landed on in particular, they will reconsider the question since the British colours wave on the summit of this islet. The rocky slopes have been transformed into terraces of flowers, and a neat European-built cottage, with broad verandah, and a roof thatched with sugar-cane leaves, contained the archives of the British Consulate. The natives looked upon this house as a perfect marvel of art; the windows, papered rooms, and above all, the staircase,--the first ever made in Fiji,

—proved a source of never-failing curiosity and admiration.

Miss Pritchard made tea in the English fashion, which I thoroughly enjoyed, after being so long compelled to take it from the hands of rude natives. A room was given up to me, and every comfort Fiji afforded was bestowed upon me. To sleep once more in a well-constructed, clean bed, under a good mosquito curtain, is a luxury that only those who have been obliged to forego for some time can fully appreciate. It was high time that I arrived at such quarters, as I began to experience symptoms of dysentery,—a disease which has proved fatal to many new-comers from Europe. However, a judicious supply of Fijian arrowroot, and a few glasses of port-wine, soon restored me to perfect health. Mr. Storck, who had been suffering from his fall and those ulcerations to which most people going to the tropics for the first time are subject, also began to get better after being a few days at Lado, so that both of us had reason to be extremely thankful for the hospitality conferred.

There being no collective name for the waters situated between Moturiki and Ovalau, and sheltered by the Yanuca (= Yanutha) islands, Mr. Pritchard, in honour of the Honourable Arthur Kinnaird, who takes a deep interest in Fiji, termed them Port Kinnaird, and endeavoured to form a settlement on the south-western parts of Ovalau. When I first visited this settlement there were about twenty-five whites, some of whom had cleared a little land; but most of them seemed to belong to that class of immigrants who arrive almost penniless, and are disappointed on not becoming transformed into capitalists on landing. I endeavoured to urge them to begin planting their land with such tropical products as the climate favours, and told them of my little cotton plantation at Somosomo. All hoped to make their fortune when Port Kinnaird should become the capital of Fiji, and their land rise in value.

The question of where the capital of Fiji is going to be on the country becoming a European colony, is a much debated one in the islands. The unfitness of Bau, the native capital, for all commercial purposes, being generally acknowledged, four places have laid claim to that distinction,—Levuka, Ga Loa, Port Kinnaird, and Suva. Levuka has always been a favourite resort of the white population, and has a central position, and a tolerably good though not large harbour, but there is no room for a town. Rocks rise from almost the water's edge, allowing space for only one or two rows of houses, the heat in which is suffocating; and unless a series of works is commenced similar to those which render Valetta a city of terraces, there is no hope of making Levuka more than a trading village. When I finally left it, in November, 1860, there were only few weather-boarded houses, belonging to the consuls and missionaries,—all the rest of the dwellings were large huts built by the natives. The finest house was that of Mr. Binner, beautifully situated on the top of a hill, and commanding a grand view of the reef and its curling surf. Closely adjoining Levuka—as London does Westminster, New York Brooklyn, or Hamburg Altona—is Totoga, a fortified place with thick walls and gateways, where the Roman Catholic missionaries and several French reside. True, this place might be incorporated with Levuka, but it is surrounded by swamps, the drainage of which would be a matter of difficulty to a young community.

Ga Loa, or Black Dusk Bay, on the southern side of Kadavu, is the next place that recommends itself to consideration. Should a steam communication be established from Brisbane, Australia, to Central America, and *viâ* Fiji, Ga Loa would recommend itself as a fit place for steamers to call at; and I have advocated its claims both in the 'Athenæum' and before the Royal Geographical Society of London, and shall speak of it again when describing our movements at Kadavu. But I do not think it well suited for the capital of Fiji. Kadavu, on which it is situated, is one of the southernmost islands, and separated by a sea of more than sixty miles from Viti Levu, the principal island, and by more than one hundred and fifty miles from the centre of Vanua Levu and Taviuni. Small canoes or open boats could not venture thither except in fair weather, and its isolation would always be against its becoming the true metropolis.

Port Kinnaird offers great advantages, independent of its central position. It is a very fine port, perfectly landlocked; and if a portion of Moturiki could be devoted to a site for a town, it would speedily rise in importance,—for Moturiki is probably the finest little island in the group. The entrance to Port Kinnaird is popularly regarded as difficult and impracticable, but a consultation of Captain Denham's survey proves ingress and egress to be easy. Port Kinnaird would doubtless become the future capital if its advantages were not totally eclipsed by Suva in Viti Levu. So convinced has every one capable of forming an opinion become that Suva will be the capital, that the land around the harbour has enormously risen of late; £20 an acre was asked in November, 1860; and £10 I saw actually refused for land a few years previously not worth more than a few pence at the utmost. Not a single house had then been built. The general conviction that Suva must become the capital seems to have been the sole cause of this sudden rise. If one were to write a puff for a land speculator, one would hardly string together a greater number of favourable conditions. There is a good harbour, with mud bottom, deep water right alongside of the shore, sheltered by a reef, and having a wide passage for the largest vessels to beat out. When once inside the passage there is clear sea-room, no outlying shoals or reefs. Suva commands the most extensive agricultural district in Fiji, through which run fine rivers (the Navua and Wai Levu or Rewa) navigable for boats for many miles inland. Suva has besides outside reef communication completely around Viti Levu, with the exception of a few miles on the southern shore and the westward, and continuing to the northward to Vanua Levu, and along the entire southern shore of that island. The convenience of inside reef communication is demonstrated in the case of parties employed in sawing. Logs are purchased at a distance of forty miles from the pits, and floated up by natives at a trifling cost. Were there no reefs, this would be an impossibility. Suva Point is a gently undulated country, free from swamps, and about three miles wide or thereabout at the base. It has on one side Suva Bay, on the other Laucala (= Lauthala) Bay; the latter first surveyed by Sir Edward Belcher,* and offering many conveniences. The point itself is open to the prevailing winds; it is thinly timbered with bread-fruit, cocoa-nut, dawa, and other trees of no great growth, and thus requires but little clearing.

A few days after my arrival at Lado, we were gratified by a visit from Mr. Cæsar Godeffroy, of Hamburg, who had been several years in the South Sea establishing a direct trade with Germany, and planting agencies in the most important groups. Messrs. Godeffroy and Co. are the first great house who have entered this comparatively new field of commercial enterprise, and there is every reason to believe their operations successful. There is a great market in the South Seas, but only those who have an intimate acquaintance with the articles required should ever be tempted to enter it. Even the comparatively few things I took out for barter taught me the value of inquiring most minutely into the exact nature of the articles here current. Knives with white handles were rejected or but slightly esteemed, though their blades were even better than those having black ones, and so with everything else.

Judging from the crowds of boats and canoes daily arriving at Lado—for every one here has either the one

* Rewa Roads are called in the Admiralty Chart Nukulau Harbour; the special chart published embraces the surveys of Sir E. Belcher.

or the other—the sudden disappearance of this Consular establishment would be felt as a serious inconvenience. The British Consul is now the sole authority that keeps order in Fiji—the natives having voluntarily made over to him the entire jurisdiction of the group, and found it preferable in their quarrels with the whites to abide by his judgment, rather than break their own heads and those of the white settlers by an appeal to the club. It was easy for them to arrive at this conclusion; meanwhile, the person who thus found himself called upon to adjust the differences of a native population about twice that of New Zealand, and a thick sprinkling of white immigrants, some of whom hold queer ideas of poetical justice, had no idle time of it; and if Mr. Pritchard had not acquired a thorough mastery over the Polynesian mind by means of his intimate acquaintance with all their customs, usages, and traditions, of which he skilfully avails himself, there would be endless fights and dissensions, to the great detriment of the native population and the interests of commerce. I have repeatedly listened to the proceedings in court, and been struck with the logical acuteness of the natives. Their mind seems indeed of a much superior cast to that of most savages; and their discussions are as much above those of the Maoris reported in the New Zealand newspapers, as the talk of men is to the prattle of children.

On the 28th of June, Cakobau (or Thakombau, as his name may be written according to English orthography), King of Fiji, and supreme Chief of Bau, paid a visit to Lado, and I was formally introduced to him. His Majesty has been described repeatedly as a man of almost gigantic dimensions. But he is only of fair proportions, and does not measure more than six feet in height. I can speak very positively on these points, having often seen him with nothing more than a few yards of native cloth on, as well as in a blue naval uniform. When dressed in uniform, people would scarcely believe that he could be the same man whose powerful build excited their attention. When one day in his company I got quite close to him, in order to take his measure without his becoming aware of the attempt. But his quick eye had detected the studies of comparative anatomy in which I was engaged, and very good-naturedly he offered to stand close to me, when it was found that he was more than two inches shorter than I am, without his shoes and socks, whilst I measure exactly six feet two inches, so that he is after all only six feet high. It is not difficult to reconcile the statements relating to his gigantic stature with what I have ascertained. People not accustomed to move much amongst natives almost in an absolute state of nudity, are generally deceived about the size of the person they see before them. Moreover, the King, previous to his conversion to Christianity, wore a large head of hair, all frizzled and curled in such a way as to stand literally on end, and covered with a piece of white native cloth,—a device which must have greatly added to his height, and induced foreigners to believe him much taller than he really is. He has of late years suffered a little from elephantiasis, but generally enjoys very good health. None of the portraits that have been published do justice to him, and he feels rather annoyed that Europeans should think him as ugly as those representations make him. His face expresses great shrewdness and good-humour; his bearing is very dignified on public occasions; and it was gratifying to see him at church behaving in a manner that no reasonable man could find the slightest fault with.

The Queen of Fiji, to whom Cakobau has been married according to Christian rites ever since he abandoned heathenism, is a rather stout, quiet woman, about five feet two inches in height. I have only seen her once *dressed*, and that at the time of our first official interview about the cession. She then wore a neat bonnet, latest Parisian fashion, a coloured silk dress, and a black mantilla trimmed with lace. I need scarcely add that the use of crinoline was not unknown even in this remote quarter of the globe. The Queen, at the interview alluded to, was rather bashful, owing to a wish expressed by the Consul that she should sit at her husband's side, instead of, as the rules of the country demanded, behind him. However, she comported herself very well indeed, but I daresay was very glad to get her clothes off as soon as the official interview was over.

Cakobau calls himself "Tui Viti," or King of Fiji, and has a perfect right to it. True Fiji is divided into a number of petty states, yet all of them acknowledge vassalage to Bau by paying either a direct tribute to it, or being tributary to states so circumstanced. It is highly probable, however, that at one time all Fijians were under one head, and formed perhaps a more compact nation than they do at present. Of course, I am aware the title "Tui Viti" has been revived only lately; owing, it is stated, to a letter which General Miller, formerly H. B. M. Consul-General at the Hawaiian, or Sandwich Islands, addressed to "Tui Viti," and which Cakobau, as the most powerful chief of the leading state, thought it right to open. But the title "Tui Viti" occurs in many ancient legends current in various groups of Polynesia, and could scarcely have originated with such close neighbours, who would rather be apt to detract than to magnify the power of a foreign nation already far above them in the exercise of various useful arts and manufactures. Old traditions further state the Fijians to have been an unwarlike people, until they had established a more intimate and frequent intercourse with the light-coloured races of the eastern groups, when sanguinary intratribal quarrels became almost their normal condition. These traditions would be favourable to the existence of a powerful monarchy in Fiji, such as legendary evidence represents it as being at one time, and also its ultimate extinction and remoulding by the growing power of petty chiefs, skilful in new practices of war acquired whilst abroad. The hypothesis advanced derives additional strength from the fact of all Fijians, though scattered over a group of more than two hundred different islands, speaking one language, having a powerfully developed sense of nationality, and feeling as one people. No ancient Roman could have pronounced the words "*Civis Romanus sum*" with greater pride or dignity than a modern Fijian calls himself a "*Kai Viti,*" a Fijian. We can scarcely conceive these general sentiments to have taken hold of the popular mind with such force, if the people had always been divided into petty states as at present.

Away from the capital and Cakobau, some of the Fijian kinglets talk very boastfully of their total independence, and wish you to believe the suzerainty of Bau merely applies to certain inferior chieftains; whilst the *social* supremacy is seldom disputed, and the court dialect is understood by all the chiefs, even those living in the remotest parts of the group, and it has therefore very properly been adopted by the Wesleyan missionaries in their translation of the Bible. Each of these states or principalities has its ambassador at Bau (*Mataki Bau*), who, however, does not constantly reside in the capital, but only when there is any business to transact, which may occasionally last for weeks or months. On arriving at Bau, he takes up his abode at the house of the Bauan "minister," if he may be called so, charged with the affairs of the district from which he comes as ambassador, and he is by his host introduced to the King of Fiji. When Bau has any business to transact abroad, the ambassador selected is invariably the minister of the affairs of the district to which he is sent, and his place at the capital is temporarily filled by a relative. The office of these diplomatic agents is hereditary in certain families, and they are appointed by the ruling chiefs. Title and office are quite as much valued as they are in Europe by ourselves,—human nature being human nature all the world over.

On the 28th of July, Mr. Pritchard and myself set out in the consular gig for Navua, Viti Levu, to pay our visit to Chief Kuruduadua. There being rather a strong south-easterly breeze, we arrived two hours after dark at Bau, thoroughly wet from salt water, and heartily glad to take shelter under the hospitable roof of Mr. Collis, a gentleman connected with the mission. Until 1854, Bau, which is the name of the metropolis, as well as the ruling state, was opposed to the missionaries, and the ovens in which the bodies of human victims were baked scarcely ever got cold. Since then, however, a great change has taken place. The King and all his court have embraced Christianity; of the heathen temples, which, by their pyramidal form, gave such a peculiar local colouring to old pictures of the place, only the foundations remain; the sacred groves in the neighbourhood are cut down; and in the great square where formerly cannibal feasts took place, a large church has been erected. Not without emotion did I land on this blood-stained soil, where probably greater iniquities were perpetrated than ever disgraced any other spot on earth. It was about eight o'clock in the evening; and instead of the wild noise that greeted former visitors, family prayer was heard from nearly every house. To bring about such a change has indeed required no slight efforts; and many valuable lives had to be sacrificed,—for although no missionary in Fiji has ever met with a violent death, yet the list of those who died in the midst of their labours is proportionally very great. The Wesleyans, to whose disinterestedness the conversion of these degraded beings is due, have, as a society, expended £75,000 on this object; and if the private donations of friends to individual missionaries and their families be added, the sum swells to the respectable amount of £80,000.

Bau is built on a small island on the east side of Viti Levu, with which it is connected by a long flat of coral, fordable at high water, and in places bare at low. The annexed sketch, taken in 1860, by Mrs. Smythe, and kindly placed at my disposal, will give a better idea of the place than any description. The island is at the back about a hundred feet high, and around the beach thickly covered with native houses, arranged in crooked streets. The top of the island, where the British flag is waving, was a mere receptacle for rubbish, until the industry of the missionaries converted it into smiling gardens and eligible sites for dwelling-houses. At my first visit the natives were just finishing their new *Bure ni sa*, —a building, one or several of which are found in every town, and which may be described as a compromise between our club-houses and town-halls. It was 125 feet long, but not quite so high as the adjoining church, which is 100 feet high, and seems a tremendous edifice for natives to erect without nails, and the use of such tools as are employed by us.

The King's residence is close to the beach, and a large native-built house, to which several out-houses are attached: one of which is inhabited by Peter, a Tonguese, who fills the office of prime minister, and seems much attached to the King. In front of the house is a fine lawn of couch-grass, and groups of iron-wood, and other native shrubs and trees,—the whole, I believe, a creation of Mrs. Collis, the wife of the resident training master at Bau, who will ever live in my memory, for

Drawn by Mrs Smythe. Vincent Brooks, lith

BAU, CAPITAL OF FIJI.

having, amongst other great acts of kindness conferred, never failed to supply me in this land of pork and yams with bread, cakes, and other acceptable presents whenever I came in that neighbourhood.

Bau is said to own its present superiority to the fortunate accident of having been the first familiar with the use of fire-arms. Charles Savage, a Swede, introduced it about the beginning of this century. But it was not only to this accident that Bau is indebted to its permanent ascendency. Like England, but on a lilliputian scale, it is a great naval power, able to send its fleets of canoes to any part rebelling against its authority, or refusing to discharge its annual tribute. The Bauans are a fine race, nearly all members of noble families or gentlefolks. Most of them are tall, well-proportioned, and often with a handsome cast of countenance. In Fiji, as in fact all over the South Sea, a man is estimated by the height of his body, and little men are regarded with contempt. Their tall figures prove a great advantage to the Bauans. This general contempt for small men arises from the fact, that throughout Polynesia the chiefs and upper classes are taller than the lower orders, and with a finer physical they combine a greater mental development. They are in every respect superior to the people whom they rule. They are as genuine an aristocracy as ever existed in any country. They know every plant, animal, rock, river, and mountain; are familiar with their history, legends, and traditions; and strict in observing every point of their complicated etiquette. They swim, row, sail, shoot, and fight better than the common people, and excel in house and canoe building. Thus they keep their place amongst a people not able to fall back upon dress and finery to lend distinction to rank, dignity to person.

We were desirous of pushing on early the next morning, but as the tide did not suit, we ran over to Viwa, a small island close to Bau, where a permanent printing-press has been established in the first stone house ever built in the group. The greater portion of the Fijian Bible has been printed at this establishment; and the edition, now exhausted, is very much esteemed by the natives. A Fijian and English Dictionary, composed by D. Hazelwood, is another great work produced here in 1850. This Dictionary is full of a mass of reliable information, and must be regarded as the best contribution the Fijian missionaries have made to science. Ethnologists, geographers, and naturalists, and philologists as a matter of course, will find here facts and observations not met with elsewhere.* Viwa is full of fruit-trees, and altogether a charming spot. The cocoa-nut palm seems to be the only plant that does not flourish. After having attained a certain height it begins to wither—the foliage looking as if boiling water had been poured over it.

We found Messrs. Martin and Baker, the two gentlemen connected with the mission of this place, absent,—they having gone to look for an eligible new station on Vanua Levu. But their wives were at home, and glad to see us safe. Through telescopes they had watched our boat on the previous evening, as long as daylight

* I believe Messrs. Trübner and Co., Paternoster Row, London, have still a few copies of this publication on hand.

lasted, fearing that we might meet with some accident in the rough sea we had to cross.

On going back to Bau, Mr. Fordham, the principal missionary, represented to Mr. Pritchard the desirableness of prohibiting the importation of firearms and gunpowder into Fiji. Fighting, he thought, might thus be prevented. Mr. Pritchard agreed with him that there was not much use for those articles, there being no wild animals, and only a few ducks and wood-pigeons to shoot, but that it would be impolitic to venture upon making any prohibitive law, waiving all considerations as to the right of doing so, when there were no officers to execute it. Even supposing that a certain pressure could be put upon the English subjects, who was to prevent the Americans, Germans, and French from selling any number of firearms, and any amount of gunpowder, to the natives? On a previous occasion, Mr. Pritchard was seriously asked by another gentleman to introduce the Maine liquor-law. No spirits of any kind should be landed or sold. This idea the Consul also refused to entertain. The law had broken down when enforced by all the power of a great state, and could scarcely be expected to work well under less favourable circumstances.

CHAPTER V.

THE WAI LEVU, OR GREAT RIVER.—CANAL DUG BY NATIVES.—MATAISUVA. —INSTITUTION FOR TRAINING NATIVE TEACHERS.—SACRED GROVES, TREES, AND STONES.—MOSQUITOES.—ISLAND OF NAIGANI.—MR. EGGERSTRÖM'S KINDNESS.—FEUDS AT NADROGA.—NUKUBALAWU.—TAGURU.— NAVUA RIVER.

THE Rewa, Wai Levu, or great river of Viti Levu, has four large mouths, and its deltas are extremely fertile, and cultivated to some extent by the natives. About eighteen miles from its mouth it receives the Wai Manu, which comes from the west, whilst the main branch takes its rise in the Namosi Valley. It was explored in 1856 by Dr. Macdonald, of H.M.S. Herald, Captain Denham, accompanied by Mr. Samuel Waterhouse, of the Wesleyan Mission, and a full account of their proceedings has been published.* Mataisuva, our next stopping-place, is built on one of the large deltas, a little below the town of Rewa. From Bau it may be reached either by sea or by going up the Wai ni ki, or Kaba mouth. The natives have shortened the latter passage more than

* "Proceedings of the Expedition for the Exploration of the Rewa river and its Tributaries, in Na Viti Levu, Fiji Islands. By John Denis Macdonald, Esq., Assistant Surgeon of H.M.S. Herald, Captain N. M. Denham," in the Journal of the Royal Geographical Society, vol. xxvii., pp. 232–268, with a Map by Arrowsmith.

twenty miles by cutting a canal, Kele Musu, across the longest of the deltas. Taking advantage of the tide setting in, we left Bau about noon and soon found ourselves in the canal, probably the greatest piece of engineering ever executed in these islands, affording a proof how thickly they must have been populated to allow such an undertaking, at a time when there was nothing but staves to dig the ground, hands to shovel it up, and baskets to carry it away. It has not been ascertained when this canal was dug; all that can be elucidated is, that it was made long ago, and for the purpose of carrying out a military stratagem. It is about two miles long, sixty feet wide, and large canoes pass without difficulty. On a subsequent occasion, our schooner, the 'Paul Jones,' finding it impossible to get from Bau to Rewa by sea on account of a heavy gale, actually made her way through this canal, by taking due advantage of the tide.

We neared Mataisuva, the mission-station, about sunset, and passing the mangrove forest, were surprised to see the immense number of Flying Foxes, or Bats (*Notopteris Macdonaldii*), rising from them. They measure nearly a yard from the extreme points of their wings. Mr. Pritchard informed me that at Samoa, the same or a very nearly allied species is a great pet with the natives of that group, and probably the only known instance of a domesticated bat.

Passing the town of Rewa, we reached Mataisuva at half-past six on the evening of the 29th of June, and were hospitably received by the Rev. W. Moore, who was then the superintendent of an institution for training native teachers. A large square piece of ground had been set aside for a number of houses surrounded by little gardens in which the teachers resided. Some of these teachers were Fijian, some Tonguese. The natives like their own countrymen best, because they always suspect the Tonguese, and with good reason, of playing into the hands of the Tonguese chiefs, whose great aim is to make themselves masters of Fiji. These teachers, after having been properly trained at this institution, are sent as residents to those parts of the country which have applied for them; and they are of very essential service in preparing the ground for the white missionaries, whose limited number is quite inadequate to the great task set before them, that of christianizing Fiji. Many parts of the group are now anxiously desiring the Gospel, but, with so few labourers in the field and only limited funds, it is impossible to do much more than is now attempted. Apart from any religious consideration, I should always support the Protestant missionary in preference to the Roman Catholic, because the latter attempts simply the conversion of the heathen, whilst the Protestant not only christianizes, but at the same time civilizes them. The quiet, well-regulated family life and cleanly habits which our Protestant missionaries set before the savage, are of inestimable value to the people whom they endeavour to raise in the scale of humanity. It is quite wrong to suppose that savages do not notice whether a man wears clean linen and is well washed or not. They do notice it, and never fail to draw comparisons in favour of those who, by means of their comfortable homes, are enabled to appear before them as good examples of cleanliness.

Though most of the white Wesleyan missionaries are perfect masters of the language, they own themselves that the native teachers they had trained generally beat them in the choice of local illustrations. Of course, there is occasionally a want of tact on the part of the latter. Thus, one of them, wishing to illustrate how wisely in everything nature had adapted the means to the end, chose the hand, and commenced by saying, " Now, when you eat a human hand, you will perceive," etc. This illustration would have sounded odd to a Christian congregation at home, but never excited any notice amongst a people just emerging from cannibalism.

The church at Mataisuva is not so large as that at Bau, but it is much better finished, and some of the beams under the roof are covered with different-coloured fibres of the cocoa-nut worked in various elegant patterns. The ridge-beams, always projecting on both ends, according to strict Fijian customs, are ornamented with white shells (*Ovulum ovum*, Swb.), and in front of the church there are some curiously-cut stems of tree-ferns. Altogether the building is a fine specimen of native architecture, and the only thing to complete it is a good tolling bell. Hitherto the congregation has been obliged to be called together by large drums, made of Tavola wood, beaten by thick and short pieces of wood,—a contrivance which may be heard for several miles around, but sounds essentially unchristian.

The Rev. William Moore, as an apt Fijian scholar, devotes some of the spare moments he can snatch to a subject hitherto much neglected, that of collecting the " *mekes*," or old songs of the natives, now fast fading away. He has also made considerable advance in translating 'The Pilgrim's Progress' into Fijian, a task which, if I mistake not, has been somewhat facilitated by Mrs. Binner's unpublished version of a portion of that book. Bunyan's great allegory has already been translated into one or two Polynesian languages, and the natives seemed to like it very much as long as they believed it to be a genuine story, but when they heard that it was only a series of "lies," their interest abated. It will be interesting to know how the Fijians receive it. They are as true believers in the genuineness of their own numerous fairy tales and doings of their gods, as the ancient Greeks were in those of their gods and demigods; — the hold which Homer had on the national mind arising, probably, quite as much from his embodying this feeling, as well as expressing it in language still the admiration of mankind.

Accompanied by Mr. Moore we went to the town of Rewa, in order to gather specimens of two new palms, one of them a fan-palm (*Pritchardia pacifica*, Seem. et Wendl.), the leaves of which are only used by chiefs, as was the case with those of the Talipot palm in Ceylon. I also collected some interesting information about the bread-fruit, of which there are no less than ten different varieties cultivated at Rewa, including the best of the group.

On our way home we fell in with a little schooner belonging to the mission, and returning from a trip up the Rewa river, where she had been sent for yams. She had not accomplished her object, as two hostile parties of natives had not allowed her to pass, and even fired at her, without however wounding or killing any one. Formerly these inter-tribal feuds were of much more frequent occurrence, and often protracted over a considerable period of time; but since firearms have become accessible to all parties, the same result followed in Fiji as in Europe upon the invention of gunpowder.

Sacred groves and trees form as prominent a feature in the paganism of the Fijians as they did in that of the Indo-Germanic nations. A fine grove still exists in the Rewa district near the mission-station of Mataisuva, and at a point of the coast termed Na Vadra Tolu (the three screw-pines), probably from three specimens of the *Pandanus odoratissimus*, still a common plant in that locality, having stood there. Leaving the mission-premises, and keeping along the sandy beach, an enormous Yevuyevu tree (*Hernandia Sonora*, Linn.) presents itself, forming a complete bower, which leads to a curious group of vegetable giants. A venerable Vutu rakaraka (*Barringtonia speciosa*, Linn.), more than sixty feet high, has thrown out several huge branches, two of which form, in connection with the stem, bold arches. The large aerial roots of epiphytical fig-trees are holding the monster in close embrace; several kinds of ferns and climbing *Aroideæ* and wax-flowers (Hoyas) interlace the struggling masses, and tend to increase the wildness of this fantastic scene. The dense foliage of surrounding Vesi, Ivi, and other fine trees ensures a constant gloom and sombreness to the place; and only through the bower, serving as an entrance, does the eye obtain a glance at the open sea, and perchance the sight of a passing canoe with its large triangular sail. It was at this lonely spot, far away from human habitations, where in the depth of night the heathen priest used to consult the gods whether it was to be war or peace. If at dawn of day blood was found on the path, more blood was to be spilt; if no such sign was discoverable, peace was the watchword. Several celebrated groves were destroyed on the introduction of Christianity, and a large one near Bau was felled the day after King Cakobau had embraced the new faith, the native carpenters trembling when they had to lay the axe on objects so long sacred to them by all the laws of "tabu." They were taught by tradition that when, once upon a time, their forefathers felled some of these trees, and repaired the next day to the spot in order to square the logs, they found the trees again in their proper position, and growing as if no sacrilegious axe had ever laid them low.

Besides these groves, there were isolated trees which were held sacred; and in days of yore European sawyers came occasionally in unpleasant contact with the Fijians when, unknowingly, they had cut them down for timber. Vesi (*Afzelia bijuga*, A. Gray) and Baka (*Ficus sp.*) seemed to have been those principally selected. The Vesi furnishes the best timber of the islands, and may, as the most valued tree, have been thought the fit residence of a god; there is nothing in its appearance that is extraordinary, our beech most nearly resembling it in look. The Baka is not famous for its timber; but its habit is as remarkable as that of the banyan-tree of India, aerial roots propping up its branches and forming a fantastic maze which no words can describe. At first living as an epiphyte on other trees, it soon acquires such dimensions that it kills its supporter, and henceforward must draw its nourishment from the soil. There are fine specimens of the Baka on the Isthmus of Kadavu; and on an islet belonging to Mr. Hennig the aerial root of the Baka formed a cabin in which Mr. Pritchard, myself, and all our boat's crew took shelter during a heavy tropical shower; and twenty persons might have found room there. The crown of this tree was one hundred and fifty-two feet in diameter, or four hundred and fifty-six feet in circumference. The horizontal branches and the large roots issuing from all parts of the stem, and more sparingly from the branches, rendered this tree a noble object, well calculated to inspire pleasure or awe. The Rev. W. Moore lamented the destruction of one of these fine trees near Rewa, committed by a sick man in hopes that it might be pleasing to the Christian God, and incline him to favour his convalescence. These sacred groves and trees were not worshipped as gods, but, as in the Odic religions of our ancestors, looked upon as places where certain gods had taken up their abode.

Sacred stones, to which the natives pay reverence, exist in Fiji; for instance, near Vuna and Bau, as well as in many other parts of Polynesia. Fully granting their being the supposed abode of certain gods and goddesses, as has been contended, we can only hope to arrive at their real meaning and primary origin, by considering them in connection with the ideas associated with or represented by other monoliths. I would particularly direct attention to their peculiar shape, of which the missionaries Williams and Turner* have published some good illustrations. Compared with certain remnants of Priapus worship, as found in Indian temples, the "Museo segreto" of Naples, and, freed from all obscenity, in the obelisks of Egypt, their nature becomes evident. More or less, these monoliths represented the generative principle and procreation; and, if the subject admitted of popular treatment, it would not be difficult to show that the Polynesian stones, their shape, the reverence paid to them, their decoration, and the results expected from their worship, are quite in accordance with a widely-spread superstition, which assumed such offensive forms in ancient Rome, and found vent in the noblest monuments of which the land of the Pharaohs can boast. Turner, after stating that he had in his possession several smooth stones from the New Hebrides, says that some of the Polynesian stone-gods were supposed to cause fecundity in pigs, rain and sunshine. A stone at Mayo, according to the Earl of Roden, was carefully wrapped up in flannel, periodically worshipped, and supplicated to send wrecks on the coast. Two large stones, lying at the bottom of a moat, are said to have given birth to Degei, the supreme god of Fiji. In all instances an addition to objects already existing was expected from these monoliths. There was a stone near Bau, which, whenever a lady of rank at the Fijian capital was confined, also gave birth to a little stone. It argues nothing that these stony offsprings were fraudulently placed there. The ideas floating in

* Williams's 'Fiji and Fijians,' p. 220. Turner's 'Nineteen Years in Polynesia,' p. 347.

the minds of the bulk of the people absolutely tended towards the unbiassed conviction that some mysterious connection existed between the large stone and the Bauan ladies. Since the introduction of Christianity to these districts, it has been found necessary to remove the large stone, leaving its numerous posterity behind, to get on as best it may.

During the rainy season, the mouth of the Rewa river is notorious for myriads of mosquitoes. On some evenings the hetacombs slain by incautious contact with the flame, actually put the candles out. Mr. Moore once contrived a room on the principle of a mosquito-curtain; but the contrivance was not found to answer, as few persons could be induced to purchase freedom from irritating bites by confinement for several hours of a hot night in an insufficiently ventilated kind of cage, which, from its very nature, could not be so large as to admit of much moving about, or the introduction of lights for reading or writing. Mosquitoes are objects to which the attention of all new-comers is irresistibly directed. Those of Somosomo never favoured us with a call until after breakfast, and very obligingly withdrew about sunset, to let us have the evening to ourselves. In other parts of the group the evening is their very time for paying visits. The moment one of their monotonous *solos* is heard, a *tutti* will immediately follow. The difference between the voices of the various species is al most as great as that observable in those of men; and a naturalist studying these insects as thoroughly as they study him should either possess an ear musically trained or else carry a fiddle, in order to determine the exact note struck up. I am persuaded that every mosquito, from the large sluggish one which annoyed us when searching for Sir John Franklin in the Arctic Circle, to the little swift one of the Equator, may be known as readily by its peculiar note as by any artificial diagnosis,—the Sydney one pre-eminently by its very deep tone.

On the 2nd of July, about noon, we left Mataisuva, and at 7 P.M. reached Naqara (the Cave), in the island of Naigani, where Mr. Eggerström, a Swedish gentleman, had taken up his abode. He was just recovering from a serious illness contracted by incautious contact with the Kau karo, or Itchwood, a poisonous tree (*Oncocarpus Vitiensis*, A. Gray = *Rhus atrum*, Forst.) peculiar to Fiji and New Caledonia, the stem of which he had been converting into a flag-staff. Mr. Eggerström received us cordially, and had tea and supper prepared. He also wished us to sleep under his hospitable roof; but the mosquitoes were so very troublesome that we could hardly finish our meal, and were obliged to beat a hasty retreat to our boat, though our kind host assured us that if we remained a little longer we should get quite as much used to their bites as he was, and feel no inconvenience. We spread the awning over our gig, and made every preparation for sleeping. As it was still early, Mr. Pritchard read, and I went again on shore, to the native village, which I found, as I had been assured, quite free from mosquitoes. The natives were very friendly; they showed me their canoes, and brought me cocoa-nuts and sugar-cane to eat; I gave them a few sticks of tobacco in return, and wanted them to dance;

but they informed me, through the interpreter, that the missionaries desire them not to dance nor practise any more their game of throwing canes, after the yams have been planted. They said they should sing instead, and forthwith commenced. I let them go on till they came to a "*meke*," or song, in which they mimicked the missionaries; I then stopped them by wishing them "good night."

Most of our crew passed the night on shore, and Mr. Pritchard and I slept in the consular gig, anchored close to the shore. Early next morning we were awoke by the arrival of a large canoe from Nadroga. The man in charge came to ask the Consul's advice about making peace with the heathens who had for several months made war upon Nadroga for becoming Christian. They had only ten towns, six of which had been taken by the heathens, and several inhabitants been baked and eaten. The Nadroga people had only captured two towns, and now feared they could not hold out much longer unless Christian natives of other districts hastened to their assistance. They were now going to Rewa and Ovalau, to ask for such assistance, and had with them a lot of tortoiseshell, to be exchanged for muskets and powder. Mr. Pritchard told them that he should visit them in about a month, and then use his influence to restore peace. I may as well add in this place, that he did so in August, with Colonel Smythe, and that they conjointly sent a messenger to the heathens, inviting their chiefs to an interview. The messenger was received with blows, and told it was fortunate that he had come by himself. If two had been dispatched, one would have been sent back to tell the tale; now, as only one had come, he should merely be half killed, and might go home to say that they neither cared for the Consul nor for Colonel Smythe, and declined all interference on their part.

We went again on shore, as Mr. Eggerström had invited us to breakfast and to inspect his establishment by daylight. Mr. Eggerström had expended a great deal of labour on his retreat, cut steps in the solid rocks, and made a large basin for bathing, and seats near the spring from which the water was supplied. He seemed to have been anxious to render his new home as pretty as possible, and paid less regard to the requirements of the crop he wished to grow. He complained that nothing would flourish, and I told him that unless he sacrificed more trees, his sweet potatoes, yams, and bananas, to say nothing about European vegetables, would be, as hitherto, a prey to snails, caterpillars, and insects, and his house never free from mosquitoes. But he said he loved the shade, and could not make up his mind to do that.

Although the place was swarming with mosquitoes the previous night, there was now not one to be seen. The sky looked very rainy, and we hesitated whether to stay or push on. We decided on adopting the latter course, but had hardly been afloat more than ten minutes when the rain began to come down in such torrents that our boat required constant baling. We took shelter at Nukubalawu, in the house of an American, Mr. Work, who, like most of the old white settlers, is better known in Fiji by his nickname, in this instance "Moses." He had a sawing-pit, which he worked with natives, one of whom had been with him for years. Though he was moving across the bay, to take up his residence on the little island inhabited by Mr. Eggerström, he made us very comfortable; and I took advantage to arrange my collection of plants, which had seriously suffered from the heavy shower that drove us to seek shelter in this place. The rain continued all day, so that we were quite unable to stir.

Leaving Nukubalawu next morning, we passed a remarkable rock on the shore of Viti Levu, which from its peculiar shape and large dimensions Mr. Pritchard and I named the "Giant's Thumb." The rain continued, and after an hour's pulling and sailing, we were obliged to land at Taguru, where we found three white men engaged in sawing and building boats. As Taguru belongs to Kuruduadua's dominions, we dispatched a messenger to Navua, the chief's residence on the coast, to inform him that we would be with him as soon as the weather permitted. Towards sunset there was a lull in the rain, and we at once resumed our way to the chief, who was not yet under missionary influence, and about whose cannibalism and despotic government we had heard so much.

A pull of about two miles westwards brought us to the Navua, one of the largest rivers in Viti Levu, and not yet explored by any scientific man. There are several extensive deltas at its mouth, composed of rich alluvial soil, and exceedingly well adapted for cotton. From information gathered, I was led to conclude that the sago-palm was a member of the Fijian flora. My inquiries commenced in the eastern part of the group, and I was always directed westward, and assured at every place that I should find the object of my search a few miles further on; but that not proving the case, I began to look upon it as a mere phantom, when at last, after a search of several hundred miles, whole groves of fine sago-palms (*Sagus Vitiensis*, Herm. Wendl.) greeted me on the banks of the Navua river. This is an interesting discovery; botanically, because no sago-palm had ever been found so far south; philologically, because the plant is here termed *Soga*, calling to mind the names of Sagu, or Sago, by which it is known in other districts peopled by the Papuan race; and commercially, because it adds an important article to the export list of these islands. The Fijians made no use of the farinaceous pith the Soga contains, though they are familiar with converting that of the *Cycas circinalis* of the district into cakes, eaten by the chiefs.

CHAPTER VI.

STAY AT NAVUA.—CHIEF KURUDUADUA'S HOUSEHOLD.—" HARRY THE JEW."
—A PRINCE AS HE WAS BORN.—MASSACRE PREVENTED.—KURUDUADUA'S
CHARACTER.—STATEMENT OF MR. HEEKES RESPECTING THE NAMUKA
OUTRAGE.—TOWN AND BURES OF NAVUA.—TATOOING.—RETURN TO LADO.

WE were soon at Navua, a town some three miles up the river, and the residence of Kuruduadua, the great chief of this district. The messenger dispatched from our last halting-place having announced our visit, we found the chieftain seated in his large house, surrounded by councillors and attendants, awaiting his guests. As he and his territory are little known to the whites, our arrival created some sensation. The ceremony of presentation is novel. On entering the house, Charles Wise, our interpreter and guide, as already schooled, addressed the chief to the effect that the Consul had come to introduce a chief from England, who had been sent to explore the country; and that we purposed doing ourselves the honour of being his guests for several days. After a few minutes' silence, the chief orator replied, in the name of Kuruduadua (it would have been against Fijian etiquette for the latter to address us personally at the first formal visit), that the stranger chief and the Consul were welcome, for their presence conferred a distinguished honour on Navua, and the neighbouring tribes should know the fact as soon as the great drum could send forth its rolling peals. As he concluded, all the men in the house clapped their hands, and exclaimed, " *Mana, mana, mana!* " At the same instant the great drum, or *lali*, was beaten lustily, and our presence in Navua was heralded throughout the district.

The chief's eyes glistened, and a proud smile of exultation gleamed over his face as we threw ourselves at full length on the clean mats spread for us. Our loquacious interpreter here began to describe a huge iron pot that was near the door, and to tell how wickedly it had been appropriated to boil the carcases of slaughtered men instead of *bêche-de-mer;* thus confirming the rumour which Macdonald had told in the Geographical Society's Journal. A rather unpleasant feeling stole over us, and we thought of friends and homes far away. Our peace of mind, however, was soon restored, when the chief proposed that we should join him in a bowl of *kava*, a beverage prepared from the root of the South Sea pepper, by being masticated by young men, and tasting like soapsuds, jalap, and magnesia! A baked pig and some half-dozen baskets of yams were next brought in by women, headed by the chief's favourite wife, all crawling on their hands and knees. Hungry as we were, the story of the big pot made us rather revolt from this frugal meal; but ascertaining that it was a real pig we beheld before us, we dined. It is a curious fact, that Fijian custom does not permit the host to partake of the meal which he provides for his guests; and the chief eyed us askance as we ate. About this time a carronade, that guarded the entrance to the house, was discharged—emphatically to demonstrate the chief's delight. Kava, or "*yaqona*," as it is called in Fiji, was masticated and drunk every half-hour. We observed that the string by which the bowl is suspended when not in use was always thrown towards the chief. The object of this is to distinguish the "great man," for if any one incautiously walked upright in his presence, the club is his fate.

Kuruduadua has ten wives, and as he himself does not exactly know the number of his children, we were left ignorant on that point. The great drums were beaten every hour of the night, in honour of the guests, but much to our annoyance, for they kept us awake some time after we retired. Our bed was made of several layers of mats, and over us was a large mosquito screen, about twenty feet long, made of the bark of the paper mulberry. As many as eight or ten natives sometimes sleep together under one of these screens. Before retiring, the Consul presented various articles, as knives, axes, prints, etc., to the chief; and the usual complimentary speeches, expressive of mutual confidence and goodwill, ensued.

On the following morning "*Harry the Jew*" presented himself—the only Englishman who has lived for any length of time in the wild and unknown regions of the interior, and has managed to throw a halo of mystery around himself. His real name is John Humphrey Danford, and he has been for so many years living with Kuruduadua and his family, cut off from all intercourse with civilization, that he seemed to have lost his reckoning, and was not quite sure whether he had been sixteen, eighteen, or twenty years in the islands. His story is full of adventure. Born in London, he was early apprenticed, first to one then to another trade, but his employers being all men with whom he "could not agree," he left them in disgust, and took to the sea. This brought him to the South Pacific, where he discovered that the captains he had to deal with were disagreeable men ; and, after exchanging from vessel to vessel, he finally ran away at Tongatabu. There, after twelve months' residence, amid many privations, partly caused by a great hurricane and its usual successor, a general famine, he perceived the Tonguese too were disagreeable people, and at once took passage in a canoe for Fiji. Arriving in this group in distress from heavy weather, the canoe was seized at the island of Kadavu, and the crew condemned to be baked in the oven—thus finding the Kadavu people more disagreeable even than the Tonguese. By strategy, however, he succeeded in making his escape to Rewa, where he remained some time with other white men. To one, Charles Pickering, a celebrity of Fiji and the hero of some capital anecdotes, he sold a pinchbeck watch that only went when carried. Whence he got this precious article, he says it is unnecessary to tell; enough for the history, that as soon as he received the price thereof from Pickering, he jumped into a boat and started off for some distant part of the islands, condemning the white men as a disagreeable set of fellows. In his wanderings he met one "Flash Bob," for whom he acted as agent in the selection and purchase of a lady-love from a native chief. This brought him once more in contact with the disagreeable whites. He now commenced a *bêche-de-mer* establishment, in conjunction with his friend Pickering, who had given him the nickname of "Harry the Jew," in consequence of the watch transaction. After some months in his new business, a quarrel arises about the purchase of Flash Bob's wife; the drying-house of the establishment is burnt down by a party of natives; Pickering, enraged that his property has been destroyed, takes everything away, leaving poor Danford once more penniless, shirtless, and friendless, on the beach. His nickname, translated into Fijian, has begun to work mischief amongst the newly-converted natives, and he is denied hospitalities the heathens would not refuse, because he "belongs to a people who have killed Christ." The brother of Chief Kuruduadua, hearing of his forlorn condition, sends him an offer to reside at Namosi, his mountain residence, which offer is hesitatingly accepted. His heart almost fails him as he toils his way into the very midst of a nation of cannibals. But iron necessity urges him on. Tired and footsore, almost in an absolute state of nudity, he reaches the town. Messengers meet him and carry him on their shoulders. The chief then gives him wives, —how many we shall not say,—a yam plantation, two gardens, houses, and dispatches bales of native cloth to the coast, to be exchanged for European dresses for him. He is also raised to the dignity of a "brother," and allotted slaves to attend upon him. Our hero—happy man!—now, for the first time in his life, finds agreeable companions in the chief and his people. In return for the dignities heaped upon him, Harry was to repair the muskets of the tribe, and to tell the chief stories about the white men and their country. Having for about a week been an errand-boy to a London apothecary, he was able to dispense pills to the sick, and thus to assume another important stand in his new life. Years had rolled on without his seeing any white faces, when one day native messengers arrived from the coast, stating that they had been sent by a foreigner, who wished to have an interview with him, and whom they described as wearing a blue coat all covered with looking-glasses. Harry had seen many extraordinary sights, but a man thus attired excited his curiosity, and he acceded to the request. To his surprise, he found the late Mr. Williams, United States Consul, whose brass buttons had been mistaken for looking-glasses. Mr. Williams had heard of the existence of some copper mines in the interior, and was desirous of purchasing them. Through Harry's intervention, that object was accomplished, and the mines passed into Mr. Williams's possession, but they have not as yet been worked, nor indeed been examined by any scientific man. Dr. Macdonald and Mr. S. Waterhouse paid a visit to Namosi when they ascended the Rewa river; and Harry, who had long ere that sown all his wild oats, and found one wife quite as much as a sensible man could manage, begged the Rev. Samuel Waterhouse to christen his natural children. But he met with a refusal, on the ground of his not being married. "Then pray marry me," was the next demand. "Impossible," replied the missionary, "your bride is not a Christian." Danford felt this refusal very deeply. Many a long year had he waited to free himself from the reproach of not living in matrimony, and when at last a fair chance seemed to present itself, he met with disappointment. The Wesleyans have shown a strict adherence to a similar policy, and they may be right from their point of view; but in consequence many of the whites have been obliged to ask the Catholic priests to discharge those duties which their Protestant brethren refused. The Catholic priests, asking few questions, have invariably christened such children, and, remembering the full significance of the formula, that in marrying we take each other "for better, for worse," united in matrimony all loving couples presenting themselves for the purpose.

We were struck with the fact, that all the young lads were in a state of absolute nudity; and, on inquiry, learned that preparations were being made to celebrate the introduction of Kuruduadua's eldest son into manhood; and that, until then, neither the young chieftain nor his playmates could assume the scanty clothing peculiar to the Fijians. Suvana, a rebellious town, consisting of about five hundred people, was destined to be sacrificed on the occasion. When the preparations for the feast were concluded, the day for the ceremony appointed, Kuruduadua and his warriors were to make a rush upon the town, and club the inhabitants indiscriminately. The bodies were to be piled into one heap, and on the top of all a living slave would lie on his back. The young chief would then mount the horrid scaffold, and scanding upright on the chest of the slave, and holding in his uplifted hands an immense club or gun, the priests invoke their gods, and commit the future warrior to their especial protection, praying he may kill all the enemies of the tribe, and never be beaten in battle; a cheer and a shout from the assembled multitude concluding the prayer. Two uncles of the boy were then to ascend the human pile, and to invest him with the *malo*, or girdle of snow-white *tapa*; the multitude again calling on their deities to make him a great conqueror, and a terror to all who breathe enmity to Navua. The *malo* for the occasion would be perhaps two hundred yards long, and six or eight inches wide. When wound round the body, the lad would hardly be perceivable, and no one but an uncle can divest him of it.

We proposed to the chief that we should be allowed to invest his son with the *malo*, which he at first refused, but to which he consented after deliberation with his people. At the appointed hour, the multitude collected in the great strangers' house, or *bure ni sa*. The lad stood upright in the midst of the assembly, guiltless of clothing, and holding a gun over his head. The Consul and I approached, and in due form wrapped him up in thirty yards of Manchester print, the priest and people chanting songs, and invoking the protection of their gods. A short address from the Consul succeeded, stirring the lad to nobler efforts for his tribe than his ancestors had known, and pointing to the path to fame that civilization opened to him. The ceremony concluded by drinking kava, and chanting historical reminiscences of the lad's ancestors,—and thus we saved the lives of five hundred men! During the whole of this ceremony, the old chief was much affected, and a few tears might be seen stealing down his cheeks. Soon however cheering up, he gave us a full account of the time when he came of age, and the number of people that were slain to celebrate the occasion.*

Kuruduadua was still a heathen. He said that our religion was good, but there were few true Christians in the group, and he hated hypocrisy, and did not profess to be better or anything else than he really was. He rather favoured than hindered the spread of the Gospel. On Sunday morning I heard him interrogating two men, whether they were Christians. On their answering in the affirmative, he reprimanded them for not attending the church service, as the drum—the substitute for bells—had left off beating for some time. We induced him to make several important concessions to civilization, to prohibit cannibalism throughout his territories, and to keep the Sunday as a day of rest, if not a holy day. To this he agreed cheerfully. Indeed he seemed most anxious to stand well with the whites, and one of the first explanations he offered after our arrival was respecting an attack upon, and plunder of some white men, who resided on Namuka, an island seven miles west of Rewa. The attack and plunder was made

* The custom of standing on corpses is mentioned by several writers on Fiji, and was probably practised throughout the group. Joseph Waterhouse, in his 'Vah-ta-ah,' p. 32, a book full of facts not found elsewhere, describing the condition of Bau previous to its conversion to Christianity, says, " Down the next lane a young chief is trying on, for the first time since he was born, a narrow slip of native calico, as an indication that he now thinks himself a man. He stands on the corpse of one who has been killed to make his stepping-stone for the ceremony of the day."

by a chief then at war with him. Long after peace had been re-established Kuruduadua became by exchange the owner of some boxes that had been taken from Namuka, by the attacking party. Danford saw the danger of purchasing property thus taken, and advised Kuraduadua to get rid of it. However, his counsels were unheeded, and when at a future time the boxes were actually found in Kuruduadua's possession, the American captain sent to punish the Namuka attack, fixed upon him, as one of the guilty party, and burned Navua, then full of valuable property of all sorts, honestly acquired from white traders. Several large 32-pound shots were knocking about the town, and served the children as playthings, whilst the ruins of fine large houses were still to be seen. Kuruduadua handed us a paper from his desk, drawn up by a white trader familiar with the whole affair, which he begged might be made known to our countrymen, in order to acquaint them with the real facts of the case.

"Ovalau, November 27th, 1856.

"Being acquainted with many circumstances connected with the attack upon Namuka, and convinced that great injustice has been done to the chief Kuruduadua, living at Navua, by his being punished as an accessory to that act, I beg to lay before you the true particulars of the case as they came under my observation.

"It has been stated that Kuruduadua was a party to the attack upon Namuka, because some of his people had been some time before driven away from that place by the whites. The facts were these:—Some canoes belonging to Kuruduadua's tribe, as was their custom when voyaging, put into Namuka to spend the night. They caught some crabs, and climbed some trees for cocoa-nuts, as they had always been accustomed to do, when the whites who had purchased permission to reside upon the island rushed out and fired upon them; the natives immediately fled, leaving one canoe behind. This canoe, with the property in it, was handed over to me by Mr. Allen Dolittle, when I was residing at Nukubalawu, to return to Kuruduadua. When I took it to the chief, he was not at all displeased at his people having been driven away, and said that if they again annoyed the white residents at Namuka he would himself club them.

" Some time after this, Tui Solia was knocked down by one of the whites on Namuka. Tui Solia was at this time at war with Kuruduadua. The latter heard, through a deserter, that Tui Solia intended to avenge the insult offered to him by plundering Namuka, and put the whites on their guard. He could not protect them there, as it was not in his territory, and he was at enmity with Tui Solia's tribe. He told the whites to remove at once to Nukubalawu, into his dominions, where he would protect them from every harm. He was evidently very anxious to secure the whites from injury. Thus, so far from being privy to the attack, he endeavoured to save the whites.

" I went at once to Namuka to warn the whites, and told them of Kuruduadua's invitation to remove for protection to Nukubalawu, and offered them the use of my boat, which they declined. I was then sent for by Mr. Saunders, to remove him from Wai Turaga to a vessel at Bau in which he had taken his passage.

" Before I returned, the attack was made on Namuka, the property plundered and the white men carried prisoners to Numulo, a small town on the mainland, which belonged to Tui Solia. As soon as I heard this, I hastened to Nukubalawu and met there Mr. A. Dolittle. Finding that nothing had been done towards the rescue of the prisoners, I sent for Kuruduadua, and giving him an axe, requested him to undertake their deliverance. He immediately complied, arranged to take a small armed party and make a sudden descent upon Numulo at early dawn. This he did. The people of the town, panic struck, fled, and the chief was thus enabled to convey the wounded prisoners and some property to Namuka, where we had gone to await the result of the expedition.

"It has been said that this chief was a party in the affair, because, at a subsequent period, some boxes, taken from Namuka, were seen in his house. They came into his possession in this manner: some time after the Namuka outrage, Kuruduadua attacked and captured a town belonging to Tui Solia, the defeat causing the latter to sue for peace. Friendly intercourse being re-established, Kuruduadua subsequently exchanged several pigs for boxes in Tui Solia's possession, and forming part of the plunder of Namuka. It is quite false that Tui Solia was at the time of the outrage under the influence of Kuruduadua; so far from that, they were enemies and at war.

"Kuruduadua has ever behaved kindly to the whites, and in this respect set a good example to other chiefs. Upon one occasion a vessel got ashore in the neighbourhood. He assembled his people, got her afloat, and made his subjects return the property they had taken,—this at a time when, in almost every other part of Fiji, the lives of the shipwrecked were taken and the vessel and cargo plundered.

"I was present at Nukubalawu, when Mr. Williams, the American Consul, and Phillips, a Rewan chief, came to inquire into the Namuka matter. Mr. Dolittle said, that after buying the island of Namuka they were entitled to protection. Phillips, the chief, then emphatically denied that the island had been sold, but said that a gun, a keg of powder, and a whale's tooth had been given as the 'yaqona' for permission to reside on the island, and that he could not sell it, as there were others who were co-owners with himself.

"JOHN HEEKES."

Navua is at present a collection of about forty houses, and built on the left bank of the river of the same name, and at the foot of a hill on which there is a private *bure ni sa* of the chief, enjoying a fine view of the flat land around, the river winding in bold curves, and high mountains in the distance. Two creeks intersect the town, over which isolated trunks of trees are thrown, the nearest approach to bridges I have seen in the country. In the two squares are several venerable Tahitian chestnut-trees (*Inocarpus edulis*, Forst.) densely covered with parasites (*Loranthus*), about a dozen species of epiphytical ferns,—one of them not larger than a moss,—wax-flowers, orchids, mosses, and lichens. There was no heathen temple (*bure kalou*), but a fine one might be seen from the top of the hill, about a mile off. I noticed three *bures ni sa*, strangers' houses, or sleeping bures. At least two of the latter are invariably found at every Fijian town or village. They may be compared to our clubs; and those frequented by the ruling chiefs do not seem visited much by the lower class of people. That at Bau, already mentioned, was the largest I saw. All along the sides are sleeping-places, covered with fine mats, and large enough for two men to sleep; and between each there is a fireplace, and stages to put the legs on. Overhead a good supply of firewood is stowed. The centre of the building is covered with loose grass, generally Co dina (*Paspalum scrobiculatum*, Linn.). There are no windows, only low doors, which may be, and are always closed towards evening, by means of thick mats, in order to keep the mosquitoes out. A large kava-bowl, and bamboo vessels filled with spring-water, seem to be the only utensils admitted. In buildings or bures like these, all the male population, married and unmarried, sleeps. The boys, until they have come of age, erect a bure of their own, often built on raised stages over the water, and approachable only by a long, narrow trunk of a tree. The women and girls sleep at home; and it is quite against Fijian etiquette for a husband to take his night's repose anywhere except at one of the public bures of his town or village, though he will go to his family soon after dawn. In the daytime the bures are generally deserted. Towards four o'clock people begin to pour in, and if any strangers arrive they will invariably take up their quarters at these places. Here politics and all events of the day are discussed, and when talking, the men, even high chiefs, will be plaiting cocoa-nut fibre into *sinnet*, so much used in the construction of houses, canoes, and arms. And a great deal these people have to talk about: the politics of the groups, independent of the new element introduced by the cession of the country to England, the never-ending intrigues of the Tonguese immigration, the endeavour of missionaries, consuls, and traders to spread Christianity and civilization, are rather complicated, and give rise to a good deal of discussion and speculation.

When evening is coming on, the bure is beginning to fill; most of the fires between the sleeping-places are lit, and the natives are leisurely stretched on the mats, their legs cocked up the stages, like Yankees in a tavern—all smoking their cigarettos, made of self-grown tobacco and dry banana leaves. Now come the kavachewers, comely-looking youngsters, carrying the large wooden bowl, a cocoa-nut shell for drinking the beverage, the bamboo water-vessel, a handful of fibre for straining the kava, and the root of the South Sea pepper from which it is prepared. No sooner have they taken their seat, and commenced chewing, taking care to throw the rope affixed to the kava-bowl toward the person highest in rank, than a leading man, perhaps a heathen priest, begins chanting a song, in which the whole assembly joins; and two young fellows beat time with little sticks, applied on a bamboo or any other sounding wood that happens to be handy. The leader of the chant does not sit motionless, but waves his body, arms, and hands in such a variety of ways, and with such extreme ease, that you fancy you can imitate him as readily as the whole assembly does. But the very first time you fail, to the great delight of your native spectators. His motions are not difficult, but you never know what they are going to be until it is too late to imitate, and he has already passed on to something else. The interest of this bye-play is thus well kept up, and the Fijians deserve full credit of having obtained hold of one of the great secrets of fixing the attention on an object, or making it, in other words, interesting. They know the art of concealing the end as long as possible. What would our novelists do without the use of this machinery! How dull would life itself be if the future was unveiled to us!

The lads, having chewed a sufficient quantity of the root, place the masticated mass into the bowl. Now water is poured on, the whole yellowish-looking fluid strained through fibres, and a cup filled. Whilst the cup-bearer is holding it to hand to the chief or highest personage present, an old man gives the toast of the evening. It is pathetic or humorous, as occasion demands, and listened to with attention; all singing and beating with sticks having ceased the moment the cup was filled. A general shout follows the conclusion of this toast, the cup is emptied in one draught, and thrown by the drinker on the mat, to be filled again and handed to the next in rank, until the whole assembly has been served.

The song becomes less and less hearty, the conversation slackens, and one by one the men drop off to sleep. Strange sight! Their pillows are made of a thick stick, have four legs, and are put just under the neck, so that the hair of the sleepers may not be deranged. They have had it only recently newly done up, washed with lime to make it frizzed like that of negroes, dyed in various colours, and arranged in many different ways. Several days must have been spent to get some of these extraordinary heads dressed. And for this reason—no other—they are ready to sleep all their lives on a pillow made of a stick of wood, and so constructed that a European could not rest his neck five minutes upon it without suffering dreadful pain. It is very fine talking about the ease of living in a state of nature, but the inconveniences to which savages put themselves in order to gratify their vanity are quite as great, if not greater, than those forced upon us by the fashions and dictates of our own society. Think of the agonies of tatooing! What would the natives give to escape it, if society would let them? But the stern laws of fashion allow of no exception. In Fiji this practice is confined to the women, the operation being performed by members of their own sex, and applied solely to the corners of the mouth, and those parts of the body covered by the scanty clothing worn by them. The skin is punctured by an instrument made of bone, or by the spines of the shaddock-tree; whilst the dye injected into the punctures is obtained chiefly from the candle-nut. No reason is given for the adoption of the custom, beyond its being commanded by Degei, their supreme god. Neglect of this divine commandment is believed to be punished after death. The men probably refrain from tatooing, because their skin, generally speaking, is so dark that the designs would not be seen, and the painful operation undergone would be mere labour thrown away.

In Polynesia tatooing seems to have attained its culminating point in the Society Islands and the Marquesas, where both men and women submitted to it; proceeding thence eastward to Samoa and Tonga, we find it restricted to the men; in Fiji to the women, and altogether ceasing in the New Hebrides. Yet, strange to add, Polynesian tradition asserts that the custom was known in Fiji before its being adopted in Samoa and Tonga. Two goddesses, Taema and Tilafainga, swam from Fiji to Samoa, and on reaching the latter group, commenced singing, "Tatoo the men, but not the women."* Hence the two were worshipped as the presiding deities by those who

* Turner's 'Nineteen Years in Polynesia,' p. 182.

followed tatooing as a trade; for a trade it was and is, quite as much as tailoring is in our own country, and requiring by far greater care and caution. The blue tracery once made cannot, like a coat or pair of trousers, be thrown aside when spoilt in the cut, but has to be worn for life, exposed to all the remarks which good and ill-natured friends may be disposed to make. A tradition, current in Tonga and Fiji, corroborates the fact of tatooing having been derived from the latter group. It is stated, that at a remote period the king of Tonga (Tui Toga) sent a mission to Fiji, in order to ascertain whether, as had been reported, the women of those islands were tatooed. On reaching the island of Ogea, in the eastern part of Fiji, the mission, with some difficulty, made the natives comprehend that they wished to find out what sex was tatooed (qia); to which the Fijians replied, "Qia na alewa" (women are tatooed). In obedience to orders, the first person met had been asked, and as a plain answer to a plain question had been obtained, the mission departed homewards. There were no other means of remembering the answer than by repeating it continually. This was done without interruption until their canoe reached the Ogea passage, where, the sea becoming rough, apprehensions about the safety of the canoe began to be entertained, and in the ensuing excitement the repetition of the precious words was neglected. Suddenly the neglect was perceived, and it was asked all round what the words were. Somebody replied, "Qia na tagane" (men are tatooed), instead of "Qia na alewa" (women are tatooed); which mistake, passing unnoticed, was repeated until the crew reached Tonga; and on being reported to the king, he exclaimed, " Oh, it is men, not women that are tatooed! well, then, I will be tatooed at once." The example set was speedily followed; hence the custom, that in Fiji the women, in Tonga the men are tatooed; hence also, adds the tradition, the name of the Ogea passage, " Qia na tagane."*

Kuruduadua accompanied us on an exploring trip down the Navua river, which we found to have several deltas, one of which is called Deuba. We passed the mouth, and went several miles westward, as far as Vanuadogo point, which is near Qamo peak. Close to one of the villages we stopped at there was a miniature temple, built of tree-fern wood, and thatched with Makita-leaves. Here parties of young men assemble for several weeks in order to practise certain tricks, which, when they are perfect in them, are exhibited before a numerous audience, but as long as they are practising nobody is supposed to go near them. On the day of the performance, the actors oil their bodies well and dress in white native cloth. The spectators, old and young, having formed a ring around them, the actors commence by chanting songs and beating time on bamboos, until they have worked themselves up to a certain pitch of excitement. Now a spirit (*Kalou Rere*) is supposed to enter them, and they pretend to be invulnerable to spear, proof against musket ball, and safe against the effects of heat or flame. By sleight of

* Another version of the tradition is given by Williams, 'Fiji and Fijians,' vol. i. p. 160, where a man, repeating the intelligence, violently strikes his foot against the stump of a tree, and in the confusion ensuing changes its tenor.

hand, they endeavour to make good their pretensions. A spearhead is softened so as not to hurt when thrown; the ball put in the musket is too small, and thus rolls out when the actors begin to dance about previous to discharging it; and the fiery oven into which a man creeps and allows himself to be covered up, has a tunnel and vent-hole, by which he has a chance of escaping. Accidents, however, will happen even in this well-regulated community. The spear unskilfully handled has been known to hurt; too much wadding put into the gun has prevented the ball from rolling out; the tunnel has been apt to fall in, and after some hours the man who allowed himself to be thrown into it, has been found to be perfectly baked. The Kalou Rere, with its high poles, streamers, evergreens, masquerading, trumpet-shells, chants, and other wild music, is the nearest approach to dramatic representation the Fijians seem to have made, and it is with them what private theatricals are with us. They are also on other occasions very fond of dressing themselves in fantastic, often very ridiculous costume; and in nearly every large assembly there are buffoons. Court fools, in many instances hunchbacks, are often attached to a chief's establishment.

Finding that Kuruduadua was a man in whom confidence could be placed, we made arrangements for going to Namosi, so as to connect the discoveries of Macdonald and Samuel Waterhouse with the southern coast of Viti Levu; but, as the weather had become extremely boisterous, and heavy rains had rendered travelling in the interior impossible, we determined to wait for more favourable weather, and return at once to Lado.

As a heavy south-east gale was blowing, the chief told us we should not be able to proceed very far, and he hoped that if on reaching the sea we should find it too boisterous, we would not mind coming back. We were out of tea, biscuit, and all the other necessaries a European requires, unable to walk about,—the heavy rain having rendered the neighbourhood of Navua a perfect swamp,—and tired of staying indoors and waiting for the weather to clear up; so we left on the morning of the 9th of July. The sea was rougher than we had expected. We had to bale constantly, and therefore effected a landing on the sandy beach, and walked to Taguru, where we had to stay two days. The boat, lightened, reached the place with difficulty. On the third day the gale and rain, which now had lasted a week, abated, and we pushed on once more. Calling at Naigani Island, we heard from Mr. Work, whom we found quite established in his new home, that the Kau karo, or itchwood, the poisonous properties of which had caused Mr. Eggerström to be ill for two months, grew on the banks of a small river of Viti Levu, nearly opposite the island. We at once made up our mind to fetch specimens, in order to ascertain the real name of the tree. We had no difficulty in finding it, and it proved to be *Oncocarpus Vitiensis*, A. Gray, or, as Foster nearly a hundred years ago called it, *Rhus atrum*. There was a considerable village about a mile and a half up the river, which we could reach in our gig. The inhabitants looked dreadfully unhealthy; most of the men had elephantiasis, and many of the children were covered with ulcers. No doubt the site of the village in a low valley in a great measure accounted for this. We were roving over the hills, when a message from Ovalau reached us with the glad tidings that Colonel Smythe had safely arrived in Levuka, and was desirous of seeing us.

Without loss of time we returned to Mr. Work's house, left it after midnight, and reached Mataisuva at eight o'clock in the morning, where we breakfasted with our kind friend Mr. Moore. There had been some trouble with the natives. An Englishman had run away with the wife of a Viwa chief, and refused to give her up. The chief, justly exasperated, threatened revenge, and would have proceeded to extremities, if Mr. Moore had not persuaded the Rewa chief, in whose territory the eloped one resided, to step in, on the grounds that the Viwa chief had no right to cause a disturbance on territories not his own. They referred the case to Mr. Pritchard, who remonstrated with the white man, telling him that if he, an Englishman, was clubbed in consequence of the provocation given, no government could possibly ask for satisfaction; and on the other hand, that if no notice were taken of his murder, the lives of the other whites would be in danger. So the woman must instantly be given up.

We had hoped to reach Lado that day, but the loss of time caused by this troublesome man delayed our departure until noon. We again passed through the Rewa river and the Kele musu canal, and towards sunset reached Kaba, where we took up our quarters at the house of Peter, a Tonguese teacher connected with the Wesleyan mission. He was a fine specimen of his race, and made us as comfortable as his means permitted. This man and a boy had been saved from drowning by our interpreter, Mr. Charles Wise, whom he welcomed with cordiality. When picked up at sea, he had been several days in the water—incredible as it may appear. His canoe had been upset, and his companions, all good swimmers, had against his entreaties separated from him, and they had all perished, being probably eaten by sharks These animals were furious in their attacks, and Peter killed several of them with his knife during the time he was in the water; they troubled him little during the night, but became very rapacious as soon as daylight was established. He was also attacked by a small sea-animal which bored regular holes into his flesh, and would have caused his death if he had not been speedily delivered. When Wise took him on board, he was perfectly exhausted, and continually cried for water. Every means were used to restore his strength; his body was oiled, and food and drink given to him.

When the moon rose we took our departure, and early next morning reached Lado Alewa, in Port Kinnaird.

CHAPTER VII.

ARRIVAL OF COLONEL SMYTHE FROM NEW ZEALAND.—THE 'PEGASUS' AND 'PAUL JONES.'—VISIT TO BAU.—QUARRELSOME DISPOSITION OF THE CHIEF OF THE FISHERMEN.—CESSION OF FIJI TO ENGLAND.—FIRST OFFICIAL INTERVIEW WITH THE KING.

THE native war in New Zealand continuing and keeping all available naval force employed, Colonel Smythe had been unable to obtain a Government vessel to take him to and about Fiji, and had therefore been compelled to charter the ' Pegasus,' an extremely slow-sailing, ill-manned ketch, commanded by a gossiping captain, who ultimately returned to New Zealand without paying even the crew, which the British Consul had been obliged to put on board. Mr. Pritchard and myself called on Colonel Smythe on the 16th of July, and regretted to hear of his long and stormy passage. He had arrived on the 5th of July, and we found him comfortably quartered at Levuka, in the house of Mr. Binner. Mrs. Smythe was making a water-colour drawing of the Levuka reef, which from Mr. Binner's house, situated as it is on the top of a hill, displays itself in all its grandeur, and together with the little islands at a distance, and the shipping of the port, forms a panorama not easily matched.

The ' Pegasus ' not having accommodation for more

than Colonel and Mrs. Smythe, Mr. Pritchard and myself chartered the 'Paul Jones,' the same little schooner which fetched me from Somosomo. She was scarcely better than an open boat, and we had to wash, dress, and take our meals on deck, the cabin being too small to hold more than two bunks, an apology for a table, and two lockers serving also as substitutes for benches. But we managed very well, and as she beat the 'Pegasus' even in short distances by whole days, we generally reached our destination long before Colonel Smythe's party did, and soon transferred our abode on shore. When I came from Somosomo she was swarming with cockroaches, to such an alarming extent that there was no staying in her; and when going to sleep we had to cover our faces, to screen at least that part of our bodies against attack. But she had since been sunk under water,—the only method here practised to free vessels from that pest,—newly painted, and done up, so that as far as her size would allow she was tolerably comfortable. Besides Mr. Storck, we had Mr. Charles Wise, the consular interpreter, on board, a half-caste who had been brought up by the late Rev. John Hunt, for whose memory he entertained a warm admiration, justly shared by all who knew that excellent man.

It was arranged with Colonel Smythe, that we should visit the principal chiefs, commencing at Bau, the capital of the group. The two vessels met at Port Kinnaird; and we finally left Lado, at that time the Consul's residence, on the 24th of July, at noon. The 'Paul Jones' anchored off Bau on the same day, but the 'Pegasus,' to give an instance of her bad sailing qualities, only arrived on the following day late at night.

There was a serious quarrel between the Chief of the Fishermen and Ratu Abel, the King's eldest son, the former having insulted the Queen, and the latter sent him a challenge in consequence. A duel was impending when we arrived, and the British Consul's persuasive powers were appealed to by various parties. Mr. Pritchard publicly asked the Chief of the Fishermen why he had offered the insult to his sovereign, but he refused to answer; Mr. Pritchard then told him he would wait for an answer, even if he had to sit up all night. The Chief, seeing that the Consul was as good as his word, and that there was no escape possible, after a silence of two hours gave the desired answer, begged the King's pardon, and all was arranged amicably. Ratu Abel was present during the whole interview, and behaved extremely well in the affair. He is a fine specimen of a Fijian prince, and will doubtless succeed his father to the throne, though some of the missionaries have been trying to persuade the King to change the law, by settling the succession upon his younger son, born after he had become converted to Christianity, and married according to our rites. But such a change would doubtless lead to endless complications and confusion, and be unjust towards a child perfectly legitimate according to the custom prevailing at the time of his birth. It is in petty interferences like these that, doubtless much to the regret of the enlightened minds composing the Board directing the truly grand machinery of the Wesleyan Society, the missionaries draw upon themselves the censure of people who fully sympathize with the noble work they have in hand, and who would do anything in their power to advance their true interest. Be it known, that interference in politics on the part of the Wesleyan missionaries is decidedly disapproved of by their Board at home, and that stringent instructions are published to that effect.

The Chief of the Fishermen, an important body in Bau, is a scheming fellow, who more than once caused mischief. On one occasion, when some British interest was involved, Mr. Pritchard, who, born and bred in Polynesia, is perfectly familiar with native modes of thought, and owes a great deal of his influence to it, wished to impress the chief with the idea that whatever plots he was hatching they were sure to be found out by those more clever than himself.* Instead of stating this in such language as one European would use to another, he said to the native, " As Chief of the Fishermen, you know all the fishes, the small as well as the big, and of course the turtle, according to your notions the king of the whole." The Chief smiled assent, flattering himself that by the turtle he himself was alluded to. To the great delight of the bystanders, the Consul continued:—" Familiar with all

* Commodore J. B. Seymour, writing to the Lords Commissioners of the Admiralty, in a letter dated, Auckland, September 2, 1861, and published in the ' Correspondence relating to the Fiji Islands,' presented to both Houses of Parliament by command of her Majesty, May, 1862, seemed also favourably impressed with Mr. Pritchard's way of dealing with the natives :—" I cannot conclude this letter," he writes, " without expressing the obligations I am under to Mr. Pritchard, whose manner with the native chiefs (being neither too deferential nor the reverse) seemed to me to be exactly what it should be. He speaks the language, and is evidently liked by all parties of Fijians ; and without his ready assistance . . . it would have been impossible to arrive at so speedy a settlement of affairs."

its habits, you are aware that at certain periods this king goes on shore to lay its eggs, and you, knowing its way, look for its footprints on the white coral sand of the beaches, and suddenly light upon what is hatching." No further amplification was required to make the chief comprehend the drift of the story. The bystanders saw at a glance that the chief had put his foot in it the moment he identified himself with the king of the fishes, and that his plots were so clumsily constructed that anybody who knew him could easily trace them out.

The public interview with King Cakobau, or Thakombau, was to take place on the 27th of July, when he would once more confirm the cession of his country made to Great Britain in 1858, through Mr. W. Pritchard. In order to place the whole subject fairly before the reader, it will be necessary to insert here the original deed of cession:—

"*Cession of Fiji to England, and Ratification of it by the Chiefs.*

"EBENEZER THAKOMBAU, by the grace of God, sovereign chief of Bau and its dependencies, Vunivalu of the armies of Fiji, and Tui Viti, etc., to all and singular to whom these presents shall come, greeting.

"Whereas we, being duly, fully, and formally recognized in our aforesaid state, rank, and sovereignty, by Great Britain, France, and the United States of America, respectively;

"And having full and exclusive sovereignty and domain in and over the islands and territories constituting, forming, and being included in the group known as Fiji, or Viti;

"And being desirous to procure for our people and subjects a good and permanent form of government, whereby our aforesaid people and subjects shall enjoy and partake of the benefits, the prosperity, and the happiness, which it is the duty and the right of all sovereigns to seek and to procure for their people and subjects;

"And being in ourselves unable to procure and provide such good and permanent government for our aforesaid people and subjects;

"And being, moreover, in ourselves unable to afford to our aforesaid people and subjects the due protection and shelter from the violence, the oppression, and the tyranny of foreign Powers, which it is the duty and the right of all sovereigns to afford to their people and subjects;

"And being heavily indebted to the President and Government of the United States of America, the liquidation of which indebtedness is pressingly urged, with menaces of severe measures against our person, and our sovereignty, and our islands and territories aforesaid, unless the aforesaid indebtedness be satisfied within a period so limited as to render a compliance with the terms of the contract forced upon us utterly impossible within the said period; this said inability not arising from lack of resources within our dominions, but from the inefficacy of any endeavours on our part under the existing state of affairs in our islands and territories aforesaid, to carry out such measures as are necessary for, and would result in, the ultimate payment of the aforesaid claims; and having maturely deliberated, well weighed, and fully considered, the probable results of the course and the measures we now propose; and being fully satisfied of the impracticability by any other course and measures to avert from our islands and territories aforesaid, and our people and subjects aforesaid, the evils certain to follow the non-payment of the sum of money demanded from us by the Government of the United States of America;

"And being confident of the immediate and progressive benefits that will result from the cession herein now made of our sovereignty, and our islands and territories aforesaid;

"Now know ye, that we do hereby, for and in consideration of certain conditions, terms, and engagements, hereinafter set forth, make over, transfer, and convey, unto Victoria, by the grace of God, Queen of the United Kingdom of Great Britain and Ireland, etc., her heirs and successors for ever, the full sovereignty and domain in and over our aforesaid islands and territories, together with the actual proprietorship and personal ownership in certain pieces or parcels of land as may hereafter be mutually agreed upon by a commission, to consist of two chiefs from Great Britain and two chiefs from Fiji; the said commission to be appointed by the representative of Great Britain in Fiji, who, in case of dispute, shall himself be umpire; the said pieces or parcels of land to be especially devoted to government purposes, and to be applied and appropriated in manner and form appertaining to Crown lands in British colonies, or as the local government of Fiji, appointed by commission from the aforesaid Victoria, Queen of the United Kingdom of Great Britain and Ireland aforesaid, may deem fit, proper, and necessary, for the use and requirements of the said local government;

"Provided always, and the cession of our sovereignty and our islands and territories is on these conditions, terms, and considerations, that is to say;

"That the aforesaid Victoria, Queen of the United Kingdom of Great Britain and Ireland aforesaid, shall permit us to retain the title and rank of Tui Viti, in so far as the aboriginal population is concerned, and shall permit us to be at the head of the department for governing the aforesaid aboriginal population, acting always under the guidance, and by the counsels, of the representative of Great Britain and head of the local government appointed by commission from the aforesaid Victoria, Queen of the United Kingdom of Great Britain and Ireland aforesaid;

"That the aforesaid Victoria, Queen of the United Kingdom of Great Britain and Ireland aforesaid, shall pay the sum of forty-five thousand dollars ($45,000) unto the President of the United States of America, being the amount of the claim demanded from us, procuring for us and for our people a full and absolute acquittance from any further liabilities to the said President or Government of the United States of America aforesaid;

"For and in consideration of which outlay, not less than two hundred thousand (200,000) acres of land, if required, shall be made over, transferred, and conveyed, in fee-simple, unto Victoria, aforesaid Queen of the United Kingdom of Great Britain and Ireland aforesaid: the selection of which said land shall be made by the commission hereinbefore named and referred to, to reimburse the immediate outlay required to liquidate the aforesaid claim of the President and Government of the United States of America;

"And we, the aforesaid Ebenezer Thakombau, by the grace of God, sovereign chief of Bau and its dependencies, Vunivalu of the armies of Fiji and Tui Viti, etc., do hereby make this cession, transfer, and conveyance, of our sovereignty, and of our islands and territories aforesaid, unto the aforesaid Victoria, by the grace of God, Queen of the United Kingdom of Great Britain and Ireland, etc., aforesaid, her heirs and successors for ever, on behalf of ourselves, our heirs and successors for ever; on behalf of our chiefs, their heirs and successors for ever; on behalf of our people and subjects, their heirs and successors for ever; hereby renouncing all right, title, and claim unto our sovereignty, islands, and territories aforesaid, in so far as herein stated;

"In witness whereof, we have hereunto set our hand and affixed our seal, this twelfth day of October, in the year of our Lord one thousand eight hundred and fifty-eight.

his
TUI VITI, ×   (L. S.)
mark.

"Signed, sealed, and ratified by the aforesaid Tui Viti, and by him formally delivered, in our presence, unto William Thomas Pritchard, Esq., Her Britannic Majesty's Consul in and for the aforesaid Fiji; the aforesaid Tui Viti, at the same time, affirming and admitting to us personally, that he the said Tui Viti fully, wholly, perfectly, and explicitly, understands and comprehends the meaning, the extent, and the purpose of the foregoing document, or deed of cession; and I, the undersigned John Smith Fordham, formerly of Sheffield,

England, but now temporarily residing at Bau, Fiji, aforesaid, do hereby solemnly affirm that I myself, fully, wholly, and explicitly translated the foregoing deed of cession unto the said Tui Viti, in the presence of the aforesaid William Thomas Pritchard, Esq., Her Britannic Majesty's Consul in and for the said Fiji, Robert Sherson Swanston, Esq., His Hawaiian Majesty's Consul in and for Fiji aforesaid, and John Binner, formerly of Leeds, England, but now resident at Levuka, Island of Ovalau, Fiji, aforesaid.

" In witness whereof, we have each and all set our respective names and seals, this twelfth (12th) day of October, in the year of our Lord one thousand eight hundred and fifty-eight aforesaid.

"JOHN SMITH FORDHAM, Wesleyan Missionary. JOHN BINNER, Wesleyan Mission Trainer. ROBERT S. SWANSTON, Hawaiian Consul, Fiji. WILLIAM T. PRITCHARD, H. B. M. Consul."

" We hereby acknowledge, ratify, and renew, the cession of Fiji to Great Britain, made on the 12th day of October, 1858, by Thakombau. In witness whereof we have hereto affixed our names this 14th day of December, 1859.

" RABICI ROKO TUI DREKETI (his × mark), of Rewa.
JIOJI NANOVO (his × mark), of Nadroga.
NA WAGA LEVU (his × mark), of Rakiraki.
TUI LEVUKA (his × mark), of Ovalau.
KOROI COKANAUTO (his × mark), of Bau.
KOROI TUBUNA (his × mark), of Tavua.
NAIBUKA KOROIKASA (his × mark), of Nakelo.
RATU ISIKELE (signed), of Viwa.
TUKANA (his × mark), of Noco.
TUBAVIVI (his × mark), of Rakiraki.
CURUICA (his × mark), of Korotuma, Ra Coast.
SESEBUALALA (his × mark), of Korotubu.
TUDRAU (his × mark), of Dravo.
SAMISONI (signed), of Viwa.
NA GALU (his × mark), of Namena.

RATIFICATION OF CESSION.

"KOROIKAIYANUYANU (his × mark), of Lasakau.
DABEA (his × mark), of Kuku, Viti Levu.
KO MAI VUNIVESI (his × mark), of Nakelo.
PITA PAULA (his × mark), of Viwa.
TUI BUA (his × mark), of Bua.
THAKOMBAU (his × mark), of Fiji.

"We hereby certify that the foregoing chiefs have signed this document with a full understanding of its meaning, in our presence, this 14th day of December, 1859.

"HT. CAMPION, Commander, R.N., H.M.S. Elk.
WILL. T. PRITCHARD, H.B.M. Consul.

"We hereby certify that we translated the foregoing document to the Chiefs who have signed, and that they thoroughly understand its meaning.

"W. COLLIS, Wesleyan Mission Training Master.
E. P. MARTIN, Wesleyan Mission Printer.

"January 16th, 1860, at Levuka.

"RITOVA (his × mark), of Macuata.
TUI CAKAU (his × mark), of Taviuni.
TUI BUA (his × mark), of Bua.

"Witness to marks:

"JOHN CAIRNS, Owner of 'Lalla Lookh,' and Merchant of Melbourne.

"TUI TAVUKI (his × mark).
TUI BUKELEVU (his × mark).
TUI YAME (his × mark).
TUI NAKASALEKA (his × mark), per Qarinivalu of Nukuraleka.
VERI LEVU (his × mark), of Yali.
RATU SAVUNOKO (his × mark), of Ono and Januiana.
TUI NACEVA (his × mark).
Witness to Tui Naceva's mark, C. J. Baird.

"Translated by us, before whom the above Chiefs made their marks, this 15th day of August, 1860:

"JAMES S. H. ROYCE; CHARLES WISE.

"I hereby ratify the above cession, Navua, Sept. 4th, 1860.

"KURUDUADUA, (his × mark).

"Witnesses to signature:

"BERTHOLD SEEMANN, Ph.D.; W. T. PRITCHARD, Consul."

Precisely at eleven o'clock on the morning of the 27th of July, the King fired a salute. When arriving at the place of meeting, the royal residence, we found the King and Queen, both dressed in European fashion, the former in a blue uniform, seated on chairs, of which several had been arranged in a semicircle for our use. There were present, besides Colonel Smythe, Mr. Pritchard and myself, Messrs. Fordham and Collis from the mission, not to mention the ladies. Ratu Abel, the King's eldest son, a fine-looking fellow, was absent, but sent for, and the chiefs and principal landholders soon dropped in, all dressed in native costume. Mr. Fordham interpreted for Colonel Smythe, Mr. Charles Wise for Mr. Pritchard. I wrote down all at the time, and the following, obtained from both sources, may be regarded as a faithful *résumé* of what was spoken:—

"It having been represented to Her Britannic Majesty," said Colonel Smythe, addressing King Cakobau, "that the King and Chiefs of Fiji are disposed to become British subjects, her Majesty has directed an inquiry to be made into the matter, and hear what King and Chiefs have to say on the subject, in order that it may be reported to her."

The King replied: "The arrangement respecting the cession entered into with Mr. Consul Pritchard is still in full force, and shall not be disturbed by any foreign Power."

"Great Britain," continued Colonel Smythe, "produces many things that Fiji does not, and *vice versâ*, so that by an exchange of products the two countries would be mutually benefited. I refer especially to cotton, which grows luxuriantly in Fiji, and is valuable in England."

The King replied: "I am fully aware of it; and in consequence of what Mr. Consul Pritchard told me at the interview at Levuka, about the desirableness of cultivating this article, I have directed it to be planted, and my commands have been carried out to some extent."

"In ceding the country," Colonel Smythe resumed, "every man will retain his own property and land, and everybody will be protected, so that a stop will be put to the fearful feuds that have decimated the population."

The King rejoined: "There may be people in the group who at present cannot fully appreciate that idea; but it is somewhat like Christianity, which, though a blessing, is looked upon with prejudiced eyes by many not familiar with its beneficial tendency."

When the chiefs and landholders were asked whether they had any observation to make, they remained mute, and at the conclusion of the whole raised shouts of approval. All then retired, and nothing more was said except what has been stated in substance above. Colonel Smythe states, in one of his official communications, as printed in the Blue-books, that the King "could not convey to Her Majesty 200,000 acres of land as consideration for the payment of these claims for him, as he does not possess them, nor does he acknowledge to have offered more than his consent that lands to this extent might be acquired by Her Majesty's Government for public purposes in Fiji." Nothing to this effect was broached during the official interview; on the contrary, the King distinctly said, that "the arrangement respecting the cession entered into with Mr. Consul Pritchard is still in full force." Nor was the Consul aware that Colonel Smythe had on any other occasion elicited information from the King that could be thus construed. It was perfectly well understood by all the leading chiefs that each and all would have to make over a certain portion of land, in payment of the debt fastened upon them by the American Government; and Bau, and King Cakobau as its representative, would have borne his share to make up the 200,000 acres. The very fact that all the chiefs, without any exception, and even those living in the remotest districts, ratified the deed of cession, proves that King Cakobau was backed by all the influence of his country, and had a perfect right to cede the sovereignty of the islands.*

* In order to place this fact beyond dispute, I have printed the names of all those chiefs who ratified the deed of cession,—this ratification being a document omitted in the Blue-book on Fiji. Some information as to the real position of Bau in Fiji will be found at pp. 74–80 of the present work.

CHAPTER VIII.

EXCURSIONS TO KOROIVAU AND NAMARA.—DEPARTURE FROM BAU.—PASSAGE THROUGH THE GREAT RIVER OF VITI LEVU.—BURETU.—APOSTATE CHRISTIANS.—REWA.—ARRIVAL AT TAVUKI, KADAVU.—WHALE SHIPS.—ATTEMPT TO ASCEND BUKE LEVU.—THE ISTHMUS OF KADAVU.—GA LOA OR BLACK DUCK BAY.—DEPARTURE FOR NAVUA,

I TOOK advantage of our stay at Bau, which lasted till the 2nd of August, to pay several visits to Namara, Koroivau, and several other parts of Viti Levu. There was a fine pyramidal temple at Namara, no longer used for religious purposes, and near it was standing an isolated Fan-palm (*Pritchardia Pacifica*, Seem. et Wendl.), both objects peculiarly Fijian. The natives here were extremely friendly, and carried us through bogs and mud when occasion required. At first, the children, on seeing our white faces, were much frightened, and some boys and girls from twelve to fourteen years old would run for their lives when we attempted to get near them or even looked hard at them. However, they soon got reconciled to our colour, or rather want of colour, and a few jew's-harps and beads, judiciously distributed, would make them as happy as kings and quite attached to us. The women were busy grating the seeds of the Ivi (*Inocarpus edulis*, Forst.), now ripe, and made into bread. The hill-sides were planted with a great number of pine-apples and cassava-root, and around nearly all the yam, banana, and sweet-potato patches I observed the cotton-trees, which had been planted by order of the King and at Mr. Pritchard's instigation. The village of Koroivau was a complete cotton garden; the trees were twelve to fourteen feet high, and formed regular avenues in the streets. In my rambles in the forest I met with some natives who were clearing pieces of ground for cultivation. They were extremely friendly, and invited me to partake of some wild yams ("Tivoli") which they had just been roasting in the hot ashes. I gladly availed myself of their offer, and found the roots like cultivated yams, and quite as good in taste. Though no smoker myself, I carried a pipe and tobacco, which passed from mouth to mouth, every one having a few puffs and then passing it on to his neighbour; and when I intimated to them that the pipe was theirs, and presented an additional stick of American tobacco, they were highly pleased, and hoped that I would soon come again to "gather leaves." In the swampy parts of the forest I found a new Aroideous plant, the Viu kana (*Cyrtosperma edulis*, Schott) under cultivation. Like the Taro, or Dalo, as it is here termed, which it somewhat resembles, its root is edible, and very much used.

We left Bau on the 2nd of August, early in the morning, our party consisting of Colonel and Mrs. Smythe, Mr. and Miss Pritchard, Mr. Collis and myself, all embarked in two boats belonging to the mission, and proceeding to Rewa by way of the river and the canal, a route, it will be remembered, which Mr. Pritchard and myself took on a former occasion. After two or three days' rain and gale, there was a temporary lull in the weather, and our trip was altogether a pleasant one. About noon we halted at Buretu, a fortified town, which has never been taken, and is therefore regarded as impregnable. If it is so, that must be owing entirely to the bravery of its inhabitants, for the low walls with which it was surrounded did not impress us with any great strength. Some years ago a good number of the Buretu people embraced Christianity, but when at a subsequent date the town rebelled against Bau, they became apostates, nor did the restoration of peace make them relinquish their pagan religion, and they had at the time of our visit, one of the finest temples in the whole group. These and similar fluctuations must be expected in all attempts to introduce a new faith, but from which Fiji has been more free than many other countries similarly operated upon. Wherever Christianity was preached in the group it took a quick and firm hold, and the ultimate conversion of the whole population is merely a matter of time and £. s. d. If the Wesleyan Society had more funds at its disposal, so as to be able to send out a greater number of efficient teachers, a very few years would see the whole of Fiji christianized, as all the real difficulties formerly in the way of the mission have now been removed. On my representing the case in this light, his Majesty the King of Hanover was graciously pleased to subscribe as his first gift, £100, towards so desirable an object, at the same time expressing his admiration for the labours of individual missionaries I named. If the Fijis should be taken by any European government, the prosperity of the country would best be advanced by placing ample funds at the disposal of the Protestant missionaries for the christianization of the natives, for which the machinery as now worked by the Wesleyans would offer the most efficient and readiest means. The Catholics would probably effect the christianizing part with a lesser outlay, but it must not be forgotten that one of the great advantages of Protestant missions is, that they civilize as well as christianize, whilst the Catholic priests, having no home, no family life to exhibit for imitation, simply christianize.

We reached Rewa, or rather Mataisuva, the mission station, about three o'clock in the afternoon, and were scarcely sheltered in safety, Colonel Smythe and his wife with Mr. Waterhouse, the chairman of the Fijian district of the Wesleyan mission, Mr. Pritchard and all the rest of us, with Mr. Moore, than a strong south-easterly gale, accompanied a heavy rain, commenced, which lasted for six days. Our vessels had been ordered to round the south-east extremity of Viti Levu, and call for us at Rewa; but this bad weather had baffled all their attempts, and the 'Paul Jones' thought it best to endeavour to come through the canal, which connects the two branches of the great river of Viti Levu,—an attempt which proved quite successful.

At Rewa, a meeting of all the chiefs and landholders was held, and the same proceedings gone through as at Bau. All expressed themselves in favour of ceding their country to England in the manner already detailed. Amongst those assembled was a son, still a boy, of Cakonauto, better known amongst the whites as Philips, a chief friendly to civilization and the whites. During his lifetime, he had accumulated a great number of European and American manufactures, curious clocks, musical boxes, etc., but on inquiry I found that all these things had become scattered. His son would ultimately succeed to the chieftainship, and was made a great deal of by his people. At present the government was in other hands. He was a comely-looking youth, of a much lighter complexion than the rest of his countrymen.

The 'Pegasus' being again late, Mr. Pritchard and I started for Kadavu (Kandavu), the largest of the southernmost islands of the group. Leaving Rewa road on the 13th of August at six P.M., we made Tavuki Bay, on the northern side of the island, at seven o'clock on the following morning, where we took up our quarters under the hospitable roof of Mr. Royce, one of the resident missionaries. In consequence of the strong southeasterly gale, the temperature was very agreeable, and during the previous week Mr. Royce observed the thermometer to go down to 62° Fahrenheit, the lowest ever observed in the group.

There were three American whaleships in the bay, taking in wood, water, and fresh provisions, commanded by Captain James Nicols Charles Nicols, and Thomas Sulivan. They had been nearly all their lives in the South Sea whaling trade, and were very well known to Mr. Pritchard when he was at Samoa. Their business had evidently been a lucrative one, and this was to be one of their last, if not their last voyage. They had hitherto taken in their supplies at Samoa or Tonga, but the natives of those two groups had become so exorbitant in their charges as to render it imperative to look for cheaper provision markets. Fiji had answered their purpose much better, and they predicted the arrival of a regular whaling fleet as soon as the great facilities here offered should have become more generally known amongst the trade. Having their families with them, they gave us several pressing invitations to come on board, which the Consul, myself, and all the missionaries gladly accepted. These vessels enjoyed the reputation of being patterns of what whaleships should be; and I must record my surprise at the scrupulous neatness, cleanliness, and even elegance prevailing. The Captain's cabins were fitted up and kept better than I have ever seen them in any vessel.

When our friends heard that we were anxious to ascend Buke Levu, the great mountain situated at the western extremity of Kadavu, they offered us one of their whale-boats for that purpose; and one of their mates, a skilful steerer, volunteered to pilot us to the foot of the mountain. Mr. Pritchard and I left Tavuki 13th of August early in the morning. It was quite fine when we started, but after an hour's pull, a gale sprang up, and after being nearly swamped in going through a narrow passage of a reef, where the water was breaking, we were compelled to postpone our attempt to a more favourable time, and land at Yawe, a town famous in Fiji for its very large specimens of pottery, made without a wheel, and taking as our crockery does, its name from the place of manufacture. We hoped that it might clear up during the night, to allow us to proceed in the morning; but the next day the rain was more heavy than it had been even during the previous one, and we had no option but to return to Tavuki. During the night our interpreter had heard that a circular letter had been received from the Tonguese chief Maafu, advising his countrymen how to act, so that the policy of England with regard to the cession of Fiji might be frustrated, and the country ultimately fall into the hands of Tonga; and also that a similar letter had been sent to Bega (Mbenga). The Tonguese teachers in the pay of the Wesleyan Society were made the agents for diffusing the burden of the message. When we got back to Tavuki Mr. Pritchard communicated what we had heard to Mr. Royce, and he sent for one of the leading Tonguese teachers, who made no secret of these machinations, and promised to procure the letter received in Kadavu. Ere two hours had elapsed he succeeded, and it is now in the Consulate. Mr. Royce pointed out the impropriety of teachers of the Christian religion allowing themselves to be used as tools in miserable political intrigues; but the Tonguese said that, however glad to be excused, they could not help themselves, and had to do what their chiefs told them. The doings of the Tonguese form an important chapter in the history of the Fijis, and will be treated under a separate heading, and I merely mention here this fact, because it has been disputed that the teachers allowed themselves to be used as political agents.

Tavuki, from being made the centre of the mission of the district, must be regarded as the capital of Kadavu, and is situated in latitude 19° 3′ 9″ south, longitude

178° 6′ 23″ east, according to observation taken by Mr. Sedmond, master of H.M.S. Harrier, 17, Captain Sir Malcolm M'Gregor. Tavuki is an open bay on the northern coast, with no deep water close to the shore, and at ebb tide one has to walk about half a mile over the coral reef before being able to reach the boats. The missionaries had endeavoured to make a pier, on which those whom the chiefs would wish to punish for any petty offences were made to work; but at the time of our visit little progress had been made, and one could almost have wished that a greater number of petty offences had been committed.

The island of Kadavu, of which so little is known, and no accurate hydrographical survey exists, is highly cultivated, notwithstanding its being so hilly, and rising on its western extremity four thousand feet high. A strong belief has sprung up that there must be gold, and old gold-diggers from the Australian colonies, judging from the formation of the quartz rocks, maintain that the island is auriferous. Quite recently Kadavu has been examined by two miners from Melbourne, who certainly did find a quartz reef, but not the precious metal they were in search of. The fact of the matter is, that neither of these parties had the means to provide themselves with proper tools for a thorough and final exploration. The discovery of gold has actually been reported from Vanua Levu. The population of Kadavu, said to number about ten thousand, is a mixture between the Fijian and Tonguese races, all of whom, with the exception of seven individuals, have nominally become Christians. The island is twenty-four miles long, stretching from east to west, and being contracted about the centre into the narrow isthmus of Yarabali, literally "Haul-across," so named from the fact of canoes and boats being dragged across it, in order to save the trouble and escape the danger of a long passage around the east and west point. Colonel Smythe and myself, in company with Mr. Royce, crossed it on the 16th of August, and found the northern portion of the isthmus a fine avenue of cocoa-nut palms, the southern more or less a mangrove swamp. A similar short cut for canoes is effected at Naceva Bay in Vanua Levu. On both sides of Yarabali there is a bay; the northern, Na Malata, is shallow and open; the southern, Ga loa, has deep water, good anchorage, and three passages through the reef outside, which acts as a natural breakwater. We found its shores full of pumice-stone, drifted here from the Tongan volcanoes. The different exploring expeditions having quite overlooked this fine bay, Mr. Pritchard made a rough survey in 1858, it being not improbable that if the much discussed communication between Sydney and Western America—the shortest route to England—should be established *viâ* Fiji, steamers would prefer calling at this southernmost bay, with plenty of sea-room outside, to running the risk of entering the labyrinth of rocks, shoals, and reefs, which render the navigation of the central parts of the group, in the absence of a complete chart, a rather difficult task.

Ga loa, or Black Duck Bay, derives its name from the largest of three islands situated in it. Ga loa island is two hundred feet high, about a mile long, and half a mile across, and full of fruit-trees. It was pointed out as the spot where, only a twelvemonth ago, a man was baked and eaten. Cannibalism in Fiji will soon number amongst the things that have been. The influence of all the whites residing in or visiting the group is steadily directed towards its extinction, and though a person who ought to have had more charity has asserted in print that *he had been told* some of the white residents were habitual partakers of human flesh, I think, for the honour of our race, such second-hand stories ought to be indignantly rejected. Antiquaries know that cannibalism of a certain form lingered in Europe long after the Reformation; that mummies, said to be Egyptian, were extensively used medicinally, and that only after it was found out patients had not partaken of the contemporaries of Thothmes I. or Rameses the Great, but of bituminized portions of their own fellow-countrymen, this precious quack medicine fell into absolute disuse. Even in our own times we may still meet in certain parts of Europe people doing what has been recorded with horror of the Fijians—that of drinking the living blood of man; but mark! with this essential difference, that the former, watching their opportunities at public executions, do it in hopes of thereby curing fits of epilepsy, whilst the latter did it to gratify revenge and exult over fallen enemies. As for a European, even of the lowest grade, coolly sitting down to a regular cannibal feast, the idea is too preposterous to have ever been allowed to disgrace the pages of a modern publication.

Taudromu, another of the islands of Ga loa Bay,

scarcely half a mile round, now belongs to an American Indian of real flesh and blood; and in former times was inhabited by Ratu-va-caki, a mighty spirit, who, with his sons, all like their father, of prepossessing appearance, and bearing poetical names,* seem to have played the same part in Fiji as the Erl-King and his daughters did in Europe. Many are the stories told of their deeds and adventures. Generally they used to go out together, but if Ratu-va-caki was disinclined, the boys, who, young rascals! had as keen an appreciation of a pretty face and a good figure as their old rake of a father, would rove about by themselves, principally moving about in heavy squalls and gales; hence their invisible canoe was termed "Loaloa;" and if, soon after stormy weather, any fine young girls suddenly died, it was proverbially said that Ratu-va-caki and his sons had carried off their souls. However, poetical justice was done at last. One day, when all were at Yanuca, near Bega, their presence, notwithstanding their having assumed human shape, was discovered by the local god, who rightly guessed their intentions. When they were performing a dance, and all the girls were admiringly watching their graceful movements, the local god caused his priest to prepare a certain mixture, which, on being sprinkled over the visitors, made their arms, legs, and other parts of their bodies assume such ridiculous shapes, that they became the laughing-stock of all, and could never think of again undertaking similar expeditions.

* The sons were called, Teketeke-ni-masi, because he, the eldest, wore a wreath of flowers over his white tapa, Tawake-i-tamana, Reaugaga, and the youngest Valu-qaiaki (or rising moon).

The meeting with the chiefs and principal landholders of Kadavu was held at Tavuki, and passed off as satisfactorily as that at Bau and Rewa, the natives expressing their eagerness to become British subjects. We purchased from the natives a good many curiosities, such as clubs, fans, spears, etc., for our ethnological collections, some of which were remarkable specimens of carving, and evidently very old. The great size and heaviness of these things made them very inconvenient objects to carry and stow away on board, crammed as we were for space. One afternoon all the children of the town and neighbourhood, wishing to show their goodwill, came in full procession, and singing, up to the mission-house, each carrying a present. Some had bundles of sugar-cane, some bunches of taro, some struggled under the weight of an enormous yam. All the presents were piled in a heap at our feet, and it was intimated that they were meant for the special gratification of Mrs. Smythe. Then all the children sat down in rows on the ground, and sang a number of songs, accompanied by grotesque gestures, and movements of body and arms, but at the same time not without meaning. One of these songs, or "mekes," described the horror of the natives when seeing for the first time a horse and a man on its back,—how they fled in wild terror, and took refuge on high rocks and trees, so that the monster might not hurt them.

Both 'Pegasus' and 'Paul Jones' left Tavuki Bay on the morning of the 17th of August, and after a few hours' sail arrived at Qalira, where we hoped to ascend Buke Levu, but the sea was so high that we found it impossible to land. We hoped for better luck at Nasau, which we reached late at night, and were in full hopes of gaining the top of the fine mountain, constantly exhibiting to us its dome-like summit. The next morning, however, was so very rainy, that we had to give up all hopes of accomplishing our object that day; and it was therefore resolved to postpone our ascent, and cross over to Viti Levu, in order to pay a visit to Kuruduadua, for the exploration of whose dominions Mr. Pritchard and I had already paved the way.

CHAPTER IX.

DEPARTURE FROM KADAVU.—ARRIVAL AT NAVUA.—A COURT OF JUSTICE.—STARTING FOR THE INTERIOR.—THE NAVUA RIVER.—ITS FINE SCENERY.—RAPIDS.—A CANOE UPSET.—TOWN OF NAGADI.—HOSPITABLE RECEPTION.—SOROMATO.—KIDNAPPING.—FAMILY PRAYERS.—HEATHEN TEMPLE.—A LARGE SNAKE TO BE COOKED.—MARCH ACROSS THE COUNTRY.—VUNIWAIVUTUKU.—A DIFFICULT ROAD.—A PURSE LOST.—NO THIEVES.—ARRIVAL AT NAMOSI.—DANFORD'S ESTABLISHMENT.—HIS USEFULNESS AS A PIONEER.

LEAVING Kadavu on Saturday the 18th of August, at noon, our schooner cast anchor off Navua early next morning, where we were hospitably received by Kuruduadua, the chief of the district. Danford, the Englishman, whose history has already been told, was also there to conduct us to his place of residence at Namosi, as had been previously arranged. We took up our quarters in the new Strangers' House (*Buri ni sa*), where there was ample room to hang up mosquito curtains and open our luggage. There had been a quarrel between an Englishman and a Tonguese, both residing at Taguru, in Kuruduadua's dominion. The Englishman had allowed his pigs to grub the fields belonging to the Tonguese, and the latter, after repeatedly remonstrating without effect, had thought it advisable to enlighten the Englishman by setting fire to his shed. Both parties appealed to the British Consul for justice, and, with

Kuruduadua's approval, the case was gone into as it would before any magistrate in England, witnesses being called to establish the truth of the various statements advanced. The result was, that the Englishman was told that, according to Fijian customs, the pigs, not the fields, were fenced in, and that he had no right to allow his animals to destroy neighbours' property; whilst his neighbour, for taking the law in his own hand, was ordered to erect, in a specified number of days, a new shed, in every way equal to the one destroyed. Kuruduadua was highly pleased with the way in which the whole had been managed; and though it was late when the case was decided, he sent for several of the leading men to give them an account of it, and they sat up the greater part of the night discussing the fairness of the proceedings.

Having made arrangements with Kuruduadua for proceeding into the interior on our previous visit, we were able to start on the morning of the 21st of August. The travelling party consisted of Colonel Smythe, Mr. Pritchard, the Rev. J. Waterhouse, Danford, Chief Kuruduadua, and a host of followers, all embarked in canoes. The weather, which, during the previous week, had been rainy, became very fine at starting. The boat in which Mr. Pritchard, Danford, and myself were seated, was always ahead, and all attempts made by the others to beat us proved failures. At one time we had a most exciting race, the rival canoes putting forth all their strength, but to no avail: we kept ahead in spite of all their efforts.

Danford and the natives were quite in their element,

and indefatigable in offering explanation. I thought I could not do better than take advantage of their local knowledge and dot down all I heard, saw, and had pointed out. " Look to the right," cried one, " there is Tamana, with a large temple at the top." " Look to the left," interpolated another, " if you wish to see Solu, a small town, just disappearing betwen those banana plantations. You have already lost it. Those bamboos, high reeds, and tall treeferns, have shut it out. Do you see the wild plantain? There! there it is! You can always know it from others by its having erect orange-coloured branches instead of nodding ones, like the cultivated species. One more sago-palm in that swamp, probably the last, as we ascend the river; it does not like rocks, and here, you see, they begin. This is the first rapid: no danger, all the canoes pass over safely. Three hawks chasing a pigeon! Now for bold scenery! The rocks are at least two hundred and fifty feet high, full of fine timber at the top. And those splendid waterfalls! Here we are at Kuburinasaumuri; cliffs on both sides, and the river full of fresh-water sharks, of which the chief killed a very large one for biting his brother. This is Na Savu drau—the hundred waterfalls. In the rainy season that number is quite correct; even now, if you count all those little streaks of silver pouring over the cliffs, you will find it not far short. On the right is the Wai-ni-kavika (the river of the Malay apples), where a mighty spirit dwells."

And thus they went on talking and pointing out all they considered interesting or worth looking at. We had gradually exchanged the low, flat land of the coast for bold river scenery, and poled and paddled against a strong current. Judging from the water-mark observable on rocks and trees, the Navua, which flows almost due south, must be navigable for large boats during the rainy season; but when we ascended there was little water, and it required no ordinary skill to get the canoes over all the rapids that presented themselves. I have never appreciated the fun of passing over rapids, where a single false stroke or inattention of the steersman may upset you, and one may congratulate himself by simply escaping with bruises.* On one or two occasions we had to drag our little flotilla over them by means of ropes. At length we arrived at one worse than any we had previously encountered. We all landed, and told our crew to put our luggage on shore; this order, however, was only partially obeyed. Colonel Smythe's people, wishing to save themselves the trouble, headed the rapid. In an instant the torrent, breaking

* I well remember the anxious faces on board a steamer going over the rapids of La Chine, on the St. Lawrence; the band playing all the time, " The Rapids are near, and the daylight is past." There were on board then nearly all the members that had assembled to attend the meeting of the American Association for the Advancement of Science, at Montreal, Canada, I, as official representative of the Linnean Society of London, amongst the number; and judging from the serious tone that prevailed, and the sudden silence when we drew near the rapids, I don't think there were many present who thanked the managing committee for having provided this passage for our special amusement. Everybody was glad when it was over, except perhaps those Canadians who, by frequent repetition, had become used to this sensation passage. The temporary gloom was, however, soon dispelled by an animated discussion as to whether the honour of taking the first steamer over La Chine—the Indians had always taken their canoes over—was due to an Englishman or American. I did not wait for the end of the discussion; but whatever countryman, he must have been a most daring and cool-headed fellow.

the rope, had swept away the canoe, dashed it with great force against a steep rock on the opposite side, smashing the outrigger, swamping the little vessel, and leaving all the luggage and provisions swimming in the water. All the natives plunged in the river, and succeeded in saving the property. Of course the clothes were saturated, the tea had been made, the sugar was dissolved, and the biscuit looked like so much bread and butter pudding. To me, who often got a wetting in crossing rivers, it was quite amusing to see Colonel Smythe and Mr. Waterhouse busy in wringing and hanging up their clothes, and I could not resist the temptation of asking them whether any mangling was done there.

Fortunately, the stores which Mr. Pritchard and I had brought were quite safe, and so we could supply most of their deficiencies. The mishap being repaired as much as possible, we pushed on, and soon arrived at Na Mato,—a place where the river was entirely blocked up by huge rocks, said to have fallen from the top of the mountain on the right-hand bank, during an earthquake some forty years ago. The natives assured us that when this catastrophe first took place, the stoppage of the river was complete; and the water rose so high that for a long time it inundated their fields, and they had to dive for their provisions. They did obtain cocoanuts, but could not get at the taro, and there was a famine in consequence.

We left our large canoes at Na Mato, and in smaller ones, which Kuruduadua had in readiness, passed a steep rocky shore, where the people of Nagadi bury their dead. Excavations are made into the rock, and the corpses laid on their back, with the head towards the west. A small species of bamboo, of which the natives make pan-flutes, was here most common, as indeed all along these rocky shores, and greatly added by its graceful feathery habit to the beauty of the scenery.

Sunset was close at hand when we reached Nagadi, a town built on the top of a high steep hill, composed of rich clayey soil. For the night, we took up our quarters at the Bure ni sa, or strangers' house, invariably found at every Fijian town or village, and reminding one of the Tambo or Tambu of South America, between which and the strangers' house of Polynesia there appears to be a connection which ethnologists do not seem to have appreciated sufficiently. Both are public establishments, where travellers have the right to pass the night, and where they obtain meat and drink.* This Bure proved extremely dirty, and was much too small for all the people assembled to welcome our party. By spreading clean mats over a portion of the floor, and putting out most of the smoking fires kindled between each of the sleeping-places, we succeeded in making ourselves comfortable. Pigs, yams, and taro, all baked on hot stones in true Polynesian style, as Captain Cook described it one hundred years ago, and a quantity of pudding, consisting of ripe bananas boiled in cocoa-nut milk, and sweetened with

* One of the meanings of the Polynesian word *tabu*, or, as the Fijians pronounce it, *tambu*, is "set apart," "reserved," etc.; and I often wondered—that is all I could do with my slight philological knowledge—whether the name of the houses "set apart" or "reserved" for travellers in the Andes, the Tambos or Tambus, was in any way connected with this word.

rasped sugar-cane, were brought in and presented to Chief Kuruduadua, who, after accepting the gift through his speaking-man, again presented it to us. We had to go through the same ceremony of accepting the food, and had also the obligation to distribute it amongst the whole travelling party. This task was accomplished satisfactorily by Danford, whom his long life amongst the mountain tribes of Viti Levu has made familiar with all their complicated ceremonies.

After supper the kava bowl was brought out. Whilst the beverage was preparing the whole assembly chanted songs; and when ready, Danford gave the toast, and the cup-bearer handed the first cocoa-nut full to the chief. As soon as our bowl was empty, another and another was prepared, until the whole company had been served. Fortunately, kava, unlike distilled spirits, does not make people quarrelsome; it has rather, like tobacco, a calming effect; and when Fijians extol the virtues of their national beverage, they often, and justly, make this observation.

When leaving Navua we had more volunteers for accompanying us than there was any occasion to employ, and we were compelled to reject the services of a good many. Amongst them was a young chief, named Soromato, or, as his companions nicknamed him, "Montemonte." I told him that I did not wish to crowd our canoe, and he must stay behind; but he declared that he had made up his mind not to leave me as long as I was in the island. I told him I would not have him on any account, and if he did not take himself on shore directly, I would pitch him in the river. He intimated that he could swim, and that his clothes would not spoil, as he wore none. It not being prudent to give in to the natives, I had no option but to carry out my threat, choosing the very moment our flotilla was under weigh. He thought it a good piece of fun, and declared he would be with me nevertheless. He was as good as his word. When we landed at Nagadi he was there already, having come by the mountain road. I had now no alternative. He proved to me most useful and attentive, and never left me until I finally embarked, when he cried bitterly on being told that it was quite out of the question he could go to Europe with me, where he would probably have to exchange a life of ease and plenty for one of toil and poverty, and not be treated as a chief but as a common man.

The tribes of which Kuruduadua was the head, had for some time been molested by their neighbours, and we found at Nagadi a party of soldiers just returned from an unsuccessful ambush. They had endeavoured to kidnap some of their enemies, and were rather disappointed at having to report ill success. I recognized several of them as having been at Navua during our first visit to that place, and they gave us some account of Kuruduadua's son, whom Mr. Pritchard and I invested with his *toga virilis*. He was in the depths of the mountains, and a message had been sent to him that he might come to pay his respects to us.

Before retiring to rest we had family prayers in English, Mr. Waterhouse officiating. Kuruduadua commanded silence, and it was very impressive, amongst a profound stillness, to hear a Christian minister offering up supplications to heaven for the conversion of the benighted beings crowding around us. They were all attention, and in their minds evidently compared the convulsive ravings of their own priests with the dignified bearing of the Christian missionary.

The next morning I paid a visit to the heathen temple at Nagadi. Unlike other temples on the coast, which are generally erected on terraced mounds, and quite free from any enclosure, this was on level ground, and surrounded by a high bamboo fence; some of the sticks used being the young shoots entire, with unexpanded leaves, and looking like so many fishing-rods. The temple itself was a mere hut, scarcely twenty-five feet long and fifteen wide. In one corner there was an enclosure of reeds, where the spirit was supposed to dwell or descend. Kava-roots and leaves, clubs, spears, and little twigs of *Waltheria Americana*, suspended from various parts of the roof, had been presented as offerings. In some old temples the various offerings have been tastefully arranged, making the interior of the building look like a great armoury. There were no images of any kind,—indeed, I never saw idols of any sort throughout Fiji. The priest and his family also lived in this place, and readily exhibited all the curiosities accumulated. Amongst the things attracting my attention was a lot of bamboo-canes tied in a bundle, which, on being struck on the ground with the opening downwards, produced a loud and hollow sound. Two single bamboos of unequal length are beaten contemporaneously with this large bundle in religious ceremonies. I gave the young priest a jew's-harp, with which he expressed himself highly pleased.

At Nagadi the river branches off in two different directions: the eastern branch is not navigable even for small canoes, but said to be about forty miles long; whilst the northern has deep water, of which we took advantage in resuming our journey the next morning. All our luggage was sent by land, on the backs of natives. The weather still continued fine, so that we fully enjoyed the beautiful scenery and rich vegetation around us. We passed Bega, where our river was joined by a small tributary stream; hence the site of the town (or koro) is termed Uci wai rua, the junction of two rivers, the rivers being the Wai Koro Luva, and the Wai ni Avu. We finally abandoned our canoes at Wai nuta, to proceed on foot to Namosi—there being no horses, mules, or any other mode of conveyance.

On stepping on shore I was shown the largest snake I ever saw in Fiji. It was only six feet long, two inches in diameter, of a light brown colour, and with a triangularly-shaped head. I was very desirous of obtaining it for my zoological collection; but the natives said that Kuruduadua had just seen it and ordered them to prepare it for his supper on his return from Namosi. As he had passed on, I could not get the order revoked; and the reptile having been put alive in a bamboo, which was corked up at the ends, the boys, much to my regret, trotted off with it.

Climbing at once commenced. The paths being very narrow we walked in single file, Kuruduadua taking the lead, and showing us the sites of the various towns which he or his fathers had taken when their victorious army gradually fought its way from the interior of Viti Levu to its southern coast. The soil appeared everywhere of the richest kind. We saw no plains of any size, but series after series of undulating ranges of no very great height, well suited for growing coffee, tea, and cotton. Now and then there was a fine bird's-eye view of the country, which Kuruduadua was always careful to point out, evidently enjoying our expressions of delight on these occasions. I saw a good many plants that interested me, and their collection ultimately isolated me and Soromato, henceforth my shadow, from the rest of the party.

I had just been speculating on the cause of the Fijian, in common with other insular floras, being poor in gay-coloured, and rich in green, white, and yellow flowers, when, lo! a look in the valley revealed bushes covered with a perfect mantle of scarlet and blue, thrown up to great advantage by the bright rays of the sun. I saw my travelling companions had made a halt near the very spot where nature had condescended to refute a deeply-rooted generalization. I clambered down the hill as fast as the condition of the ground would admit, and for awhile lost sight of the gay display by intervening objects. A few more steps and I stood before a startling sight—Colonel Smythe's artillery uniform hung up to dry in the sun!

In detailing the violent emotions I had passed through, my companions enjoyed a good laugh at my expense, and invited me to cool myself by sitting down to a cup of hot tea, pork, and yams, all spread out picnic fashion on the grass, and in the shade of some fine cocoa-nut palms. The village where I met with this mortification rejoiced in the name of Vuniwaivutuku, and consisted of about thirty houses, some of which were neatly fenced in with *Dracænas*. The place where we had squatted down was in front of the Buri ni sa, an old and not very large building, surrounded by a good many erect stones, indicating the number of dead bodies eaten under its hospitable roof. The grass-plot in front, and several fine leaf plants, gave an air of neatness to the whole; whilst the extensive view it commanded over the whole valley, proved the situation a well-chosen one for a strangers' house. Kuruduadua informed us that there were two roads from here to Namosi, and that he should take us the longest, and bring us back the shortest, so that we might see as much as possible of his territory. He told us the road would be rather a rough one, and, without any exaggeration, it proved quite equal to the worst roads I traversed in South America. Now we had to climb perpendicular rocks, now creep underneath low bowers formed by reeds, now again wade through rivers and rivulets, or pass over swampy ground. Our clothes were torn by brambles, our hands and faces cut by sharp-edged leaves of grasses; indeed, one was forcibly reminded of the flight of the mechanics through the forests, which Puck relates with roguish delight in the 'Midsummer Night's Dream:'

"For briers and thorns at their apparel snatch;
Some sleeves; some hats; from yielders all things catch."

On proceeding, Colonel Smythe discovered that he had left his purse at Nagadi, having placed it last night under his mat, and forgotten to put it in his pocket before starting. "Make yourself perfectly easy about it," said Kuruduadua, when this loss was communicated to him, "I allow no thieving here; I club all thieves: they don't do that at Rewa or Bau. A man shall go back for it at once, and in a short time the purse will be brought." A messenger was sent accordingly, and, sure enough, when it was brought not a coin was missing.

Covered with mud and very tired, we reached towards sunset the town of Namosi, where Danford many years ago took up his residence. The beauty of its situation had not been exaggerated, and the accompanying sketch, for which I am indebted to Dr. Macdonald, will give some conception of it. It is built in a lovely valley, very much reminding me of Ischl. High mountains are rising on every side of an extremely fruitful valley, through which the Wai dina is winding its serpentine course, and passing many miles of fertile country, ultimately discharges its waters into the sea at Rewa. The temperature being considerably lower than that of the coast, a European is filled with a thrill of delight as he begins to breathe the air so much resembling that to which his constitution is best accustomed; and it requires no prophetic soul to predict that if ever the Fijis become a European colony, Namosi will be a favourite resort during the hot season, and the surrounding hills a mass of coffee and tea plantations.

We went straight to Danford's house, one of the largest in the town, built close to the rocky banks of the river, and surrounded by a neat bamboo fence, enclosing fine cocoa-nut, bread-fruit, orange, and Tahitian chestnut-trees, which diffused an agreeable shade over the extensive courtyard, whilst gay-coloured dra-

VALLEY OF ONO-BALEAGA, INTERIOR OF VITI LEVU

cænas and croton shrubs gave quite a finish to the place. Danford evidently enjoyed our surprise at finding everything so clean and comfortable, and new mats and even calico curtains. It was the best kept native-built house I had visited in Fiji. Afterwards, when having seen more of us, he told us how much annoyed he had been by certain remarks the whites on the coast had made to his disadvantage. Those people, who should be nameless, had insulted him by asking him point-blank how cannibal food tasted, and how he could think of forsaking the Christian religion and assisting in heathen rites. He had nothing to oppose of these accusations but silent contempt, and his well-fingered Bible was a good proof of his real disposition. In his own way he had evidently done a great deal of good; was the direct means of abolishing many abominable practices; and without this pioneer we should never have been able to reach this little-known region of the world. He was very fond of reading, and had accumulated a good many books, mostly presents from consuls, missionaries, or captains and officers of ships. I increased it by a copy of Shakspeare, after which he had a hankering. The natives often came to look at his picture books, and the 'Illustrated London News' was a source of endless delight to them.

CHAPTER X.

POPULAR IDEAS RESPECTING THE INTERIOR OF VITI LEVU.—MALACHITE AND ANTIMONY.—ASCENT OF VOMA PEAK.—VISIT TO A HEATHEN TEMPLE.—"SPIRIT FOWLS."—OFFICIAL MEETING WITH KURUDUADUA AND HIS SUBJECTS.—A REBELLION TO BE SUPPRESSED.—PRESENTATION OF FOOD.—"THE OLDEST INHABITANTS."—A COURT-FOOL AND HIS TRICKS.— MR. WATERHOUSE PREACHING.— DEPARTURE OF COLONEL SMYTHE, AND MESSRS. PRITCHARD AND WATERHOUSE, FOR NAGROGA.

To the north of Namosi there is a good deal of unexplored country, and we tried hard to get some information about its general features. A popular belief, current amongst the white settlers in Fiji, affirms that there is a large table-land and an inland lake in Viti Levu. Nothing could be learnt of this table-land, but the natives had heard of a lake on which canoes were. Not far from Namosi, still in sight of the town, exists a mountain, which the late Mr. Williams, American Consul for Fiji, bought for its rich veins of copper ore. After Mr. Williams's death a number of specimens from this mountain were found in his possession, of which his executor gave me several. They proved to be malachite, closely resembling the Australasian, and next to that of the Ural, considered the best. Nothing has as yet been done to work these mines. The natives also informed us of the existence of ore cf antimony about ten miles from Namosi, and at a place called Umbi, where it is said to occur in large veins in the side of a hill. Macdonald and S. Waterhouse also heard of and saw quantities brought down by the natives in bamboos, and concluded that it must be plentiful. The black sand so frequently found on the banks of the Rewa river, and attracted by a magnet, has also been washed down from these mountains. Danford at one time fancied he had discovered gold in the neighbourhood, and in 1856 he took the 'Herald's' officers to the Wai ni Ura. The rocks were spangled with iron pyrites, which made their appearance wherever the surface was broken: gold was nowhere to be seen.

Directly on our arrival we made preparations for ascending Voma, the highest peak in Viti Levu, perhaps in the whole Fijis, and never trodden by the foot of white man. The natives represented to us the impossibility of getting to the summit, but we told them that we must at least make the attempt. To this proposal they agreed, and on the morning of the 24th of August we commenced our task, guided by Natove, a famous warrior and petty chief, who proved an excellent hand in cutting openings through the forest when we got higher up.

On leaving Namosi our path led through numerous taro, banana, and yam plantations, and close to an altar made of sticks and native cloth, on which food for the spirits of the dead was placed: some of the yams were actually sprouting again. The mass of Fijians will have it that these offerings are consumed by the spirits of their departed friends and relations, supposed to have great supernatural influence; but if not eaten by animals, the food is often stolen by the more enlightened class of their own countrymen, and even some foreigners occasionally do not disdain to help themselves freely.

The ascent of Voma was steep, and made us very warm indeed. Our native attendants found it equally so, though not encumbered with any clothing like ourselves; and to cool themselves they thought it no additional exertion to climb up a tree and catch the breeze. In former times, there had been a town some considerable distance up the mountain, traces of which were still visible; and hence, though there was a thick wood, the actual virgin forest did not commence until we had attained the height of about 2500 feet above the sea. When entering that region we found the trees altogether different from those of the lowlands, and densely covered with mosses, lichens, and deep orange-coloured orchids (*Dendrobium Mohlianum*, Rchb. fil.). Some of the ferns were of antediluvian dimensions. A species of *Cinnamomum*, producing a superior kind of cassia-bark, and used by the natives for scenting cocoa-nut oil, and as a powerful sudorific, was met with in considerable quantities. The absence of all large animals, and the limited number of birds, impart an air of solemnity to these upland forests. Not a sound is heard: all is silence —repose!

We had to pass over some awkward places, and to climb several almost perpendicular rocks, rendered slippery by water trickling down. However, at half-past ten, two hours and a half after starting, Colonel Smythe, Mr. Pritchard and myself, reached the summit: Danford having stopped half-way, and Mr. Waterhouse remained behind at Namosi to scatter a little seed of truth amongst the numerous heathens pouring into the town for to-morrow's grand meeting.*

Immediately trees were cut down, and compass bearings taken of all prominent parts, by which means an important step was made to reform the geography of Viti Levu.† A great part of Fiji lay like a map at our feet; there were the islands of Moturiki, Batiki, Gau, Bega, Ovalau, and a host of smaller ones; even Kadavu was looming at the distance. We had hoped to have a

* "Before a large company of chiefs and people," says Mr. Waterhouse, in his published journal of this tour, "I gave an account of the Great Creator, and of the original state and subsequent fall of man. They loudly applauded Adam's cleverness in blaming the woman, and Eve's in accusing the serpent. I was afterwards requested to tell them about Noah and the Flood, with which demand I complied. Before I left the house, the chief said to those present, 'These missionaries are our true friends: they want us to live in peace and quietness, and to cultivate the soil; but you slaves can't understand these matters.' Many referred in glowing terms to the visit of my brother Samuel, and Kuruduadua gave a vivid description of his visit to the house of the Rev. William Moore."

† Dr. Macdonald and Mr. Samuel Waterhouse were, it is well known, the first who penetrated up the Wai dina, or great river of Viti Levu, to Namosi, and from data which they furnished was constructed the map published in the Journal of the Royal Geographical Society, vol. xxvii. Having nothing to go upon but the compass and dead reckoning, the position of Namosi, as well as the source of the Wai dina, has been placed too far west, as our route to Namosi lay almost due north. The compass bearings taken on the top of Voma Peak would have corrected errors found in recent maps; but the southern coast seems to be so far out that they cannot be made available at present. I subjoin them:—East end of Moturiki, N.E. by E.; centre of Batiki, N.E. by E. ¾ E.; west end of Gau, E. by N. ¼ N.; centre of Nukulau Island (Rewa), E. ¾ S.; east end of Bega, S.; centre of Yanuca, S. by W. ½ W.; Gamo Peak, S. by W. ½ W.; extreme sea horizon to the west, S.W. by W.; town of Namosi, N.N.W.; extreme sea horizon on the north was the west end of Ovalau.

glimpse of Bega; but that we should be able to see nearly two-thirds of the whole group was a pleasure for which we were unprepared, and which amply repaid the exertion made in the ascent. A fire was kindled to let the people of Namosi know of our success, and after collecting specimens of the vegetation, and partaking of some refreshment, we descended, and reached Namosi about five P.M., the boys carrying baskets full of rare and new plants.

In the evening we paid a visit to a Bure Kalou (heathen temple). Though not surrounded by a fence, it was situated and similar to that at Nagadi, small and insignificant in comparison with some of the temples near the coast. Danford introduced us to the priest, who kept up a roaring fire, which made the inside too hot for us to stay longer than a few minutes. We were told that the *Kalou* (=Spirit, God), for whom two-thirds of the whole building were set apart by a screen of bamboo, liked heat; but I presume the only spirit fond of a good fire was the priest himself, as he was rather an old man. Hearing from Danford that one of our party, disliking pork, had not eaten meat for several days, he very good-naturedly let us have several fowls presented to the temple. Danford dubbed them spirit-fowls, and Mr. Pritchard turned them into excellent curry, for which the materials were fetched fresh from the bush.

When retiring to the house, Danford occupied the greater part of the evening by telling us one of the best Fijian stories, one of the chiefs helping him out when memory failed. It was that of the Princess Vilivilitabua and the Vasu-ki-lagi. One of our party took down the outline of it, but unfortunately lost it, and I shall not spoil a good story by giving it imperfectly.

Chief Kuruduadua had proposed to have the official meeting at Namosi, in preference to Navua, his usual place of residence on the coast, and summoned all his tribes, their petty chiefs and landholders for the 25th of August. On our arrival, Namosi was already crowded with visitors, and parties of men, women, and children, generally bringing loads of provisions and property with them, continued to flock in from all directions during the whole of the following day. The meeting took place in the open air, and in the public square or Rara, which is situate on the banks of the river, and before the great Bure ni sa, or strangers' house, a building about ninety feet long, and built on a mound. The weather was beautiful, and the birds were singing sweetly in the numerous shaddock-trees lining the banks.

When we arrived, the people, with the exception of the women, were squatted on the ground at a respectful distance from the seats placed for our accommodation. None of the influence which civilization and missionary teaching have had on the Fijians were here perceptible. Every native appeared in primitive style, and a stranger sight it has never been my fortune to witness. Every man seemed to have used his utmost efforts to make himself look as singular as he possibly could. Their dresses were merely narrow strips of bark cloth. Some faces were quite black, some only half; again, others half black and half red, or striped in various ways. Nothing could be more curious than the endless variety displayed in the shape and colour of the wigs, and doing-up of the head; a European peruquier might have taken a lesson with advantage. Chief Kuruduadua had taken his seat on the steps leading to the principal entrance of the great Bure. He wore a turban of snow-white tapa, and a purple girdle of the same material, from which were suspended two trains of native cloth, several yards long. On his left were his brothers and councillors, amongst whom was seen his friend Danford. When we had taken our seats, the people welcomed us by clapping of hands, whereupon mutual explanations were at once entered into.

Through Mr. Waterhouse, Colonel Smythe addressed to the chief a speech similar to that delivered at Bau and other places, the purport of which has already been given. Mr. Waterhouse spoke in the Bauan (court) dialect, and Kuruduadua replied in the same, that he and his people had made up their minds to "lean upon England," as he expressed it, in the manner agreed upon with Mr. Consul Pritchard. Colonel Smythe approved of their determination as judicious, there being no country more able to protect them than mighty England. He also recommended the cultivation of cotton. On being questioned about the ownership of land, Kuruduadua replied that he considered himself the sole proprietor of all the land, the boundaries and principal tribes of which were specified; that his late brother had sold some land to Mr. Williams, deceased, and he himself some to several Englishmen, all these transactions being acknowledged as valid.

An expression of mutual goodwill concluded the

business. During the whole time the people behaved with great dignity; none spoke except those who carried on the discussion. When their foreign affairs were satisfactorily concluded, the chief, quitting his seat, begged us to remain, in order to see how they managed their internal politics. This invitation we gladly accepted by taking up our position near the entrance of the Bure, where we had a better view of the whole assembly.

It appears that one of the numerous tribes subject to Kuruduadua had rebelled against his authority, and it had been determined by the councillors that stringent measures should be put in force against it. The principal and most renowned speaker of the Government, a man about fifty, now came, staff in hand, out of the great Bure into which Kuruduadua had retired, and explained to the people at large the policy about to be pursued. He moved freely about the circle formed by his audience, and his speech was listened to with profound attention, eliciting now and then exclamations equivalent to "hear, hear!" The drift of his argument was that the rebels must be put down and peace restored, in order that they might have plenty when the white men came to their country, from whom Fiji already derived such benefits. When he had finished, other speakers got up, all in favour of the government measure, and much applauded by the multitude. One old chief was much cheered on saying, "I am no speaker, but know how to fight; and there (pointing with his hand) is the road to the enemy's stronghold."

All business matters having been disposed of, it only

remained to enact the closing scene by a great banquet. The women now appeared on the stage. All the young girls had collected in a group, some two hundred yards off, in a grove of palm-trees, each carrying a basket-full of taro. According to their fashion, they wore nothing save a girdle of hibiscus-fibres, about six inches wide, dyed black, red, yellow, white, or brown, and put on in such a coquettish way, that one thought it must come off every moment. The girls (a hundred and fifty-four) walked in single file, and all those wearing girdles of the same colour kept together. When arriving in front of the Bure, young men received the baskets and emptied their contents in a heap, leaves having been spread out to keep them from coming in contact with the ground. We counted as many as two thousand taros, after which the baskets came in so fast that we lost count. The girls, after performing their part, walked away in the same order as they came. Several young men now brought seven large hogs, roasted entire, which were placed on the top of the taro heap. The whole pile of food was then presented to the visitors. The largest pig, and I am almost afraid to say how many hundred taros—ready to be eaten—fell to our share. It took twenty men to take our share home, for the food was not supposed to be consumed on the spot, everybody being at liberty to do what he liked with his lot, and I saw but very few not taking their portion away with them.

There was a man present at this meeting, Ro Tui Kuku, who had seen five generations of the reigning chief's family, and could not have been less than a hundred and twenty years old; and there was another man, sharing the same house with him, who had seen four generations of the same family: excellent proofs of the fine physical constitution of the natives, and the healthiness of these mountains. Ro Tui Kuku was quite childish, and when we spoke to him and presented him with a little American tobacco, he said that he must be off home. He had great-great-grand-children living, the eldest of whom was about ten years old.

Another personage attracted our attention. He was the court fool of the occasion, and had dressed himself in a very fantastic manner. The fools attached to the courts of South Sea chiefs are very often hunchbacks, the natives being fully sensible of the great fund of humour which that class of people generally possess, as a set-off, it would almost appear, for the physical deformity which so often exposes them to unmerited ridicule, and which is now considered in Europe an essential condition of the most comic figure the popular mind has conceived. But the Namosi fool was an exception to this rule. He was in every respect a fine fellow, more than six feet high. On his head he wore a contrivance made of sticks and feathers resembling the shovel-bonnets ladies used to wear some years ago, and his face and body were painted in a very ludicrous manner. He talked in a feigned voice, imitating a woman, and probably gave utterance to many witticisms and good jokes, as he kept his countrymen in roars of laughter whenever he opened his mouth. When the meeting broke up, we had to recross the river in order to get to Danford's house; a strong Tonguese belonging to the mission performed, St. Christopher-like, the office of carrying our party across. Not being in a particular hurry to get over, I was waiting until all had crossed, when this fool came up to me with an offer to take me to the opposite bank. I thought he might be up to some tricks, and was rather on my guard. He landed me safely, but I soon found that I had been sold nevertheless,—my white dress looking as if printed on. The colours he had on his back had come off, and made me look almost as comic as the fool himself. The natives thought it an excellent joke, and when they saw me laughing as much as they did, their merriment knew no bounds.

On the following day (Sunday, August 26th) Mr. Waterhouse, making the most of his opportunity, once more addressed the people;* in the afternoon, he, Colonel Smythe, and Consul Pritchard left Namosi for Navua, whilst I thought it best to remain behind in order to explore the neighbourhood, and get a more intimate acquaintance with these singular people. Kuruduadua again led the way, and this time took his visitors the shorter of the two roads leading to Vuniwaivutuka. They shot down the river rapidly, and on Monday, about four P.M., reached the 'Pegasus,' and put at once to sea. On the 30th of August they found themselves at Nadroga.

* "On Sunday I preached on 'God now commandeth all men everywhere to repent,' to a congregation of about three hundred male adults, all heathens, who listened very attentively and respectfully. Now and then one or another would say aloud, 'Very good;' or, 'It's true.' When I had concluded, I requested the audience to maintain perfect quietness for a few moments whilst I engaged in prayer to the true and only God. They granted the favour, and not an individual made the slightest disturbance. As I was leaving, one of the chiefs thanked me publicly for my instruction."—*Waterhouse, in Wesleyan Missionary Notices.*

As the difference between the heathen and Christian population, mentioned in a previous chapter, had not yet been satisfactorily settled, they found the country in rather a disturbed condition. The conflict between barbarism and an incipient civilization was still going on.

"The people were glad to see a missionary," says Mr. Waterhouse. "I was sorry to find that some of our native agents had not maintained neutrality between the Christians and heathens, which, they were obliged to confess, was not only against orders, but had proved to be, so far as they were personally concerned, bad policy. Since my visit in 1851 the bones of those human beings who had been eaten had been collected together and buried. The evening was spent in examining and instructing the schoolmasters and Scripture-readers. Mr. Moore has done a noble work in preparing so many agents for these benighted parts.

"Though in some danger, yet I felt it my duty to sleep on shore to encourage my native colleagues to abide by their post of honour. Only last Tuesday a man was killed by a 'kidnapper.' There is no safety in going outside of the house after dark. In some cases the kidnappers enter the house, close or surround the doors, dispatch the inmates, and make their escape. In the event of an occurrence of this sort, I suggested that, instead of allowing the intruders to kill us, we should close in on them and bind them.

"Colonel Smythe sent a native messenger to request the heathen Chief to pay him a friendly visit. The man performed his errand, and delivered his message. The enemy then clubbed him, and sent him back with the remark, that if two had been sent, one would have been killed and eaten, and the other allowed to return and report the fate of his comrade. Under these circumstances they only *half-killed* him, and sent the other half of the poor man to tell a very sad tale and show his wounds. A present seemed to go far towards healing the sores inflicted by a pine-apple club."

Mr. Pritchard did not think it advisable to go further than Nadroga, whilst Colonel Smythe proceeded to Vuda, Ba, Vatia, Na Vatu, and thence to Naduri on Vanua Levu, and returned to Levuka on the 22nd of September. Everywhere the chiefs acquiesced in the cession of their country to England.

It will be remembered that I was still at Namosi; and I must beg the reader to return with me to that place.

CHAPTER XI.

FIJIAN CANNIBALISM.—THE GREAT CAULDRON.—NAULUMATUA AND HIS APPETITE FOR HUMAN FLESH.—BOKOLA.—VEGETABLES EATEN WITH CANNIBAL FOOD.—THE OMINOUS TARO.—APPROXIMATE NUMBER OF BODIES EATEN AT NAMOSI.—OVENS FOR BAKING DEAD MEN.—SUSPENSION OF THE BONES.—NOT ALL FIJIANS CANNIBALS.—EFFORTS OF THE LIBERAL PARTY TO SUPPRESS ANTHROPOPHAGISM.—AIDED BY EUROPEANS.—REAL SIGNIFICANCE OF EATING MAN ONLY PARTLY UNDERSTOOD.—CONCESSIONS TO HUMANITY.—ABOLITION OF CANNIBALISM THROUGHOUT KURUDUADUA'S DOMINIONS.

WHEN, in August, 1856, Dr. Macdonald, of H.M.S. Herald, then under the command of Captain Denham, and the Rev. Samuel Waterhouse, a brother of the gentleman who accompanied us, paid a visit to Kuruduadua's dominions, cannibalism was still one of the recognized institutions of the state. "A few days ago," says Dr. Macdonald, "a large canoe from Navua went out on its first voyage, when a fleet of the enemy from Serua attacked it, and succeeded in killing one man, who fell overboard. The Serua people now dispersed, and the canoe, on returning, landed a detachment with directions to surprise the enemy on coming ashore. They fell in with a party of seven, four of whom were killed, two fled, and one was taken prisoner. The latter was almost immediately boiled alive in a large cauldron. Kuruduadua, the perpetrator of this cruelty, addressed him, in short terms, to the effect that, as he had so wickedly cut to pieces a living man of his (Kuruduadua's) people, he should be served as the case deserved. The unfortunate man was then thrust headforemost into the boiling pot. The greater part of the slain was eaten at Navua, but parcels of the revolting food were distributed amongst the chief's dominions in the mountains. On the morning of the 30th of August, after a little parley with the chief, Naulumatua, the knee of a dead body, already cooked, was brought to our bure. The bones had been removed by an incision made on one side, and the whole was carefully wrapped up in banana leaves, so as to be warmed up each day in order to preserve it. Of six parcels of human flesh which we knew had been sent to Namosi, this was all we had an opportunity of seeing. One leg was said to have been deposited at the grove of Viriulu, the deceased king and father of Kuruduadua.* Mr. Waterhouse spoke to the chief very impressively on the subject, pointing out all the evils which follow in the wake of cannibalism. I saw very distinctly that this savage was quite ashamed of himself; but I saw also that, if he did feel inclined for the tempting morsel, there was now very little chance of seeing him in the act; but for my own part, I am quite satisfied, and do not now desire further ocular demonstration of the existence of cannibalism in Fiji. We have now every reason to believe that the portion of the last bokola (dead body), which Naulumatua asserted had been placed upon the rock where the remains

* We are told this king's name was " Ratuibuna," but perhaps he went by two names.—B. S.

of the last chief were laid, was eaten on the sly by this cannibal, whose morbid taste for human flesh was acknowledged by all the people in the town. . . . Tobi, one of our party, happened to stumble into the chief's house, and he distinctly saw a human hand hanging in the smoke over the fireplace. Now, although the distribution of all the other parts had been accurately detailed to us, no mention was made of this, so that the dissimulation of Naulumatua was clear enough. Most probably, had we approached the spot, the inviting morsel would have been quickly conveyed out of the way. Mr. Waterhouse was informed that the chief continued to eat his portion at intervals throughout the day, until it was all demolished; but an old favourite of the town helps him out with it." Thus far Macdonald.

Naulumatua was the half-brother of Kuruduadua, and only died a short time previous to our visit, and the court was still in mourning for him, which was the reason of our not having either dance or song. His head-wife took me to his grave, and lamenting his death, said that he might still be alive if he had only abstained from eating human flesh, and that both she and Danford had done all in their power to convince him that he was ruining his constitution systematically by that indulgence. For it appears that human flesh is extremely difficult to digest, and that even the strongest and most healthy men suffer from confined bowels for two or three days after a cannibal feast. Probably, in order to assist the process of digestion, "bokola," as dead men's flesh is technically termed, is always eaten with an addition of vegetables, which it may be ethnologically important to notice; since, thanks to a powerful movement amongst the natives, the influence of commerce, Christian teaching, and the presence of a British Consul, Fijian cannibalism survives only in a few localities, and is daily becoming more and more a matter of history.

There are principally three kinds which, in Fijian estimation, ought to accompany bokola,—the leaves of the Malawaci (*Trophis anthropophagorum*, Seem.), the Tudauo (*Omalanthus pedicellatus*, Bth.), and the Boro-dina (*Solanum anthropophagorum*, Seem.). The two former are middle-sized trees, growing wild in many parts of the group; but the Boro-dina is cultivated, and there are generally several large bushes of it near every Bure-ni-sa (or strangers' house), where the bodies of those slain in battle are always taken. The Boro dina is a bushy shrub, seldom higher than six feet, with a dark, glossy foliage, and berries of the shape, size, and colour of tomatoes. This fruit has a faint aromatic smell, and is occasionally prepared like tomato sauce. The leaves of these three plants are wrapped around the bokola, as those of the taro are around pork, and baked with it on heated stones. Salt is not forgotten.

Besides these three plants, some kinds of yams and taro are deemed fit accompaniments of a dish of bokola. The yams are hung up in the Bure-ni-sa for a certain time, having previously been covered with turmeric, to preserve them, it would seem, from rapid decay: our own sailors effecting the same end by whitewashing the yams when taking them on board. A peculiar kind of taro (*Caladium esculentum*, Schott, var.), called "*Kurilagi*," was pointed out as having been eaten with a whole tribe of people. The story sounds strange, but as a number of natives were present when it was told, several of whom corroborated the various statements, or corrected the proper names that occurred, its truth appears unimpeachable. In the interior of Viti Levu, about three miles N.N.E. from Namosi, there dwelt a tribe, known by the name of Kai-na-loca, who in days of yore gave great offence to the ruling chief of the Namosi district, and, as a punishment of their misdeeds, the whole tribe was condemned to die. Every year the inmates of *one* house were baked and eaten, fire was set to the empty dwelling, and its foundation planted with *kurilagi*. In the following year, as soon as this taro was ripe, it became the signal for the destruction of the next house and its inhabitants, and the planting of a fresh field of taro. Thus, house after house, family after family, disappeared, until Ratuibuna, the father of the present chief Kuruduadua, pardoned the remaining few, and allowed them to die a natural death. In 1860, only one old woman, living at Cagina, was the sole survivor of the Na-loca people. Picture the feelings of these unfortunate wretches, as they watched the growth of the ominous taro! Throughout the dominions of the powerful chief whose authority they had insulted, their lives were forfeited, and to escape into territories where they were strangers would, in those days, only have been to hasten the awful doom awaiting them in their own country. Nothing remained save to watch, watch, watch, the rapid development of the kurilagi. As leaf after leaf unfolded, the tubers increased in size and substance, how their hearts must have trembled, their courage forsaken them! And when at last the foliage began to turn yellow, and the taro was ripe, what agonies they must have undergone! what torture could have equalled theirs?

How many dead bodies have been eaten at Namosi, it is impossible to guess; but as for every corpse brought into the town a stone was placed near one of the bures, you get some faint idea of the number. I counted no less than four hundred around the Great Bure alone, and the natives said a lot of these stones—of which the larger ones indicated chiefs—had been washed away, when, some time ago, the river overflowed its banks.

On some of the Tavola (*Terminalia*) trees standing about the Great Bure, I noticed certain incisions, and as Macdonald, on ascending the Rewa river, had noticed similar ones at the town at Naitasiri, and was told that they were " a register of the number of dead bodies (bokolas) brought to the spot to be offered up at the bure before they were cooked and eaten," I inquired repeatedly after their meaning, and was assured by various persons that, at Namosi at least, they were entirely the work of children. As the bark of the Tavola-trees is as smooth as our beech, I carved my name on the largest of them; a much condemned habit of our race, but which, in remote corners of the earth, I have not always been able to resist.

There are ovens in the public square for baking dead bodies, and the pots in which human flesh is boiled or steamed are not devoted to any other culinary purpose. Another curious circumstance is, that whilst the natives eat every other kind of food with their fingers, human flesh is eaten with forks, having three or four prongs, and generally made of the hard wood of a species of *Casuarina*. Every one of these forks is known by its particular, often obscene, name, and they are handed down as heirlooms from generation to generation; indeed they are so much valued, that it required no slight persuasion and a handsome equivalent to obtain specimens of them for our ethnological collection.

It is customary to suspend some of the bones of those human beings that have been eaten in the trees before the Bure-ni-sa; and we saw several of these trophies, on some of which was growing a beautiful little fern (*Hemionitis lanceolata*, Hook.), not previously seen, and only gathered afterwards on the very summit of Buke Levu.*

It would be a mistake to suppose that all Fijians, not converted to Christianity, are cannibals. There were whole towns, as for instance Nakelo, on the Rewa river, which made a bold stand against this practice, declaring that it was *tabu*, forbidden to them by their gods, to indulge in it. The common people throughout the group, as well as women of all classes, were by custom debarred from it. Cannibalism was thus restricted to the chiefs and gentry, and again amongst them there is a number, who for want of a better appellation may be called the Liberal party, and who never

* Mr. Waterhouse speaks of "grinning skulls looking down on us;" but I never saw any skulls at this place, though carefully examining all the trees, nor do I know for certain whether that part of the body is ever suspended in trees.

eat human flesh, nor go near the bures when any dead bodies have been brought in, and who abominate the practice as much as any white man does, attributing to it those fearful skin diseases with which their children are so often visited. But their opponents, the Conservatives, maintain that in order to strike terror in the enemy and lower classes, it is absolutely necessary for great chiefs and gentry—a duty they owe to society— to eat human flesh. The feeling which the common people have regarding it seems somewhat akin to the horror inspired by that part of our nursery tales when the giants come home, and begin to smell the children concealed. The same enlightened party also objects to the killing of women, urging that it is just as cowardly to kill a woman as a baby. But here again those who advocate inhumanity are triumphant, arguing that if the women are killed the men will fret, and thus suffer an almost direct punishment; and further, that as whenever there is a quarrel a woman is sure to be at the bottom of it, justice demands that her sex, having caused the bloodshed, should not escape scot-free.

It is owing to this powerful ferment, which had penetrated the whole Fijian community, that cannibalism was so speedily abolished in all districts where Christian missionaries or European consuls were able to aid the good cause by supplying the combatants with fresh arguments, and backing them up with all the advantages derived from their position as respected foreigners. There may have been, and I dare say there are to this day, individual natives, who, like Naulumatua, have a morbid appetite for human flesh, sufficient opportunity to gratify it to an alarming extent, and who could no more break themselves of the habit, though death stared them in the face, than any confirmed drunkard can of his vice. But as a general rule *bokola* was not regarded in the shape of food; and when some of the chiefs told foreigners, who again and again would attack them about a custom intimately connected with the whole fabric of their society, and not to be abolished by a single resolution, that they indulged in eating it because their country furnished nothing but pork, being destitute of beef and all other kinds of meat, they simply wished to offer some excuse which might satisfy their inquisitors for the moment.

Fijians always regarded eating a man as the very acme of revenge, and to this day the greatest insult one can offer is to say to a person, " I will eat you." In any transaction where the national honour had to be avenged, it was incumbent upon the king and principal chiefs —in fact, a duty they owed to their exalted station— to avenge the insult offered to the country by eating the perpetrators of it. I am convinced however that there was a religious as well as a political aspect of this custom, which awaits future investigation. Count Streletzki, whose powers of observation have given him an insight into savage life few travellers have attained in so eminent a degree, fully agreed with me when some time ago this subject was the topic of conversation between us. There is a certain degree of religious awe associated with cannibalism where a national institution, a mysterious hallow akin to a sacrifice to a supreme being, with which only the select few, the tabu class, the priests, chiefs, and higher orders, were deemed fit to be connected. The cannibal forks obtained at Namosi tended to confirm this belief. There was the greatest reluctance to part with them, even for a handsome equivalent, and when parted with displaying them was objected to. This I thought at first very natural, as they were said to be heirlooms, and the owners did not like to expose themselves to the odium of having trafficked in things like them. But when afterwards they were shown to parties who could know nothing of the transactions, their faces always assumed a serious aspect, and they were most anxious that I should put the forks out of sight, especially that of children. My handling them seemed to give as much pain as if I had gone into a Christian church and used the chalice for drinking water.

When visiting Navua for the first time in June, Mr. Pritchard and I did not fail, as soon as we had succeeded in gaining Kuruduadua's confidence, to interpose the influence acquired in favour of humanity. The chief being a pagan, it was useless to employ any Biblical arguments, and we had therefore simply reason to fall back upon. One of the first concessions he conceded was, that as has already been detailed, no one should be clubbed on his son coming to manhood—a whole town having originally been singled out for that horrible purpose. It took him several days to consider our proposition with his leading men; and there were long and warm discussions as to the propriety of yielding to our request. We were kept well informed of the progress of the question through Danford, who, to his praise be it said, did all he could to bring about an issue favourable to humanity. At last Kuruduadua informed us, that having duly considered our request with his councillors, they had agreed to allow the Consul and myself to put on the scanty clothing, the assumption of which marked the transition from boyhood to manhood. We lost no time to break through a custom which will now never be repeated in the district, since the son of a governing chief dispensed with it.

The "large cauldron" which Macdonald mentions,* but did not see himself, stood close to the door of the chief's house. Our attention was drawn to it by our interpreter, Mr. Charles Wise; and the very thought was agonizing to be so near the awful vessel in which perhaps many a human being had been boiled. It was one of those large iron pots used by traders for curing *bêche-de-mer*, or sea-slugs, so plentiful on the reefs of Fiji, and a valuable article in the Chinese markets. It was large enough for cooking two men entire. At the mere sight of it my imagination ran riot, and a scene presented itself similar to that in the last act of Halévy's 'Jewess,' where the boiling cauldron is ready to receive the victim of Christian intolerance. The nineteenth century must be freed from so shocking a spectacle, and Mr. Pritchard and myself let Kuruduadua have no peace until he agreed to abolish and prohibit cannibalism throughout his dominions. A few months earlier he would have met with a most determined opposition in promulgating such a law, for his half-brother at Namosi,

* Journal of the Royal Geographical Society of London, vol. xxvii. p. 253.

then alive, would never have agreed to it; but our visit happened just at the right time in order to crown our endeavours with success.

When in August we saw the cauldron again, it was quite rusty, and had evidently not been used. Weeds were growing around it, and a creeper was trying to cover by its foliage this remnant of past errors and crimes. Kuruduadua had evidently kept the promise made us, caused presents of human flesh sent to him to be buried, and given strict orders that even in the fight impending the bodies of the slain enemies should be left to be buried by their friends, and on no consideration be removed by his own people.

Batinisavu, who succeeded the cannibal Naulumatua as governor of Namosi, belonged to the party always opposed to anthropophagism. He was quite a young man; had, according to all accounts, never tasted human flesh; and there is every reason to believe, great friends as he was with Danford, that as long as he holds the post no *bokola* will be seen at Namosi. The widows of the late governor paid me repeated visits, and said there would be no more cannibalism at Namosi, since Kuruduadua's orders were very strict. Soromato, the young chief who had attached himself to me, asked Danford one day whether he remembered a conversation they had years ago, when he was a very young boy, and in which he told him of a vow he had made never to kill a woman when able to wield a club, or eat human flesh, when old enough to do so. Danford said he well remembered it, as it struck him as very singular that a mere child should feel so strongly on these subjects as to make a solemn vow. "Well," Soromato replied, "I still adhere to that determination, and shall do so as long as I live."

I quote this as a specimen of the way in which a certain party of heathen, untaught Fijians, endeavour to bring about the same reform in their customs, which, from different points of view, and with different means, their best friends have for years laboured to effect.

CHAPTER XII.

STAY AT NAMOSI PROLONGED.—THE GOVERNOR'S ATTENTION.—"CROWN JEWELS."—THE CLERK OF THE WEATHER.—SORCERERS.—FIJIAN FAMILY LIFE.—STORY-TELLERS POPULAR.—A FIJIAN TALE.

THE people were highly pleased when they heard of my resolution to stay some time longer with them, and treated me with great cordiality. Batinisavu,* one of the younger brothers of Kuruduadua, who is the governor of Namosi, was never tired of showing me attention, and shooting ducks and fowls for me, or making different kinds of puddings, on the excellence of which he prided himself. Chiefs always make it a point to excel in everything they undertake; and this is no doubt one of the reasons why they maintain their ascendency over the people. They build canoes, houses, or temples, in a style and with a finish to which the lower order cannot come up; in agriculture they take the lead; in fighting, rowing, pulling, racing, and all manly exercises, they are patterns for imitation; in the history, legendary lore, and traditions of the country, they carry off the palm; they know every rock, river, plant, and animal, by its local name, and can give some account of everything connected with them. If to all this be added

* Batinisavu,—literally, the edge of a waterfall.

that their physical development is much superior to that of the lower classes, that they are not only taller and better made, but generally possessed of much handsomer features, we need not wonder that some travellers have thought them a different race from the rest of their countrymen; and that in their own land they have been able to resist all democratic levelling, and remain to this day as genuine an aristocracy as ever existed, because in every respect a superior class.

The widow of the late governor of Namosi asked me to see the " crown jewels" in her charge. They were kept in a wooden box, and carefully wrapt up in soft pieces of native cloth and cocoa-nut fibre. There were among them a large whale's tooth, highly polished, and quite brown from repeated greasing, a necklace made of pieces of whales' teeth, the first that ever came to these mountains, and a fine cannibal fork in the shape of a club, and bearing the ominous name of "strike twice," *i.e.* first the man and then his dead body. The woman told me a lot of other crown property had been burnt when, some years ago, the Americans destroyed Navua; among it, she assured me, was a short club which would kill a man on the spot, and was never known to miss when thrown by the hand of the supreme chief. Whales' teeth are with the Fijians what diamonds are with us, and in former days there was no favour a chief would refuse if a number of these were offered. The European and American traders soon found this out, and did not fail to bring quantities whenever they touched at these islands. The consequence has been that on the coast and amongst the christianized population whales' teeth have suffered considerable depreciation, though they have not as yet entirely been reduced to their proper value. In the interior of the great island they maintain their old importance, and Kuruduadua, on seeing us handling some money, expressed his astonishment that we should prefer coins to whale's teeth. We told him not many years would elapse before he changed that opinion, but he thought that time would probably never come.

During my stay, one of the days was rainy, preventing me from making an excursion. On expressing my regret to that effect, a man was brought to me who may be called the " clerk of the weather." He professed to exercise a direct meteorological influence, and said that by burning certain leaves and offering prayers only known to himself, he could make the sun shine or rain come down, and that he was willing to exercise his influence on my behalf if paid handsomely. I told him that I had no objection to give him a butcher's knife if he could let me have fine weather until my return to the coast, but if he failed to do so he must give me something. He was perfectly willing to risk the chance of getting the knife, but would not hear of a present to me in case of failure; however, he left to catch eels for me. When returning, the clouds had dispersed and the sun was shining brilliantly, and he did not fail to inform me that "he had been and done it." I must further do him the justice to say that I did not experience any bad weather until I fairly reached the coast, and that no sooner had I set my foot in Navua than rain came down in regular torrents. This man has probably been a close observer of the weather, and discovered those delicate local indications of a coming change, with which people in all countries living much in the open air are familiar, and he very likely does not commence operations until he is pretty sure of success.

As one of my objects in Fiji was to find out "all about the leaves," I was anxious to be initiated in an art productive of such astonishing results. A little inquiry, however, convinced me that an initiation would make me rather an object of fear than respect. The adepts in the art of Vaka-drau-ni-kau-taka (literally, to effect with leaves) are in fact regular sorcerers, whose craft I thought it prudent not to join. Not satisfied with causing rain and sunshine, they exercise a direct and much more criminal influence over life and death, by working upon the superstitious fears of the natives to such an excess that it causes serious illness, if not death. They are identical with the disease-makers of Tanna, though not enjoying such a prominent position, and accomplish what European impostors effected, and in some districts still effect, by praying to death people silly enough to make themselves nervous about any influence these rogues pretend to exercise. If a Fijian wishes to cause the destruction of an individual by other means than open violence or secret poison, the case is put in the hands of one of these sorcerers, care being taken to let this fact be generally and widely known. The sorcerer now proceeds to obtain any article that has once been in the possession of the person to be operated upon. These articles are then burnt with certain leaves, and if the reputation of the sorcerer be sufficiently powerful, in nine cases out of ten the nervous fears of the individual to be punished will bring on disease, if not death; a similar process is applied to discover thieves. In order to comprehend the working of this abominable system, and the mischief and extortion to which it gives rise, one must take into consideration the absolute helplessness of the Fijian, in fact the Polynesian generally, when anybody has acquired a moral ascendency over him. A certain white settler being very much annoyed by a native, told him in as powerful language as he could muster, that he wished him dead, and that he had no doubt he would die within a twelvemonth. The native professed to treat this prophecy with derision; nevertheless on calling about a year afterwards, the foreigner was informed that the native had fretted so much that he died. The words spoken in anger had thus had a fatal result, and the white man in confiding them to me seemed truly sorry for what he had done.

The inhabitants of Namosi on being asked for their name, will never give it when anybody else is present to answer the question. I inquired for the reason, but they could give no other explanation except that it was their custom. It probably offends their dignity. They feel in this respect more acutely than ourselves, who deem it polite always to apologize when having to ask a person's name, and generally endeavour to find it out in a less direct way.

The family life of the Fijian, especially in places like Namosi, where not modified by Christian teaching, is very curious. The men sleep, as has already been observed, at the Bure-ni-sa, or strangers' house, those of about the same age generally keeping together, whilst the boys, until they have been admitted publicly into the society of adults, have a sleeping bure to themselves. It is quite against Fijian ideas of delicacy, that a man ever remains under the same roof with his wife or wives at night. In the morning he goes home, and if not employed in the field, remains with his family the better part of the day, absenting himself as evening approaches. Rendezvous between husband and wife, of which no further explanation can be given, are arranged in the depths of the forest, unknown to any but the two. After childbirth, husband and wife keep apart for three, even four years, so that no other baby may interfere with the time considered necessary for suckling children, in order to make them healthy and strong. This in a great measure explains the existence of polygamy, and the difficulties the missionaries had to contend with in fighting against its abolition. The relatives of a woman take it as a public insult if any child should be born before the customary three or four years have elapsed, and they consider themselves in duty bound to avenge it in an equally public manner. I heard of a white man, who being asked how many brothers and sisters he had, frankly replied, "Ten!" "But that could not be," was the rejoinder of the natives; "one mother could scarcely have so many children." When told that these children were born at annual intervals, and that such occurrences were common in Europe, they were very much shocked, and thought it explained sufficiently why so many white people were "mere shrimps." Adultery is one of the crimes generally punished with death; and Kuruduadua himself had not long ago one of his nephews clubbed for taking undue liberties with one of his wives. What is called amongst us the "social evil," and thought to be an unnatural excrescence of our artificial state of society, is not unknown amongst these barbarous races. There being no streets, nymphs of a certain description waylay travellers on the high roads—a direct refutation of the Mormon argument, that "polygamy is the only cure for this corruption of our great cities."

Fijians have been charged with want of natural affection; and the strangulation of widows on the death of their husbands, and the killing of parents when beset with the infirmities of old-age by the hands of their own children, have been advanced as proofs thereof. Yet these facts are perhaps the best arguments that human nature is not different in the Fijis than elsewhere. Affection for the departed—of course, mistaken affection—prompted their relatives or friends to dispatch widows at the time of their husbands' burial; and the widows themselves have been known to seek death by their own hands, if their relatives refused to fulfil that duty which custom imposed upon them. Even widowers, in the depth of their grief, have frequently terminated their existence, when deprived of a dearly beloved wife. On the death of a near relative people will cut off joints of their fingers in order to demonstrate their grief, and they will mourn for a long time for their lost ones. The sentiment of friendship is strongly developed, and there is scarcely a man who has not a bosom friend, to whom he is bound by the strongest ties of affection. The birth of a child is a perfect jubilee, and it is truly touching to see how parents are attached to their children, and children to their parents. Under such circumstances, the greatness of the sacrifice that children are sometimes called upon by their infirm old parents to terminate their sufferings by putting them to death, becomes evident. It is a cruel slander of the native character to put any other construction on this singular, though mistaken proof of filial affection. In a country where food is abundant, clothing scarcely required, and property as a general rule in the possession of the whole family rather than that of its head, children need not wait "for dead men's shoes," in order to become well off, and we may, therefore, quite believe them when declaring that it is with aching heart and at the repeated entreaties of their parents that they are induced to commit what we justly consider a crime. The two old men present at our meeting at Namosi, were living proofs that children however, even in these wild parts, will not always be induced to lay hands on their parents.

I told a native who sometimes called at Danford's house, and seemed to be a most respectable man, a belief had been spread in our country that the Fijians were almost without natural affection. He replied, there might be some amongst his countrymen, as well as the whites, who had not much feeling; but those who denied the Fijians natural affection, either understood them very little, or else represented them in such black colours for some purposes of their own. "When leaving home," he continued, "all my thoughts are with my family, and I am never so happy as when I am under my own roof, and have my wife and children around me. When a few days ago my youngest boy was ill, I sat up with him three nights, and it would have broken my heart had he died." The man was a savage, a heathen, yet could any Christian parent have spoken more warmly or naturally? Fortunately, affection is wisely placed by Providence beyond the reach or influence of any system, right or wrong. Like a beautiful flower, it springs up freely in any soil congenial to its growth. If the Fijians were only half as black as they have been painted, they would long ere this have been numbered amongst the extinct races; for no society, however primitive, can possibly continue to exist, if the evil passions—the destructive elements—preponderate over the good. The best vindication of their national character is their national existence; the best proof of their living a life as free from vice and corrupting practices as any heathen can be expected to live, is a physical development on an average far above that of which our own race, with all its advantages of civilization, can ever hope to boast.

In the evenings, Batinisavu or other men would come and entertain me with some of those innumerable stories, in which the natives may be said to photograph themselves, show in what direction their fancy wanders, and which no travellers, worthy of the name, should omit writing down. The supernatural element plays a prominent part in all Fijian stories, and whilst possessing a decidedly local colouring, they forcibly remind one of our own nursery tales. The natives are very fond of them, and a good story-teller can never starve. Danford informed me that the "Arabian Nights" have been a source of income to him. "Aladdin, or the Wonderful Lamp," is paid for at the rate of two fat pigs, equivalent to about eight dollars; and the "Forty Thieves" meets with a similar success whenever that charming tale is told, several friends clubbing together in order to make up a purse for the story-tellers. What a source of pleasure one would open to these islanders, by translating for them the "Arabian Nights" or Grimm's "Household Stories."

Chief Batinisavu was always careful to inform me that he did not tell stories for pay, and in printing one of those he told me I must do him also the justice to add that it was a very long one. Taking up several hours in telling, I can merely give the pith of the whole, and have to leave out those details which, without ample explanation and local knowledge, would be quite unintelligible and uninteresting to the generality of readers.

THE STORY OF ROKOUA, AS TOLD BY BATINISAVU, GOVERNOR OF NAMOSI.

"Once upon a time there dwelt at Rewa a powerful god, whose name was Ravovonicakaugawa,* and along with him his friend the God of the Winds, from Wairua.† Ravovonicakaugawa was leading a solitary life, and had

* Ravovonicakaugawa, *i.e.* a long way off.

† This god was and is supposed to reside at a little brook in the lovely valley of Namosi, on Viti Levu, pointed out to us when we visited the interior of the island in September, 1860. When the Rewa people come to the Namosi valley, they never fail to make sacrificial offerings at Wairua (which is both the name of the locality and its god). Even some of those that have become Christians continue this practice.

long been thinking of taking a wife to himself. At last his mind seemed to be made up. 'Put mast and sail in the canoe,' he said, 'and let us take some women from Rokoua, the God of Naicobocobo.'* 'When do you think of starting?' inquired his friend. 'I shall go in broad daylight,' was the reply, 'or do you think I am a coward to choose the night for my work!' All things being ready, the two friends set sail, and anchored towards sunset off Naicobocobo. There they waited one, two, three days, without, contrary to Fijian customs, any friendly communication from the shore reaching them, for Rokoua, probably guessing their intentions, had strictly forbidden his people to take any food to the canoe. Rokoua's repugnance, however, was not shared by his household. His daughter, the lovely Naiogabui,† who diffused so sweet and powerful a perfume that, if the wind blew from the east, the perfume could be perceived in the west, and if it blew from the west, it could be perceived in the east—in consequence of which, and on account of her great personal beauty, all the young men fell in love with her. Naiogabui ordered one of her female slaves to cook a yam, and take it to the foreign canoe, and at the same time inform its owner that she would be with him at the first opportunity. To give a further proof of her affection, she ordered all the women in Naicobocobo to have a day's fishing. This order having been promptly executed, and the fish cooked, Naiogabui herself swam off with it during the night, and presented it to the Rewa God.

* Naicobocobo, on the western extremity of Vanua Levu, the supposed starting-point of departed spirits for Bulu, the future place of abode.

† Naiogabui, *i.e.* one who smells sweetly.

"Ravovonicakaugawa was charmed with the princess, and ready to start with her at once. She, however, begged him to wait another night, to enable Naimilamila, one of Rokoua's young wives, to accompany them. Naimilamila was a native of Naicobocobo, and against her will united to Rokoua, who had no affection whatever for her, and kept her exclusively to scratch his head or play with his locks, hence her name. Dissatisfied with her sad lot, she had concocted with her step-daughter a plan for escape, and was making active preparations to carry it into execution. On the night agreed upon, Naimilamila was true to her engagement. 'Who are you?' asked the god as she stepped on the deck. 'I am Rokoua's wife,' she rejoined, 'get your canoe under weigh. My lord may follow closely on my heels, and Naiogabui will be with us immediately.' Almost directly after a splash in the water was heard. 'There she comes,' cried Naimilamila, 'make sail;' and instantly the canoe, with Ravovonicakaugawa, his friend, and the two women, departed for Rewa.

"Next morning, when Rokoua discovered the elopement, he determined to pursue the fugitives, and for that purpose embarked in the 'Vatutulali,' a canoe deriving its name from his large drum, the sound of which was so powerful that it could be heard all over Fiji. His club and spear were put on board, both of which were of such gigantic dimensions and weight, that it took ten men to lift either of them. Rokoua soon reached Nukuilailai, where he took the spear out, and making a kind of bridge of it, walked over it on shore. Taking spear and club in his hand, he musingly walked along. 'It will never do to be at once discovered,' he said to himself; 'I must disguise myself. But what shape shall I assume? That of a hog or a dog? As a hog, I should not be allowed to come near the door; and, as a dog, I should have to fetch the bones thrown outside. Neither will answer my purpose. I shall therefore assume the shape of a woman.' Continuing his walk along the beach, he met an old woman, carrying a basket of taro and puddings, ready cooked, and, without letting her be at all aware of it, he exchanged figures with her. He then inquired whither she was going, and, being informed to the house of the God of Rewa, he took the basket from her, and, leaving club and spear on the beach, proceeded to his destination. His disguise was so complete, that even his own daughter did not recognise him. 'Who is that?' she asked, as he was about to enter. 'It is I,' replied Rokoua, in a feigned voice; 'I have come from Monisa with food.' 'Come in, old lady,' said Naiogabui, 'and sit down.' Rokoua accordingly entered, and took care to sit like a Fijian woman would do, so that his disguise might not be discovered. 'Are you going back to-night?' he was asked. 'No,' the disguised god replied; 'there is no occasion for that.' Finding it very close in the house, Rokoua proposed a walk and a bath, to which both Naiogabui and Naimilamila agreed. When getting the women to that spot of the beach where club and spear had been left, he threw off his disguise, and exclaimed, 'You little knew who I was; I am Rokoua, your lord and master,' and, at the same time taking hold of their hands, he dragged the runaways to the canoe, and departed homewards.

"When the Rewa god found his women gone, he again started for Naicobocobo, where, as he wore no disguise, he was instantly recognized, his canoe taken and dragged on shore by Rokoua's men, while he himself and his faithful friend, who again accompanied him, were seized and made pig-drivers. They were kept in this degrading position a long time, until a great festival took place in Vanua Levu, which Rokoua and his party attended. Arrived at the destination, the Rewa god and his friend were left in charge of the two canoes that had carried the party thither, whilst all the others went on shore to enjoy themselves; but as both friends were liked by all the women, they were kept amply supplied with food and other good things during the festival. Nevertheless Ravovonicakaugawa was very much cast down, and taking a kava-root (Yaqona), he offered it as a sacrifice, and despairingly exclaimed, 'Have none of the mighty gods of Rewa pity on my misfortune?' His friend's body became instantly possessed by a god, and began to tremble violently. 'What do you want?' asked the god within. 'A gale to frighten my oppressors out of their wits.' 'It shall be granted,' replied the god, and departed.

"The festival being over, Rokoua's party embarked for Naicobocobo. But it had hardly set sail when a strong northerly gale sprang up, which nearly destroyed the canoes, and terribly frightened those on board. Still they reached Naicobocobo, where the Rewa god prayed for an easterly wind to carry him home. All Rokoua's men having landed, and left the women behind to carry the luggage and goods on shore, the desired wind sprang up, and the two canoes, with sails set, started for Rewa, where they safely arived, and the goods and other property were landed and distributed as presents among the people.

"But Rokoua was not to be beaten thus. Although his two canoes had been taken, there was still the one captured from Ravovonicakaugawa on his second visit to Naicobocobo. That was launched without delay, and the fugitives pursued. Arriving at Nukuilailai, Rokoua laid his spear on the deck of the canoe and walked over it on shore, as he had done on a previous occasion. Landed, he dropped his heavy club, thereby causing so loud a noise that it woke all the people on Viti Levu. This noise did not escape the quick ear of Naimilamila. 'Be on your guard,' she said to her new lord, 'Rokoua is coming; I heard his club fall; he can assume any shape he pleases; be a dog, or a pig, or a woman; he can command even solid rocks to split open and admit him, so be on your guard.' Rokoua meanwhile met a young girl from Nadoi on the road, carrying shrimps, landcrabs, and taro to the house of the god of Rewa, and without hesitation he asumed her shape, and she took his without being herself aware of it. Arriving with his basket at his destination, Naiogabui asked, 'Who is there?' To which Rokoua replied, 'It is me; I am from Nadoi, bringing food for your husband.' The supposed messenger was asked into the house, and sitting down, he imprudently assumed a position not proper to Fijian women. This, and the shape of his limbs, was noticed by Naiogabui, who whispered the discovery made into her husband's ear. Ravovonicakaugawa stole out of the house, assembled his people, recalled to their minds the indignities heaped upon him by Rokoua, and having worked them up to a high pitch of excitement, he informed them that the offender was now in their power. All rushed to arms, and entering the house they demanded the young girl from Nadoi. 'There she sits,' replied Naiogabui, pointing to her father; and no sooner had the words been spoken, than a heavy blow with a club felled Rokoua to the ground. A general onset followed, in which the head of the victim was beaten to atoms. This was the end of Rokoua."

CHAPTER XIII.

DEPARTURE FROM NAMOSI.—VUNIWAIVUTUKA.—THE "VELI."—MODE OF TATOOING THE MOUTH.—PASSING DOWN THE NAVUA RIVER.—NAGADI CLEARED OUT BY ITS VASU.—OUR CANOE CAPSIZED.—RETURN TO THE 'PAUL JONES.'—KURUDUADUA'S CHARACTER.—LEAVING NAVUA.—BEGA. —MR. STORCK'S ILLNESS.—RETURN TO KADAVU.—ASCENT OF BUKE LEVU.—REWA.—IMMIGRANTS FROM NEW ZEALAND.—MR. MOORE'S POWERFUL SERMON.—ARRIVAL AT LADO.—OFFICE DRUDGERY.

WHEN, on the 2nd of September, I left Namosi, there were great lamentations. The women and children cried bitterly, and Batinisavu, the Governor of the place, with several young chiefs, made up their minds to see me safe to the coast. I had witnessed a similar scene after the departure of Colonel Smythe and Mr. Pritchard, and heard chiefs and people regret that they were gone, and would probably never come again. I had been amongst them much longer, and they had got used, and, in some instances, quite attached to me. Cannibals though they be, they have many good qualities; and some of the greatest crimes laid to their door may be explained, as singular, though mistaken demonstrations of a deep natural affection.

We took the same road as that by which Mr. Pritchard and his party had returned, and in the afternoon reached Vuniwaivutuka, where we made preparations for staying the night. Directly on our arrival, some of the leading men came up to the Bure-ni-sa we were stopping at, to present a root of kava to Batinisavu, as a token of respect and goodwill, and making, in presenting it, a neat little speech, to which the Namosi Governor replied in equally friendly terms. Batinisavu struck me as a man very far above the rest of his countrymen. There was something quiet and dignified about him; and though he always went without any hesitation through all the ceremonies his station imposed, he often apologized to me by saying it was " Vaka Viti "—Fijian usage—which he could not set aside.

The bures are, in Fiji, what club-houses are with us: everybody goes there, and all the news finds its way thither. The great topic of that day's conversation was the discovery of an adultery in a neighbouring village. The friends of the woman took up the case. The bure to which the adulterer belonged resisted their attack, and the consequence was a series of broken heads. The chief offender escaped, but his father was caught and punished for his son's transgressions. The husband of the seduced wife had his taro-fields destroyed, and was told that such a fool as he did not deserve to possess them. Batinisavu strongly censured the whole proceedings. He asked, where was their justice? to punish the poor old father for his son's wickedness, was simply cruel, and to destroy the crops of an already injured man, worthy of such mountaineers and fools as they were.

No one can be long in this region of " taboo " and " tatoo " without perceiving what rich stores of human fancy and ideas, shortly to be lost or mutilated for ever, are here offered. Attention is constantly directed to them, and you have as little chance of remaining ignorant of the great deeds of Degei, Rokoua, and the Vasu-ki-lagi, as you have in the East of the stories of successful magicians, spell-bound princesses, and mighty treasures concealed in obscure caverns. In Kuruduadua's dominion I could hardly turn without hearing of the doings of the *Veli*, and the greater part of the evening at this place was again devoted to them. My curiosity had already been so much excited that I determined, come what might, to write their natural history in the very localities most frequented by them. By inquiry and frequent cross-examination, I found the Veli to be a class of spirits in figure approaching to the German *gnome*, in habits of life the fairy of England. They have been in the country from time immemorial, and live in hollow Kowrie-pines and Kabea-trees. They are of diminutive size, and rather disproportionately large about the upper part of their body. Their hair is thick, and prolonged behind in a pig-tail. Some have wings, others have not. Their complexion rather resembles that of the white race than the Fijian. They have great and petty chiefs; are polygamists, and bear names like the Fijians. They also resemble the latter in wearing native cloth or tapa, which however is much finer and whiter than the ordinary sort. They are friendly disposed, and possess no other bad quality than that of stealing iron tools from the natives. They sing sweetly, and occasionally gratify the Fijians by giving them a song. They feed on the fruit of the Tankua (*Ptychosperma*) and Boia (*Scitaminearum gen. nov.*), which they term emphatically their cocoa-nut and their plantain; and men imprudent enough to cut down these plants, have received a sound beating from the enraged Veli. They drink kava made, not of the cultivated *Macropiper methysticum*, but of a pepper growing wild in the woods, and vernacularly termed Yaqoyaqona (*Macropiper puberulum*, Benth.). The Fijians have no long stories about them, as they have about their gods. All the accounts of the Veli relate to isolated facts,— to their abode, their having been seen, heard to sing, caught in a theft, and found to beat the destroyers of their peculiar trees; but they are so numerous that it is no wonder the Fijians should consider the evidence sufficient to establish their real existence.

The women about this place, as well as about Nagadi, were tatooed around the whole mouth, not merely around the corners, as is customary on the coast. The reader may smile at this observation, but after living awhile amongst natives in an almost absolute state of nudity, the eye readily detects these minute differences, and the mind begins to comprehend why, on paying compliments, these people dwell with such emphasis on this or that part of the body, when a European, under similar circumstances, would record his admiration for a becoming toilet, whole or in part. In narrating travels in barbarous countries, the disadvantage of the people not wearing clothes is acutely felt. In order to convey, at least, some notion of what the personages encountered were like, one is compelled to notice their arms, legs, and other parts of their body, a fact for which one is not always inclined.

The next morning we left Vuniwaivutuka; and after a smart walk of about an hour and a half, we came to a branch of the Navua river, where Batinisavu had a raft of bamboos prepared. It seemed a very rickety contrivance; nevertheless it was strong, and there was no chance of capsizing in passing over rapids. But I found it impossible to keep my collections dry, so four of the boys took them on their backs to Navua. We then passed down the river rapidly, and about noon reached the town of Nagadi, where we had stopped a night on a previous occasion. There we intended to exchange our raft for a large canoe, but this intention was frustrated. On that very day the "Vasu" to Nagadi had taken away all the canoes, and other articles of the town that took his fancy. A "Vasu" is a mighty personage in Fiji. He is simply a nephew, but, according to the usage of the country, he holds all the movable property of his uncle at his absolute disposal, and can at any moment take whatever he chooses. There are *vasus* not only to families, but to towns and states, and it is considered shabby to resist their exactions. Some *vasus* have even sold the land belonging to their uncles, but Fijians say that is going a little *too* far, and exceeds the proper limits of the system. If therefore the uncles wish to keep anything to themselves, they must not let their nephews see it. I remember Batinisavu, having a grasping nephew, and several American hatchets given him, begged Danford to keep them at his house, so that the vasu might not get wind of their existence. Of course the Vasus are expected to make some return, and the Vasu to the town of Nagadi, living on the sea-coast, where salt is abundant, had presented the people, whom he had cleared out of almost everything, with a supply of that useful article, for they assured us they had no canoes left to get across the river, and should have to commence that very day to build new ones. On pushing down the river, we overtook the flotilla, heavily laden with goods of all descriptions, and had no difficulty in getting the loan of a canoe to Navua. We had little reason to congratulate ourselves on this change. At the next rapid we could not bale faster than the water came in at the stern; the outrigger lost its balance, and in another moment the canoe was capsized.* Soromato, my faithful friend, by a desperate dash saved a bundle of my clothes, including cloak, and succeeded in getting them on shore dry. Having been up to my neck in water, I felt very thankful to Soromato. The natives kindled a fire on a gravelly spot, and two of the boys had to chew kava, which, in the absence of a proper bowl and straining fibres, was made in large leaves and squeezed through ferns.

The canoe being baled out, and put again in proper order, we continued our voyage, and without any further mishap reached Navua. Kuruduadua met us close to the town; he had been all day busy in the field, and said he had a great number of people staying with him to assist in his agricultural labours. When we stepped on shore, supper was just being presented to them. It was an immense heap of provisions, and though there were probably two hundred visitors, there must have been ample for all.

* In our Plate representing Koro Basabasaga will be seen a good specimen of a Fijian river-canoe with its outrigger.

As the houses were crowded, I was very glad to learn that the 'Paul Jones,' with Mr. Pritchard on board, had arrived from Nadroga, and was then anchored at the mouth of the river. Two of the crew soon after made their appearance in the dingy belonging to the schooner, and I availed myself of the chance to get on board. On paddling down the river we encountered several heavy showers; the clerk of the weather at Namosi had only guaranteed sunshine until I should have fairly reached the coast, and now I was again in the region of salt water, mangrove-trees, and sago swamps. We took shelter under a thick tree, and with my umbrella-parasol I kept myself tolerably dry. The people living on the high banks under which we had halted, soon espied us, and invited us to come into their houses. When we refused on account of its getting too late to reach the schooner, they brought some hot yams and taro, and one of the boys was sent up a cocoa-nut palm, slippery though the trunk was, to knock down some nuts for drinking. We gave them some sticks of tobacco, of which they were very glad, and all parted with mutual expressions of goodwill.

I took leave of Batinisavu, the Namosi Governor, at Navua, and shall always remember his kindness. Kuruduadua came on board that night, and Danford accompanied him. Though he had publicly declared in favour of the cession of Fiji to England, he had not as yet formally signed the deed of cession. As he is one of the most powerful chiefs, it was important to have his signature, and in the evening he affixed his mark to that document; Mr. Charles Wise having once more carefully translated the import of the paper, and I attesting the chief's signature.

Whilst sitting in the little cabin of the schooner, Kuruduadua asked about a variety of subjects, and generally exclaimed, " Ah! ye white men are superior people. We are ignorant savages!" He was much pleased with that volume of Wilkes's 'Narrative of the United States Exploring Expedition' relating to Fiji and Tonga. Indeed, all the natives who saw it were enraptured with that beautiful publication. So faithful are the representations of places and persons, that the natives instantly recognized them. The portraits of Tanoa, the father of King Cakobau, and that of the Queen of Rewa, pleased them mightily. They always exclaimed, " They live! They can see! They speak!" I wish the artist had been there to hear the praise lavished upon his productions.

Kuruduadua left very late, and Danford went with him. Always making it a point to speak of people as I find them, I have nothing to say except what is in their favour. Both of them had been of the greatest service to us, and behaved well. Kuruduadua we found an intelligent, straightforward man, quite ready to listen to reason, prepared to come up to any obligations he had taken upon himself, and detesting all half-measures, all sham. Of Danford I have already spoken. He has been a pioneer, whose services in that direction I should not be inclined to undervalue, and without whom one of the most interesting episodes of my life would probably not have occurred.

We finally left the Navua river on the 5th of September, and stood over to Bega (=Mbenga), an oval-shaped island, about five miles long by three wide, subject to Rewa, and in some measure to Kuruduadua. No sooner had we cast anchor than Mr. Don, an Englishman, came to the Consul, complaining that the natives, under pressure from the Tonguese, wished to compel him to let them have back again the land which he had bought, as they had given the island of Bega to the Tonguese. Mr. Pritchard went to the man who represented himself as the principal chief, and told him that Mr. Don totally rejected the offer of ten fat pigs, or any other equivalent for the land he had acquired, and if they had given their island to the Tonguese, it was by no means binding, Maafu, the Tonguese chief, having publicly renounced all claims on and in Fiji; and, until her Britannic Majesty's pleasure was known, the cession of Fiji to England was valid, and could not be ignored. Two Tonguese present tried to argue the point, but were signally defeated by one no novice in native tactics.

One of our reasons for making Bega was to obtain some oil of the Dilo (*Calophyllum inophyllum*, Linn.), an excellent liniment for rheumatism, pains in the joints, bruises, etc., and enjoying a high reputation throughout the South Sea. Mr. Storck, my able assistant, had—after quite recovering from his fall in Somosomo—committed the imprudence, whilst paying a visit to his friend Peter, the King's councillor, at Bau, to sleep a night between two open doors on a matted floor of· a new house, in consequence of which he had gradually become so stiff as ultimately to be unable to move even his hands. We had to dress him, put him to bed, and even feed him, his appetite being good all the while; and he, poor fellow, was so helpless that at one time he was falling in the sea, and only saved by the presence of mind of one on board. All the Fijian doctors recommended the external application of Dilo oil; and for some calico we obtained two gourd-flasks full, with which the patient was rubbed several times a day. Fortunately our voyage was drawing to a close; and I am happy to add, the greater comfort and change of food at Ovalau soon restored him to perfect health.

We intended to proceed from Bega direct to Ovalau, but towards evening the weather became so fine—every sign of rain having disappeared—that the idea struck us to run over once more to Kadavu, and ascend if possible Buke Levu, the great mountain. The passage between Bega and Kadavu being an open sea, and we having a good pilot on board, in the person of Mr. Charles Wise, the consular interpreter, we left Bega just whilst the sun was gilding the feathery tops of the cocoa-nut palms, and diffusing a bright hue over the white coral beaches.

Sailing all night, daybreak disclosed the bold outline of Buke Levu, a mountain 3800 feet high, situate on the north-west point of Kadavu, and deriving its name from a certain resemblance to the hillocks (Buke) on which yams are planted; hence Buke Levu, the "large yam hill." No white man had ever ascended it, and, though laid down in the latest maps, its very name was not recorded. It will be remembered that we had made two distinct efforts to reach its summit, but were baffled by gales and rain. We now were about to make the third. On bringing our little schooner to anchor off the town of Taulalia, heavy showers overtook us, and we began to despair of ever attaining our object, when about nine o'clock it suddenly cleared up. The natives, who had been watching from the beach, could not understand our hesitation in not landing at once, and in proof of their friendly disposition, brought out their women and children; and, moreover, carried green boughs, as the soldiers do in Macbeth, when "Birnam wood removes to Dunsinane."

BUKE LEVU, SEEN FROM THE SOUTH.

On learning our object in coming to their town, fifteen men and boys cheerfully volunteered to accompany us. The ascent commenced the moment we left Taulalia, and passing over cultivated grounds where the people were busy with their crops of sugar-canes, yams, taros, and plantains, we reached in about a quarter of an hour a village, where another party of natives joined us, and where we saw some fine plants of the different kinds of kava, for which Kadavu is renowned. A narrow path, often winding along precipices and through rivulets, led to about 1500 feet elevation, where it gradually faded away, and the isolated patches of cultivation noticed up to this height, as well as the wood which had re-occupied ground at one time cleared and the masses of reeds gave place to an undisturbed virgin forest, through which we had to cut our way. We had taken the precaution of bringing a strong rope, sixty feet long, which, made fast to trees, proved extremely useful in dragging ourselves up almost perpendicular rocks, in the rainy season occupied by waterfalls, and even at this time of the year very slippery. On some of these were found a number of delicate ferns (*Hymenophyllum*), and quite a new species of land-shell (*Bulimus Seemanni*, Dohr.), fully two inches long, and of a bright salmon-colour.

In order to save time, we had directed one of our men to push ahead and prepare a camp-kettle full of tea— of all beverages the best when one is tired and heated. When at last, after great exertion and frequent stopping to examine objects of interest, we reached the top, he and half-a-dozen others were already there, but they had omitted to bring either matches, firesticks, or water; and even the cocoa-nuts, packed up with the rest of the day's provisions, were too old for drinking. Being extremely thirsty, we could not touch food, hungry though we were. The natives declared the nearest water to be more than 1000 feet down, and, as they had not the proper wood, it was impossible for them to kindle fire by friction. However, a man must have read 'Robinson Crusoe' to little purpose, if his resources fail him in moments like these. We were determined not to let our explorations come to a sudden stop for want of something to drink. Mr. Pritchard left me the option between procuring fire or water; to guard against lame excuses on the part of the natives, it being thought necessary that one of us should go with them in search of a spring. Knowing what a hard job it was to make fire by rubbing, without pausing, two pieces of wood together, especially in the tropics, I declared in favour of getting the water. My companion, who did not seem to relish descending so many feet and climbing up again, was evidently pleased with his lot. In spite of all the natives were saying about making the wood answer, he resolutely began rubbing away. Great exertions were required; hat, jacket, vest, and necktie discarded, to obtain greater freedom of action. At last came the reward. The wood began to smoke, sparks appeared, went out again, reappeared, and, brought in contact with a piece of bark-cloth cut off the tail of a boy's dress, soon produced a flame.

All this time I had been sitting on an old stump, feigning to be quite insensible to certain broad hints about the desirableness of looking after the execution of my part of the contract. When the first flame had appeared I at last bestirred myself, and to the surprise of the fire-kindler, instead of going a long way for water, climbed up a neighbouring tree on which I had noticed an epiphytical plant (*Astelia montana*, Seem.), the leaves of which, acting as a kind of rain-gauge, were filled with pure water: by merely emptying these the necessary supply was obtained. Ere long, tea was ready, and re-

A FINE VIEW. 215

lished all the more from recalling to mind the long established connection between cups, slips, and lips.

After all hands had partaken of refreshment, a number of trees were felled in order to gain, if possible, a view, the top of Buke Levu being densely wooded. No sooner had this been accomplished than, to our joy, the clouds which up to this time had been interposed between us and the region below, dispersed, disclosing a great part of Kadavu and the sea. Our little schooner was snugly lying at anchor, flying the British colours; but we listened in vain for the signal guns which the men had been directed to fire as soon as they should perceive the smoke of our fire, intensified at intervals by throwing heaps of green leaves upon it. We afterwards learned that it had been found impossible to distinguish between smoke and clouds. A large native canoe, with its white triangular sail, was seen approaching the shore, and the blasts of the conch shells could be heard distinctly, though we were nearly 4000 feet high; otherwise there was a deep silence, only occasionally broken by the dogs, which have become naturalized in these wilds, as the domestic fowls have in other parts of the group. The vegetation encountered was similar to that of Voma Peak in Viti Levu; there were the same bright orange-coloured orchids (*Dendrobium Mohlianum*, Reichb. fil.) and the epiphytical ferns, but also several new species of plants. The *Cinnamomum* furnishing a superior kind of Cassia-bark was here as plentiful as in Great Fiji; a kind of Gummi Guttæ (*Clusia sessilis*, Forst.) also engaged our attention. Buke Levu is evidently an extinct volcano; and hot springs at its foot, near the town of Nasau, ascertained by Colonel Smythe to be 144° Fahrenheit, may possibly stand in some connection with its former activity. The outward look of the summit is very much like the cone of Vesuvius, as it was when I ascended it in 1861; but we did not discover any large crater, simply an insignificant swamp.

Having left on one of the trees a well-corked bottle containing the record of our visit,—that of the first white men who ever ascended the mountain,—we commenced the descent, which presented in some parts serious difficulties, but, thanks to our rope, we overcame them all; only one of the lads had a rather serious tumble, by which he sprained his ankle. Before we were more than halfway down it was completely dark, when the natives lit bundles of reeds and the stems of a weed (*Erigeron albidum*, A. Gray), both of which make excellent torches. On arriving at the first grove of cocoa-nut palms a general halt was made, and heaps of nuts were brought down from the trees and emptied of their contents with astonishing rapidity. It was past nine o'clock, just twelve hours after we started, when we reached Taulalia, where the whole village was assembled at and about the house of the Wesleyan teacher, a Fijian by birth, and our native companions had to give a most circumstantial account of our day's proceedings.

We slept at the house of the teacher, which we found clean and comfortable. Early next morning all who had accompanied us had to sit in a row,—and a nice long row it was,—and every one received a butcher's knife, which elicited much clapping of hands, in proof that the gift was accepted: money would not have pleased half as much, as its use is not understood. All payments are made in kind,—a most irksome and cumbrous way, compelling you to carry a whole heap of things to defray the current expenses of a cruise; articles regarded as small change, and making one look like a pedlar, you are supposed to have always about you. In one pocket you carry pipes and tobacco—in great demand, but held rather cheap; in another, fish-hooks, jews'-harps, and beads, the spare room to be filled with scissors and knives of various descriptions. On board are kept your gold and bank-notes, represented by bales of Manchester print, especially navy blue; flannel jackets and woollen blankets,—killing the natives faster than brandy and the so-called vices of civilization,—and American hatchets, price five dollars apiece. The inconvenience and expense of paying for everything by articles of barter is increased by some of the goods not proving acceptable in all towns, and the natives refusing certain things because they happen to differ in some unimportant trifle from those generally in use. Fashion here, as elsewhere, rules supreme: knives with white handles instead of black would be objected to, though their blades might be first-rate; and I learned to my cost that it is absolutely useless to lay in stock at Sydney or Melbourne unless one obtains exact information regarding the articles in demand.

On leaving Taulalia, September the 7th, we steered eastward, passing Yawe, the famous pottery manufactory, in order to bid farewell to Mr. Royce, the principal missionary at Tavuki, under whose hospitable roof we had previously stayed. Wishing to economize time, we left Tavuki at sunset for Ovalau; we had put to sea scarcely an hour when the weather became squally and very thick, compelling us to take in all canvas except the foresail. We should have fared ill if it had not been for the presence of the consular interpreter, Mr. Charles Wise, who combines with a perfect knowledge of the Fijian language, customs, and manners, the advantage of being one of the best pilots in the group, the more appreciated amongst the maze of more than two hundred islands, of which as yet no reliable chart has been prepared, though the labours of Wilkes, Belcher, Kellett, and Denham, have already done a great deal towards that desirable end. After an anxious night amongst reefs and shoals, we found ourselves off Rewa, and, as the wind had now become a gale, the rain was coming down in torrents, and the sea was very high, we took shelter in Laucala (=Lauthala) Bay, anchoring opposite the premises of Mr. Pickering, an old settler in Fiji. The occupier was absent, but his people made us comfortable.

A small schooner had just arrived from New Zealand with sixteen immigrants on board. The captain called on the Consul, and brought a file of colonial newspapers containing the latest European news. Vessels often making Fiji a week after leaving Auckland, we generally had our latest intelligence *viâ* New Zealand. The captain was going to return immediately, taking oranges, pine-apples, and yams with him, and intending to come back with a fresh number of immigrants. Those that he had brought this time had found shelter at the houses of the various white settlers about here. Mr. Pritchard and I called on several, to see what we could do for them. In comparison to New Zealand they found it rather warm in the group, while we, on the contrary, were quite chilly, and glad to have thick clothes on. They had not brought any mosquito curtains, and, like all new-comers, had suffered dreadfully during the first night from irritating bites, to guard against which in future the ladies were busy converting their light muslin dresses into defences against them.

In the evening a boat took us over to the mission-station of Mataisuva, where Mr. and Mrs. Moore gave us, as usual, a hearty welcome. The weather still continuing boisterous, we were easily persuaded to remain, especially as the next day was a Sunday, and Mr. Moore, for the benefit of the new arrivals, was to have service in English. Sunday morning proved very fine, and when drums were beaten—why does not some kind-hearted person present this fine church with a good tolling-bell?—boats and canoes poured in from all directions, and there was a large congregation, a gratifying sight after looking so long upon dark faces. Mr. Moore, a powerful and eloquent speaker, preached an extemporary sermon, admirably adapted to those he was addressing. Its tenor was that every man ought to do his duty in the position it had pleased Providence to place him in. Amongst his hearers there were probably very few who belonged to the denomination of which he is so bright an ornament, but in these out-of-the-way places all sensible people refrain from troubling their heads about the nice distinctions into which our

Protestant Church has unhappily been split, and all Christians who are not Catholics never raise much objection to forming part of a congregation, the members of which may more or less differ from them in minor points of discipline or doctrine.

Leaving Rewa roads on the morning of the 10th of September, we reached Port Kinnaird, Ovalau, on the following day, where our little schooner was refitted, and we made every preparation for another, my last, cruise in the group. Mr. Pritchard's work, which even in ordinary times was more than he could get through without the greatest efforts, and sitting up late or even whole nights, had accumulated to an alarming extent. The clerks he engaged proved worse than useless, though the pay which he could offer was three times what they would have got in England. After my departure he fortunately obtained the co-operation of Mr. Swanston as vice-consul, who, shortly after his installation in office, wrote me a letter, dated Levuka, July 9, 1861, a passage of which I shall take the liberty to quote, as it gives some insight into consular duties in this group:—

"There were urgent entreaties from missionaries and white residents at Rewa, and all along the coast of Viti Levu, to Mr. Pritchard, to visit them. Complaints from whites to windward against Tonga movements generally; and Mr. Henry complains in particular against Maafu, and seeks consular intervention. All this, etc., keeps Mr. Pritchard cruising about, and the office drudgery falls on me, and I have more than I can attend to; to wit:—

"Naval court yesterday.—Seamen complain against 'Caroline's' going to sea unseaworthy. Merchants and others put in claims against the master; he drunk and disorderly on the beach; have to put him under arrest. My constable gets intoxicated. Consular officer has to attend to it. Harvie, a Brit. subject, dead. Mr. Pritchard hands me in papers connected with the affairs, which he brought from Gau and Koro, whither he had to go last week on official business. Claims against this estate; counter-claims, disputes, and trouble to me. Old T—— lodges a complaint against S——; accuses him of violating the person of his daughter; Levuka in a state of excitement about it. Binner in great distress about disputed land title of his. Wilson's agent here, with chiefs from Na Lavu Lavu, to complete land titles. Clarke and Hazelman, ditto, ditto, from Na Viti Levu. Order from Hort, Bros., to seize schooner 'Kate,' unlawfully kept out of their possession. Complaints from Bob Somebody that Davies has kicked him out without paying him his wages. Claims against Maafu for debts due four years ago; American citizen connected with the affair; have to refer to the U. S. Consul; go into the affair to-morrow if business permits. Maafu here to ascertain why a certain Fijiman, sentenced some time since to three years' hard labour, is allowed to be at large; crime, killing a Tongaman. He offers, and insists upon his right, to enforce the punishment if the Fiji chiefs cannot. Wilson's agent lodges complaint against Bothe, for inducing natives at Wai Levu to give to him logs belonging to the company of which Wilson is the acting partner.

"And all these in two days; and so the wheel goes: every case has to be examined into, evidence heard, judgment given, papers in connection made out, often in duplicate, and so on, and so on. I am tired. I have been at it all day; it is now midnight; so good-bye.

"Yours very truly,
"Robert S. Swanston."

CHAPTER XIV.

VOYAGE AROUND VANUA LEVU.—DEPARTURE FROM LADO.—EAST COAST OF VITI LEVU.—NANANU ISLAND.—THE FIJIAN MOUNT OLYMPUS.—BUA.—NAICOBOCOBO.—NUKUBATI.—NADURI.—INTERVIEW WITH THE CHIEF.—DISCONTENT OF HIS SUBJECTS.—BECHE-DE-MER TRADE.—MUA I UDU AND ITS SUPERSTITIONS.—NA CEVA BAY.—ARRIVAL AT WAIKAVA.—VISIT TO MY COTTON PLANTATION.—MEETING AT WAIKAVA.—DEPARTURE.

OUR schooner, which had been so much shattered during the stormy passage from Kadavu to Rewa as to require a thorough refitting, again left Lado on the 10th of October. Mr. Pritchard had agreed to meet Colonel Smythe on the 17th of that month at Waikava, a town of Cakaudrove in Vanua Levu, and to bring thither all the most influential chiefs of that island. We stood over to the east coast of Viti Levu, and made it near Tova Peak, the bold cone-shaped outline of which could be seen from Lado in fine weather. The shores looked charming; grassy slopes alternating with groves of trees, rivulets, and inhabited valleys. Towards 4 P.M. we anchored off Nananu Levu (erroneously called Annan in the charts), close to the most northerly point of Viti Levu, and near another small island bearing the name of Nananu-gata. Like the adjacent coast, it is covered with grass, isolated screw-pines, and ironwood, and would seem well adapted for sheep and cattle. Poli-

NA VATU. 223

tically it is under Viwa, which again is tributary to Bau. There may be about one hundred inhabitants, who lived in a town defended by a deep ditch and high earthen mounds. On the top of the island were extensive plantations of Kawai (*Dioscorea aculeata*, Linn.), and in the valleys thousands of bread-fruit trees. The people did not seem to take much notice of us, and altogether behaved colder than any we had yet come in contact with.

Remaining at anchor all night, our voyage was continued early next morning to Bua, Sandalwood Bay. The north-eastern portion of Viti Levu, now fast fading away, is called Rakiraki, and famous in mythology as the site of Na Vatu, the Fijian Mount Olympus, and the abode of the supreme god Degei (=Ndengei). It has been supposed that this portion of Viti was the first

NA VATU, FROM THE NORTH.

to be inhabited, because all the tribes of the islands acknowledge Degei as their chief god, and own their knowledge of him to be derived from Rakiraki. There is nothing very remarkable either in the shape or character of the mountain, and, as far as our present information goes, we are unable to account for the distinction it enjoys. The accompanying sketch, obligingly furnished by Mrs. Smythe, will help to bear me out.

About noon on the 11th of October we were off Bua, no longer teeming with sandalwood as in days of yore. Our object was to invite Tui Bua, or King of Bua, to attend the meeting at Waikava. Our schooner not going close in, we went on shore in the dingy. The town of Bua is built on the banks of a river, the mouth of which for about a mile and a half is densely covered with mangroves. The district is low, the soil a rich alluvial clay. Bua has proved so unhealthy to Europeans that the white missionaries, after several deaths had thinned their ranks, were compelled to relinquish it, and fill their places with Tonguese teachers. This circumstance is the more to be regretted as Bua was a most complete station. The church is a very neat building, and has a good tolling-bell, instead of those hideous wooden drums used in other parts for calling the congregation together; the dwelling-houses are also highly finished. We found the principal one inhabited by the Tonguese teacher, who, together with his wife, was scenting cocoa-nut oil by adding rasped sandalwood and the white odoriferous flowers of the Bua (*Fagræa Berteriana*, A. Gray), a tree from which the place probably derives its name. They were very attentive to us, and loaded us with baskets full of kavikas, or Malay-apples, and cocoa-nuts, several bottles of goat's milk, and a fine log of sandalwood, now in the Kew Museum. The houses had been stripped of most of their European furniture, the church was rather in want of repair, and the whole had that desolate appearance which all places built by Europeans, but abandoned by them to natives, invariably possess. After visiting the graves of those Christian pioneers who had here laid down their lives in a noble cause, I felt quite melancholy, and was glad to return on board.

Tui Bua, the chief, being absent, and not expected back for some days, we made sail without delay. When evening came on we anchored off Bau lailai, and next morning rounded Naicobocobo (=Naithombothombo), the west point of Vanua Levu, which is rocky and thickly-wooded, and supposed to be a general starting-point (Cibicibi) for Bulu, the future abode of departed spirits. It is erroneously called Dimba Dimba by Wilkes and all those who copied him. On the 12th of October we anchored off Nukubati, a sandy little island, full of cocoa-nut trees and breadfruit, a great many of which had been cut down or otherwise injured by the Tonguese to revenge themselves on the Chief Ritova, whose private property the island is, and who had been driven from power by them to make room for a chief more willing to comply with their extortions than Ritova had shown himself to be. I went on shore and saw a party of women making pottery, which they did without a wheel, and extremely well.

On the 14th we ran down to Macuata (=Mathuata),

—not Mocuata or Mudwater, as sometimes written,—a small, stony isle, densely covered with ironwood, and at present uninhabited. This isle has conferred its name on the whole northern coast of Vanua Levu, and was the head-quarters of three branches of the ruling Macuata family, until about twenty-five years ago dissensions amongst its members broke out, which led to the total extinction of one of the branches, and proved to the others that a house divided against itself cannot stand. The whole coast had been subjugated by Tongamen; Ritova, the head of the most powerful branch, and the legitimate king of the district, was in exile; whilst Bete, who represented the weaker and subordinate portion of the family, resided at Naduri, and was a mere puppet in the hands of the artful Tongamen.

When making Nukubati we met a canoe going to Naduri, and sent a message by it to Bete, said to be attending some festival inland, that we were going to call at his town on the following day in order to make a communication to him. We had scarcely dropped anchor off Naduri when Bete's spokesman arrived in a large canoe. The first thing he delivered was a whale's tooth, dark as mahogany from age and repeated greasing, such as Fijians hold to be of the highest value. It was offered to the consul as a *soro*, or acknowledgment of submission and atonement from the chief. Mr. Pritchard hesitated about accepting it; but as its rejection would have been a direct insult, he thought it better to take the tooth, and thus prevent any misunderstanding and long explanations, both parties being fully aware of the real meaning of the token.

We found Bete sitting in his house surrounded by councillors. Mr. Pritchard informed him that his presence was required at Waikava at the meeting of chiefs, and his absence might prove disadvantageous to himself; but his mind seemed to be made up, and he gave us to understand that he did not mean to go, as the time was too short. His Tonguese advisers had probably induced him to act in this way.

I went some distance up a rivulet to bathe, and on my return met a number of Naduri people, who complained bitterly of the way in which they were ground down by the Tonguese, and how wretchedly poor they were in in comparison with formerly, when *bêche-de-mer* traders visited the coast, and they were kept well supplied with foreign articles of barter in exchange for the sea-slugs they collected. They said there could be no revival of this lucrative trade until their old chief Ritova was restored to power, as Bete was so weak, and so little respected, that he could not get the requisite number of hands together to make up a cargo. They were most anxious to know when Ritova was likely to come back, and asked repeatedly, but I turned off the conversation. There were a great number of sail-mats in Bete's house, and the people assured me that they were some of the tribute which the Tonguese extorted from them.

The sea-slugs, or *bêche-de-mer* (several species of the genus *Holothuria*), collectively termed "Dri" by the natives,* are found in great abundance on the reefs, espe-

* The different species bear the following native names :—1. Dri votovoto; 2. Dri alewa; 3. Dri batibuli; 4. Dri tarasea; 5. Dri damu; 6. Dri valadakawa; 7. Dri daidairo; 8. Dri lokoloko ni qio, etc.
To show the profits of the *bêche-de-mer* trade, I extract from Wilkes, of

cially on the northern shores of Viti Levu and Vanua Levu. In July, 1862, they figured, perhaps for the first time in Europe, in the bill of fare at a grand dinner given in London at Freemasons' Tavern by the Acclimatization Society. A highly profitable trade in them was carried on, principally by the Americans, until a few years ago, through the political troubles caused by the invasion of the Tonga islanders, it became impossible to collect sufficient for filling a vessel fitted out on purpose. As peace has now been re-established, this trade will probably revive. As soon as a ship was full it sailed direct to Manila, where merchants were eager to purchase its cargo for the Chinese markets: a cargo of tea, sugar, and silks, was then taken in for the homeward voyage. Notwithstanding that no insurance of the vessels engaged could be effected, on account of the bad charts of Fiji, the profits realized were very great. A whole cargo, which cost $1200, brought $12,000; and another, which cost $3500, brought $27,000. As for nearly ten years no sea-slugs have been collected, any enterprising shipowner dispatching vessels there would be able to collect a rich cargo in a very short space of time.

the United States Exploring Expedition, the following costs and returns of five cargoes obtained by an American, Captain Eagleston:—

1st voyage,	617 piculs,	cost	$1,100,	sales	$ 8,021
2nd ,,	700 ,,	,,	$1,200,	,,	$17,500
3rd ,,	1,080 ,,	,,	$3,396,	,,	$15,120
4th ,,	840 ,,	,,	$1,200,	,,	$12,600
5th ,,	1,200 ,,	,,	$3,500,	,,	$27,000

A further profit also arises from the investment of the proceeds in Canton or Manila. This same trader obtained also 4488 pounds of tortoise-shell at a cost of $5700, which sold in the United States for $29,050 net.

Resuming our voyage, we found ourselves, October 15th, off Namuka, where we sent on shore for water. The crew, on returning, brought an armful of gardenias, a species quite new to science (*Gardenia Vitiensis*, Seem.), with beautiful white flowers, emitting a delicious scent; and the young leaves of the shrub being enveloped in a thick coating of greenish gum, which, as they expand, gradually dissolves. There is a strange connection between Namuka and Bau: both having, or rather having had, the same local gods, the people possess mutual rights similar to those of the *Vasus*, visitors being allowed to take whatever articles they choose. The advocates of the rights of women will also be glad to learn that the softer sex of Namuka can take their seats among the men!

On the 16th we rounded Mua i Udu, as the eastern extremity of Vanua Levu is termed, where, until lately, an old screw-pine stood, to which a strange superstition attached: a man who could hit any part of this tree between the root and the crown with a whale's tooth, made sure that at his death all his wives would be strangled. On their way to Naicobocobo the spirits of the dead are supposed to do the same thing for the same purpose, there being a screw-pine at Takiveleyava. Ratu Mara, a chief well known in the annals of Fiji as a frequent disturber of the public peace, vainly tried to hit the tree at Udu; enraged at his continued failures, he cut it down. But what use is it to wrangle at fate? Ratu Mara ended his restless career at Bau, where, for repeated treacheries, the king thought fit to hang him, and all his wives escaped the fearful doom of strangulation.

Having rounded Mua i Udu, we came in sight of Rabe and Taviuni, the wind being favourable all the while. At night we anchored in Na Ceva (=Natheva) Bay, partly to avoid rocks and reefs, partly because we could not keep our crew awake. The bay derives its name from Na Ceva (*i.e.* the south-east wind, to which it is open); Natava is therefore an erroneous spelling. In Wilkes's, and other charts founded upon his survey, it is not made deep enough, and the isthmus separating it from the southern shores of Vanua Levu, about ten miles too wide. The isthmus is scarcely more than a mile and a half across, and canoes are dragged from one side to the other, as is the case in Kadavu, though its surface is hilly. Colonel Smythe made an excursion to it from Waikava; and in the chart Mr. Arrowsmith has constructed for him, this error of long standing has been corrected, as it is in the map accompanying this work.

On the following morning we called at Rabe, a fine island, of which the Tonguese have made desperate attempts to obtain permanent possession, and towards the afternoon we reached Waikava, where the missionaries from Taviuni had now established themselves, and where the official meeting with the principal chiefs of Vanua Levu was to be held. We found Colonel Smythe's vessel, the 'Pegasus,' at anchor, just returned from Lakeba, where, under pressure from the Tonguese, the chiefs had behaved rather rudely.

On the following day I ran over to Somosomo, where, in the beginning of June, I had established an experimental cotton plantation. It took me nearly a whole day to cross the strait of Somosomo, there being almost a perfect calm. I found the plantation in the best order. To my great joy, there were ripe pods, and I could gather the produce of the very seeds only set three months ago. Mr. Coxon was glad to see me again, and availed himself of my invitation to go for a few days to Cakaudrove, as the eastern extremity of Vanua Levu is more particularly called.

Shortly after my arrival, Ritova, the deposed chief of Macuata, called on me. I told him to leave off blacking his face, as it set foreigners against him, and was regarded as a demonstration of heathenism, though it might not be intended as such. Golea, or rather Ratu Golea, the chief of Somosomo, also dropped in. He had cut his hair short, and was so much altered for the worse, that I did not know him until recognizing him by his melodious voice. He had now about thirty wives; and Eleanor, the Queen, had quite recently given birth to a fine boy, who would be " *Vasu* " to Bau, and about whom the natives were in ecstasy.

The Fijians are not so prepossessing in appearance as those lazy and handsome fellows the Tonga men, who flock over here in great shoals; but whilst the Tonguese lose, the Fijians gain by a closer acquaintance. There is a manliness about them that is extremely winning; and I quite agree with Macdonald, that if their likenesses could be accurately taken, they would form quite a contrast to the ill representations of these islanders extant. Ratu Vakaruru, whose portrait is given in the frontispiece, is one of the finest Fijians living; but I cannot say that the copy I had made of Macdonald's unpublished drawing does justice to him. Their language, so far as euphony goes, yields to none I have heard in any quarter of the globe, and to my ear it sounds as pleasing as Spanish or Italian. They are certainly not an idle people, and though not working like our own labourers, from six to six, they are great cultivators of the soil, skilful fishermen, and able builders and managers of canoes. Far from living under an absolute despotism, as is erroneously supposed, all the different States of which Fiji is composed have institutions hallowed by age and tradition, fundamentally almost identical with those cherished by the most advanced nations. The real power of the State resides in the landholders or gentry, who, at the death of a ruler, proceed to elect a new one in his stead from amongst the members of the royal family. Generally the son, but not unfrequently the brother, or even a more distant relation of the deceased, is elevated to the chieftainship, and loyally supported in his dignity as long as he carries out the policy of those who have set him up. If this "House of Commons," as by a stretch of language it may be called, finds its wishes and aims disregarded, the members avail themselves of the privilege of refusing supplies, which, in the total absence of money, consist in yams, taro, pigs, fowls, native cloth, canoes (the naval estimates!), and all the other requirements of a great Fijian establishment. The intractable chief who has attempted to play the despot is thus generally brought to a proper sense of his condition. Of course, chiefs who, by strong family connections, can afford to set the "Commons" at defiance, will occasionally do so; then new expedients have to be resorted to, and the trial of strength which follows provides one of the elements of political activity. Europeans might fancy that a barbarous people would readily adopt the more simple process of getting rid of an intractable chief by knocking him on the head; and certainly that would be the solution adopted if usage had not provided a law for his protection, according to which he cannot be killed by any one inferior to him in birth. We have here the English law, that a peer cannot be tried except by his own peers, in its rudest embryonic form. It would be "*taboo*" for any commoner or serf to lay violent hands on a chief; and, however obnoxious he might have been to the community, the taboo-breaker would not go unpunished. Outsiders might suppose that amongst a people destitute of all written law much confusion existed in regard to the application of this peculiar code of polity and customs. Never would a greater mistake be committed. All their usages are as firmly established, and as strictly adhered to, both in letter and spirit, as if they had been engraven on tablets of stone. The early white settlers soon found this out, and often owed the preservation of their lives to a thorough knowledge of this system. Thus, an Englishman, of the name of Pickering, once fell into the hands of a hostile tribe long on the look-out for his body. He soon became aware that they were making preparations for a cannibal feast, of which he was to be the principal dish, though these preparations would not have been noticed by any one less versed in their peculiar customs. He knew that before they proceeded to kill him a bowl of kava would have to be made, that a prayer would have to be said over the beverage when ready, and that the person saying the prayer could not be the one eaten. Pretending utter unconsciousness of what was going on around him, he eagerly watched the moment when the preparation of the kava was advanced to the stage at which the prayer had to be said, and suddenly, to the utter dismay of his enemies, he pronounced the well-known formula. No one would now have dared to take his life, and he had the keen satisfaction of partaking of the refreshments provided for his own funeral. Another old settler, American by birth, had also the misfortune of being an object of hatred to a tribe opposed to, and at war with, the chief under whom he lived; and, as ill-luck would have it, he met a strong party of his enemies making straightway for his boat. They were about to open fire upon him, when, with a coolness deserving all praise, he exclaimed:—" Don't shoot! I am a herald of peace, charged with carrying the token of surrender to your chief, and put a stop to further hostilities." The stratagem succeeded, and the self-styled herald effected his escape.

I returned to Waikava on Saturday, October 20th, and on Monday following the official meeting was held. The chapel had been granted for that purpose. Mr. Carey, the resident missionary, interpreted the official business. Neither Bete nor Tui Bua had made their appearance; Ratu Golea dropped in when all was over; the only three chiefs therefore present were, Ritova, Bonaveidogo, and Tui Cakau, the king of Cakaudrove. After all business relating to the cession had been disposed of, Mr. Pritchard was occupied several hours in settling disputes between native and British subjects.

Waikava, sometimes called Fawn Harbour, derives its name from a little fish (Kava), which at a certain season of the year, enters the river (Wai), on which, the native town is situated. Tui Cakau, the King, had almost promised the missionaries that on their removal from Wairiki he would follow them with his whole court to Waikava; but he had not done so as yet, and fears were entertained that he would not consider the promise binding. Jetro, the old Manila man, whom I met at Korovono, was now here, employed as a Scripture-reader. Only one of the missionary houses being finished, we had to sleep in the chapel, where large screens of bark-cloth ensured the necessary privacy. Several heathen priests, on becoming Christian, have proved highly useful to the mission, and at this place there was one who occasionally, when praying rather more fervently than most people are wont to do, would suddenly begin to tremble and shake, as he used to do in his heathen state, and had no slight difficulty in checking himself in his old propensity.

After the meeting the 'Pegasus' returned to Levuka, where she arrived on the 26th of October, and as there was no further occasion for her, she returned to New Zealand, Colonel Smythe remaining behind. The 'Paul Jones' left a few hours after her the anchorage of Waikava, steering for Matei in Taviuni; the Consul having determined to arrange, if possible, some terms between Ritova and those who had driven him from his land and estates, and thus try to heal a sore of old standing. But in order to understand the real difficulties of this case, it will be necessary to sketch the history of the Tonguese in Fiji, so far as I have been able to trace it from all the sources accessible.

CHAPTER XV.

HISTORY OF THE TONGAMEN IN FIJI.—THEIR PHYSICAL SUPERIORITY OVER THE FIJIANS.—THEIR ARROGANCE.—CAPTAIN CROKER'S DEFEAT.—EARLY INTERCOURSE BETWEEN TONGA AND FIJI.—INCREASE OF TONGUESE IMMIGRATION.—CHIEF MAAFU.—KING GEORGE OF TONGA VISITS FIJI.—CONQUEST OF KABA AND BABE.—ARRIVAL OF BRITISH CONSUL.—CESSION OF FIJI.—MAAFU'S ATTEMPTED CONQUEST.—RITOVA AND BETE.—MAAFU'S AMBITION CURBED.—PEACE RESTORED.—RITOVA INSTALLED IN HIS ESTATES.—TONGUESE INTRIGUES RENEWED.—BETE'S DEATH.—COMMODORE SEYMOUR'S VISIT.—TERMINATION OF THE WARS BETWEEN FIJIANS AND TONGANS.

ONE of the many reasons which induced the King and Chiefs of Fiji to tender a formal cession of their beautiful island to the British Crown, and to ratify it with alacrity, was to escape from the insupportable exactions and tyrannies of the Tonguese. The Tonguese, or Friendly Islanders, may well be called the flower of the Polynesian race; and Commander Wilkes was only stating a truism when saying, that there were few spots on the whole face of the earth where one could behold so many handsome people together. They are tall men, with fine intelligent features, dark, often curly, hair, and of a light-brown complexion. They are far beyond the Fijians in good looks. This physical superiority, which, independent of the difference of race, the Tonguese enjoy over the Fijians, may partly result from the different treatment to which the women are subjected amongst

these two nations. Whilst in Tonga the women have been treated from time immemorial with all the consideration demanded by their weaker and more delicate constitution, not being allowed to perform any hard work, the women of Fiji are little better than beasts of burden, having to carry heavy loads, do actual field-work, go out fishing, and besides, attend to all the domestic arrangements devolving upon their sex in other countries. Indeed, their position is almost identical with that enjoyed, or rather endured, by their poor Indian sisters in North and South America. They have to work hard, and cheerfully go through all the drudgery forced upon them by the lords of creation. I remember an eccentric friend of mine once remonstrating with a Fijian who allowed his wife to carry a large bundle of sugar-cane, whilst he leisurely walked by her side. He thought the remonstrance simply a piece of impertinence, and did not see why an inferior being should not be made to contribute to the comfort of a superior.*

The Tonguese may also be called the Anglo-Saxons of the South Seas. Originally sprung from Samoa, at least their leading chiefs indisputably, they have overrun Tonga; and finding that group also too small, they established colonies in Fiji, and of late made desperate attempts to conquer the whole group. The unqualified praise given to their good looks by all voyagers has made them rather conceited, and their success in war haughty and arrogant in the extreme. It is intelligible

* The accompanying plate, representing Koro Basabasaga, on the Wai Levu, or great river of Viti Levu, gives a good idea of the treatment; the man walking leisurely along, whilst the woman is carrying a heavy load of sugar-cane.

that they should entertain a feeling of superiority over the native races whom they subdued; but in consequence of an unlucky affair, almost forgotten in England, they look down upon all Europeans, and boast of having beaten a British man-of-war. In 1840, Captain Croker, of H.M.S. Favourite, visited the Tongan Islands, and was persuaded to take part with a body of native Christians against the heathens that opposed them, then shut up in several native forts at Bea. Carronades were brought within 106 yards of the principal fort, and all hopes of a peaceable arrangement having vanished,—

"The command was given to make the attack, the captain leading the way. The sergeant of marines was ordered to scale the barricade and to fire. The attack was soon answered by the cannon at the entrances [of the fort], and by a volley of musketry; and the captain and several of his men were wounded. Notwithstanding his wound, Captain Croker exerted himself to the utmost to enter the stockade; but failing in the attempt, and becoming faint from the loss of blood, he retired to a little distance, and while leaning against a tree for support, was shot through the heart, and dropped lifeless on the ground. His men continued the attack, but at great disadvantage: the enemy was screened by their defences; while the English, on the open ground, were exposed to the hot fire of the enemy. This sad affair ended in the death of two officers, besides the captain. The first lieutenant and nineteen men were dangerously wounded. It was with great difficulty that the survivors contrived to carry off their dead and wounded."*

The officer who succeeded Captain Croker in command saw the absolute folly of losing any more men, and relinquished all thoughts of renewing the attack. One or two carronades had fallen into the hands of the

* 'Tonga and the Friendly Islands.' By S. S. Farmer. London, 1855. Page 325 *et seq.*

Tonguese. As the case stood, the British Government did not deem it just to ask for any reparation, and simply demanded the guns left behind. However, the Tonguese were not slow in taking advantage of this turn of affairs, and quite ignoring that it was their own government as much as the foreigners who were repulsed, they have magnified the catastrophe into a grand victory, and become so arrogant, that Captain Cook, could he pay them another visit, would never dream of confirming the name of the "Friendly Islanders" which he gave them, in total ignorance of the fact, related by Mariner, that they had laid two plots to take his life, not carried out because no agreement could be arrived at respecting the details of the projected murder.*

Ethnologists have long been watching the spread of the Tonguese over the South Sea, and Viti has become a field of high interest, as the light-coloured Tonguese, a genuine Polynesian people, have here met face to face powerful representatives of the dark-coloured Papuan race. There seems to have been an intercourse between Fiji and Tonga from time immemorial, distinctly spoken of in the story of the Vasu ki Lagi and the Princess Vilivili-tabua, and other ancient Fijian legends, as, for instance, that about the spread of the practice of tatooing. Independent of this legendary evidence, there are other proofs of an early intercourse. The Tonga islands not furnishing any large timber, it was necessary to go to Fiji for materials for canoes. Fine mats and native cloth, printed in choice patterns, were bartered away for permission to cut timber and build canoes. The eastern

* Mariner's 'Tonga,' vol. ii. pp. 64, 65.

parts of Fiji, Lakeba, and the adjacent islands, being the most accessible from their proximity to Tonga, were those chiefly visited; and as it took considerable time to construct the larger canoes, a strong influx of Tonguese blood was soon perceptible in the population of those districts. Not unfrequently it happened that parties going or coming were drifted by the prevailing winds on the shores of Kadavu, and hence the mixed race inhabiting that fine island is accounted for. Lakeba and Cakaudrove were formerly intimately connected, and the latter being the high-road to Bua, the Tonguese seem to have become introduced to the locality, where, above all others, the famous Sandal-wood (Yasi), so highly valued both in Tonga and Samoa for scenting cocoa-nut oil, grew in abundance.* They were not long before they made regular trading voyages to Bua, bringing with them printed tapa, fine mats, and large pearl-shells, skilfully inlaid with pieces of whales'-teeth. Having often to wait two or three months before a cargo of sandal-wood could be got ready, a close intimacy naturally sprang up between the trading parties, intermarriages took place, and thus another district received a mixed population.

Up to this period the Tonguese had been peaceful traders, glad to exchange their manufactures for natural products denied to their own islands. Gradually they adopted a different line of policy. Being men of athletic frames, of courage and daring, they were often

---

* Cakaudrove (= Thakaundrove) has been corrupted by the Tonguese into "Tacownove," and in some old charts is applied to the whole of Vanua Levu.

asked to assist in the feuds in which chiefs friendly to them were engaged, receiving canoes and other property in return for their services. From being mere mercenaries, they gradually began to act on their own responsibility, readily avenging every outrage from time to time committed against any of their countrymen on the smaller islands of the eastern group, where they could calculate the exact number of their possible opponents.*

With the constantly increasing influx of Tongan immigration, chiefs came over, who undertook the management of their countrymen, and among them Tui Hala Fatai, mentioned by Mariner, and Tuboi Tutai, spoken of as Tuboi Totai by Wilkes. About 1848, Maafu, another of their chiefs, and destined to exercise a vast influence on Fijian affairs, made his appearance. Married to one of the highest ladies of his native country, descended from the ancient royal line (Finau), gifted with great personal advantages, and possessing as comprehensive and ambitious a mind as rarely falls to the lot of a Polynesian, Maafu began to prove a dangerous rival to King George, the chief seated on the throne of his ancestors. He had already shown his disposition in a sandal-wood expedition to the New Hebrides, which originated with Messrs. Henry and Scott.

"About December, 1842, two vessels under British colours, the 'Sophia' and 'Sultana,' and a third which was said to have carried the flag of Tahiti, arrived [at Tonga] to raise a party for the purpose of forcibly cutting sandal-wood at the New

* Compare Mariner's 'Tonga,' vol. i. p. 72–76.

Hebrides. A brother of the late King Josiah, Maafu, engaged with the leader of the expedition (Henry) to furnish sixty men. They touched at Lakeba to reinforce their numbers, but could not procure volunteers, and continued their course to Eromango. Here the party, armed with muskets, were landed, and a quantity of sandal-wood cut and embarked. The natives continued friendly for the first few days, but at the end of that time, some of them having stolen three axes, a disturbance took place, when one of the supposed thieves was shot by the Tongans. The fire was returned by arrows, which wounded a Tongan, who afterwards died. In consequence of this affray they left Eromango, and proceeded to Vate, or Sandwich Island, where he and his men were again landed, armed, and directed to cut wood, the white men remaining on board of their vessels. Before long they had a battle with the natives, who, having no muskets, were defeated with a loss of twenty-six killed, none of the intruders being injured. A fort was afterwards stormed and taken, when several more were killed; the remainder retreating to an island, where they hid themselves in a cave, whither they were pursued by Maafu and his party. After firing into the cave, which seemed to have no effect, the besiegers, pulling down some neighbouring houses, piled the materials in a heap at its mouth, and, setting fire to it, suffocated them all."*

King George, the present ruler of Tonga†, having subdued a rebellion in which Maafu took a prominent part, deemed it prudent to send Maafu to Fiji, ostensibly for the purpose of keeping his countrymen in order, but really to get him out of the way. At the same time a hint, perhaps more than a hint, was thrown out that no objections would be made if Maafu did in Fiji what King George had done in Tonga, make himself master of the whole group. Maafu's first ex-

* Eskine, 'Western Pacific,' p. 143. Behaving, in fact, as barbarously to them as a few years later a French General did to an Algerian tribe.

† Farmer's 'Tonga and the Friendly Islands,' p. 398.

ploit took place at Lomolomo. Two Fijian chiefs fighting against each other, Maafu's assistance was solicited, and readily given to the weaker party, to which a Tonguese teacher of Christianity was attached. After the stronger party had been defeated by the combined efforts of its Fijian and Tonguese opponents, the native conquerors found themselves so heavily indebted to their foreign ally, and so much in his power, that they became easy victims to his intrigues to usurp their authority altogether. Maafu never espoused a cause on its own merits. The principle upon which, in this instance, and in almost every other, he seems to have acted, was to assist the weaker party against the stronger, and after its defeat turn round upon his allies, with whose weaknesses he had become perfectly acquainted during their familiar intercourse.* The quarrel at Lomolomo made him master of the whole grouplet of Vanua Balavu, and having thus obtained a solid footing, his rise was rapid. Elated with success, he used to challenge any chiefs to try their courage and skill in a canoe of equal size, and with an equal number of men to his own; but no one, not even Ratu Mara, justly looked upon as the most able sailor and commander of Fiji, could be induced to accept the challenge. The second opportunity that presented itself to Maafu for extending his power was offered by interfering at Matuka. There again two chiefs were quarrelling, and the party to which the Tonguese teacher belonged, was

* Even in Tonga his conduct was identically the same. Compare Farmer's detailed account of the rebellion in which he took part. 'Tonga and the Friendly Islands,' p. 398.

again the weaker. In a fight between the hostile parties the Christian chapel and the house of the teacher caught fire, and were totally destroyed. Maafu at once set off to avenge the injury done to his countryman, took the side of the weaker party, defeated the stronger; and then, turning round upon his friends, displaced their rightful chief by one of his own creatures. A similar affray took place at Muala, where Maafu, by hook or by crook, was again victorious.

In March, 1855, King George of Tonga availed himself of the opportunity presented by the missionary vessel 'John Wesley,' to pay a visit of state to Cakobau, the supreme chief of Bau, and titular King of Fiji. Cakobau was at that particular time in considerable trouble. Kaba, an important place in the neighbourhood of his capital, was in open rebellion against him, headed by Ratu Mara; and as he had but recently lost much of his influence by renouncing heathenism, he felt himself scarcely strong enough to put down Kaba single-handed. In an evil hour he was persuaded to apply to King George for assistance, and the latter readily complied, on being presented with a canoe fifteen fathoms long for the promise of assistance. A large fleet of canoes, and a strong reinforcement of warriors, soon arrived from King George's dominions. By the combined forces of Bau and Tonga, Kaba, to Fijian notions an impregnable fortress, was taken (April 7th, 1855*), and the authority of Cakobau re-established.

Maafu and his countrymen had prominently distinguished themselves on this occasion, and their exploits

* J. Waterhouse, 'Vah-tah-ah,' pp. 111-121.

were the subject of comment in the remotest parts of the group. Bau acquitted itself handsomely of the debt it owed, by presenting King George with the 'Cakobau,' a schooner of eighty tons built in the United States. The example set by Bau, of putting down rebellion at home by foreign assistance, was speedily followed by another Fijian state. Rabe (= Rambeh), an island of considerable size, had disputed the authority of the ruling chief of Cakaudrove, Tui Cakau; and King George having proffered assistance, it was readily accepted by Tui Cakau. Rabe fell, and the Tonguese were in the habit of calling it their own, until, in 1860, Maafu, in the name of King George, received payment for the assistance rendered.

The conquest of Kaba and Rabe had conferred upon Maafu and his followers such a high prestige that the Fijian chiefs began to tremble for their own safety, and the impolicy of calling in foreigners to suppress rebellion at home seemed to dawn upon the more far-seeing among them. Maafu was not slow in perceiving the advantage he had gained, and his favourite plan of subduing the whole of Fiji appeared now to have arrived at maturity. By cunning intriguing and a bold system of warfare, he hoped to carry it into execution. Returning to Lomolomo, he set about building a schooner of thirty-five tons, which should at once place him at an advantage with enemies who had to rely solely upon canoes. Nor did he fail to make other preparations for conquest, and he would have commenced hostile operations without delay, if it had not been for the unexpected arrival of H. B. M. Consul, Mr. W. Pritchard, who landed in

Fiji on the 10th of September, 1858, to take up his permanent abode in this important group. Bau was again in trouble. For various outrages asserted to have been committed against the life and property of American citizens, the Government of the United States demanded indemnity from Cakobau, as supreme chief of Bau and titular King of Fiji. The corvette 'Vandalia,' Captain Sinclair, had been sent to enforce the claim, and as the sum of 45,000 dollars was altogether beyond the means of the Fijian King to pay, overtures were made to Mr. Pritchard for the cession of Fiji to Great Britain, on condition that this sum, which the natives were going to refund by assigning the proprietorship of 200,000 acres of land, be liquidated. In November, 1858, Mr. Pritchard departed home to lay this offer before her Britannic Majesty's Government, and no sooner had he left the group than Maafu commenced operations.

Ritova and Bete, chiefs of the Macuata coast of Vanua Levu, were fighting out some old family feuds. Bete, being worsted, concluded an alliance with Tui Bua, another chief of importance on the south-western coast of Vanua Levu, who owed Ritova a grudge for a defeat in a former war. But even thus strengthened, Bete was unable to cope with his rival. Maafu saw that here was his chance. Friendly messages were dispatched to Ritova, who, delighted with the moral support of so powerful a chief, forwarded sail-mats and other valuable presents. At the same time Maafu sent messages equally friendly, but more sincere, to Tui Bua, and through the Tonguese residing there prompted him to apply for assistance against Ritova. This idea was no sooner suggested than carried into effect, and Maafu became the declared ally of Tui Bua and Bete. This new combination could not but excite deep apprehensions at Bau, as tending to derange that political balance which that power deemed it necessary to uphold in order to maintain its supremacy. Maafu, duly informed of the cloud gathering in that quarter, repaired straightway to the capital, and almost succeeded in dispelling it. He made out that he had sent only a few men under the charge of his officer Wai-ni-golo, and he even endeavoured to persuade Cakobau to aid him by dispatching canoes to the scene of action, as the whole affair when terminated would add fresh lustre to the supremacy of Bau. Cakobau however contented himself with ordering one canoe to accompany the expedition, more to watch proceedings and furnish correct reports than to take any active share in the operations. On leaving Bau, Maafu gave out that he was going direct to Bua, to see how his men were getting on, instead of which he proceeded to Lomolomo for reinforcement. Wai-ni-golo, the Tonguese officer previously sent to Bua, had orders to provoke a direct quarrel with Ritova; he obeyed them by taking two villages and putting most of the inhabitants to death. By the time this was accomplished Maafu and the reinforcement arrived at Bua, where Tui Bua was taken on board the Tonguese schooner, and the whole party proceeded to the Macuata coast. The combined forces now took town after town, until they reached Nukubati, Ritova's stronghold, which, after considerable resistance, fell into their hands. Ritova, nothing daunted, retreated to the mountains at the back of Nukubati, where he was regularly besieged. But fate was against him. Chief Bonaveidogo, one of his followers, at this critical time went over to Maafu's side, to save his life and that of his vassals; and Ritova, finding further resistance on the Macuata coast hopeless, escaped with the remnant still firm to him across the mountains to Solevu, where Tui Wai Nunu, a chief friendly to him, resided.

Solevu (Sualib, of Wilkes) is a little district on the southern side of Vanua Levu, between Bua and Cakaudrove, which acknowledged a sort of vassalage to Bau, but was otherwise independent. In order to humour Tui Bua, who was eager to annex this district to his territories, Maafu had promised to subject it for him, and with that view had already left in it a detachment of men. By Ritova's retreat to this very district, a fine opportunity of killing two birds with one stone presented itself. Rounding the western parts of Vanua Levu, the allied forces appeared before the town of Solevu, which, being strongly fortified, held out against the invaders three whole months. At the end of that time, the besieged were in extreme want of fresh water, the besiegers having diverted a rivulet supplying the town from its course, and all the wells being dry. Unable to hold out any longer, Solevu surrendered. When Ritova and Tui Wai Nunu heard this news, they perceived it was hopeless to prolong the struggle. Meanwhile Maafu had caused it to be known that he had promised Mr. Swanston, the acting British Consul, to spare Ritova's life, if he were taken. Ritova therefore thought it advisable to give himself up, and for some time he was a prisoner under the immediate eye of the victorious chief. But Maafu's followers were most unwilling to see this promise kept; they pressed him hard to get rid of a man at once so bold and so dangerous. Maafu, on one side assailed by his unruly mob, on the other bound by a promise which he deemed it prudent not to treat lightly, solved the dilemma by allowing Ritova to escape to Cakaudrove, and in order to blind his vassals and allies, he pretended to be enraged at his escape, and dispatched men in pursuit of the fugitive.

Maafu now proceeded to dispose of the conquered territories. Solevu was annexed to Tui Bua's dominion; the western part of Macuata was placed under Bete, the eastern under Bonaveidogo, with the express understanding that each of the favoured parties had to pay a stipulated tribute. In this distribution, the claims of Bau on Solevu had been altogether disregarded. If anything had been wanting to open the eyes of Cakobau, it was furnished by these high-handed proceedings, which sounded like scorn to a proud people, who had been led to believe that whatever was done in this war would tend towards extending and consolidating the authority of the supreme power in Fiji. More humiliation was in store for Bau. In order to avoid as long as possible a direct contest with that state, Maafu retired to Lomolomo to direct his operations. Bau was to be got between two fires. A strong fleet of canoes was dispatched to Bega, an island, through Rewa, subject to Bau, and which, overawed by the superior force suddenly appearing, gave itself up to the Tonguese; whilst Tui

Bua was directed to get up a quarrel at Rakiraki, the north-eastern district of Viti Levu, subject to Bau through Viwa. Everything was thus progressing favourably, and a few months more would have brought about the overthrow of Bau, making Maafu virtually master of all Fiji. At this critical moment Mr. Pritchard returned from England with intimation that her Britannic Majesty's Government had taken the cession into favourable consideration. Soon after his arrival, a meeting of Fijian chiefs took place at the British Consulate, in Levuka, with the view of ratifying the cession made by Cakobau, and they availed themselves of the opportunity to appeal to Mr. Pritchard to check Maafu's grasping career. They founded this appeal upon the fact that Fiji was already ceded to the Queen of Great Britain, and that Maafu, as a foreigner, was taking the country from her. After a tedious discussion of five hours, Maafu consented to renounce all political claims on and in Fiji, and the lands conquered, by signing an instrument to that effect, in the presence of all the chiefs assembled, her Britannic Majesty's Consul, and Commander Campion, of her Majesty's ship Elk.*

"Know all men by these presents,—1. That I, Maafu, a Chief of and in Tonga, do hereby expressly and definitely state, that I am in Fiji by the orders of George, King of Tonga, as his representative, and that I am here solely to manage and control the Tonguese in Fiji. 2. That I have, hold, exercise, and enjoy no position nor claim as a chief of or in Fiji. 3. That all Tonguese claims in or to Fiji are hereby renounced. 4. That no Tonguese in Fiji shall exact or demand anything whatever

* The English version of this document is here subjoined; one of the copies of it I brought home is now in the library of the British Museum.

from any Fijian, under any circumstances whatever, but they shall enjoy the privileges and rights accorded to other nations in Fiji. 5. That the lands and districts of Fiji which have been offered by various chiefs to me are not accepted, and are not mine, nor are they Tonguese, but solely and wholly Fijian. 6. That the cession of Fiji to England is hereby acknowledged. In witness whereof I have hereto set my name, this 14th day of December, 1859. MAAFU.

"We hereby certify that the foregoing Chief Maafu signed the above document in our presence, this 14th day of December, 1859.—WILLIAM T. PRITCHARD, Consul; H. CAMPION, Commander R.N., H.M.S. Elk.

We hereby certify that we translated the foregoing document to Maafu, a Chief of Tonga, who has signed, and that he thoroughly understands its meaning.—W. COLLIS, Wesleyan Training Master; E. P. MARTIN, Wesleyan Mission Printer."

The peace of the group, which, to the serious disadvantage of trade, had been so long interrupted, was thus at length re-established; but the wounds inflicted by the war were not so easily healed. The Tonguese did not content themselves with merely taking a place. They plundered and set fire to the dwellings, cut down the fruit-trees, filled up the wells, ravished the women, and put to death as many of the fighting-men as their ferocity prompted them; even those who had given themselves up as prisoners were often mercilessly murdered in cold blood. When Maafu and his hordes had been at a place, it was as if a host of locusts had descended. Not only had every vestige of provisions, pigs, fowls, yams, and taros been devoured or carried off, but the plantations themselves had been ruthlessly destroyed, forcing the poor natives to seek such wild roots as would enable them to eke out their miserable existence. Yet, after all their provisions, tools, native cloth, canoes, and other moveables had either been carried off or destroyed, they had to set to work making cocoa-nut oil, sail-mats, and other articles for their conquerors. The intensity with which a Fijian hates a Tonguese need therefore cause no surprise. Yet there were not wanting people who applauded what had been done, and who were rather displeased to see the policy pursued by the invaders brought to such a sudden conclusion. Maafu knew full well that he stood in need of such friends, and he had set early about making them. He had three different bodies to interest in his conquest,—his own immediate followers, the foreign traders, and the Wesleyan missionaries. The Tonguese were easily attached to his cause by giving them unlimited license to rob and plunder the country, and ravish the women; the foreign traders he made his supporters, by running up heavy bills for powder, shot, and general stores, which stood no chance of being paid, unless it was in contributions in cocoa-nut oil, tortoiseshell, and *bêche-de-mer*, extorted from the conquered places; whilst the Wesleyan missionaries were kept quiet by Maafu making it the first condition, in arranging articles of peace, that the conquered should renounce heathenism and become Christians. The thousands of converts thus added to their flock, completely blinded the missionaries to the danger they were incurring in coquetting with so unscrupulous an adventurer. It was only after Macuata had been reduced, and public opinion had severely condemned the massacre of prisoners at Natakala and Naduri by Jamisi, one of Maafu's officers, that they saw the necessity of protesting against the unsanctioned use which had been made of their name.

I shall probably be accused, by those versed in Fijian affairs, of an undue partiality for the Wesleyan missionaries, by viewing their conduct in the light I do, and endeavouring to separate the doings of the missionaries from those of the barbarous hordes who overran the country. I admit that the latter is a matter of no slight difficulty. Christianity had early taken root in Tonga; and when, in 1835, the Wesleyans in that group determined to extend their operations to Fiji, they naturally fixed upon Lakeba, and those parts where a strong population of Tonguese was already established, and where they could use a language familiar to them until Fijian had been learnt. Tongamen were found extremely well qualified for acting as pioneers in teaching the rudiments of the Christian faith; and during the whole period that the Wesleyans have been labouring for the conversion of Fiji, they have employed a large number of them. They were spread over the whole country, and, unfortunately, became in Maafu's hand, ready instruments for the execution of his plans. They supplied him with reliable information about the quarrels, weaknesses, and resources of the different territories, were never tired of praising their great chief, and ever ready to prompt the Fijian rulers to apply to him in cases of dispute and war. All these facts cannot be gainsaid; and those must be strangely ignorant of the working of the Polynesian mind, who fancy that doctrines of so recent a growth as those of Christianity would ever induce a native of subordinate position to remain indifferent to the wishes and orders of his chief. When King George visited Fiji, it was in the 'John Wesley,' and it was on board of that vessel the arrangement relative to the subjugation of Kaba was concluded. Finally, nothing was said by the missionaries whilst Maafu achieved his conquest, and it was only after great atrocities had been committed that a letter of remonstrance was addressed to him.*

Yet, notwithstanding these facts, occasionally urged with great vehemence, I dismiss, as utterly unfounded, the idea that the missionaries concocted the whole plan with the Tonguese. A calm review of all the information on hand, rather leads to the conclusion that Maafu was leading the missionaries to believe that he was advancing their interest, when indeed he only abused their

* The following is a copy of a letter sent to Maafu, extracted from the records of the Wesleyan Missionary Society at Sydney, by the Rev. J. Eggleston:—

"There is something, Sir, which I wish to tell you, *i. e.* our hatred of the deed performed at Nabekavu amongst the people of Natakala. It was of no use whatever. If it was not done by your orders, please inform me, that I may defend your character. There is another subject which I desire also to make known. It is extensively reported that this war is the work of the missionaries. If this be true, tell me now which of us has sanctioned the hostile proceedings. Was it me, or whom? Please inform me, for it will be published prejudicially all over the world. If we are belied, be kind enough to vindicate us in your letter to me. Tell your people also to announce you (as the author), and not to announce us. I do not wish to prevent your approach to Ulumatua and Wai Nunu. Please yourself about this; for yours is its goodness, and yours is its evil. But command your warriors to announce you; do not let them announce us, as we do not sanction it in the least. It is also rumoured that it is our advice that Mara, Ritova, Tui Levuka, and another be put to death. If you seize these, do not deliver them *to be killed*, lest it be said that it is by our advice. We have not come to make known a message of death; our work allotted to us is preaching. But if a man disturb the country, let his chief bring him to a trial.—30*th July*, 1859."

[I have not seen the answer to this letter, if there was one.—*B. S.*]

name in order to advance his own; and they perceived too late that they had been made the dupes of an unscrupulous and ambitious man.

At the height of his power, Maafu is supposed to have had no less than three thousand fighting-men of his own nation, independent of his Fijian allies, and after the signing of the document of the 14th of December, 1859, had placed a curb on his ambition, the number remaining was still sufficiently great to cause uneasiness to the natives. On the part of Mr. Pritchard it required extreme watchfulness, lest the bloodshed which had so seriously diminished the population and injured the prosperity of the islands should be renewed. Maafu exhibited little inclination to return to Tonga; there was still hope that, in case England should reject the proffered cession, the conquest of the whole group by Tonguese arms might become a reality. He therefore enjoined his partisans to remain quite passive until the danger was past, and not commit any rash act. A characteristic letter to that effect was sent in the middle of 1860 to Bega and Kadavu, the contents of which became a public secret. But men, who had so long been accustomed to behave with all the insolence of conquerors, who regarded Fiji in no other light save a fair field for lust and plunder, and would not disdain to plant the battle-axe in the public squares, and insultingly demand either an ample supply of animal and vegetable food or the heads of so many Fijians—such men were not easily kept quiet. Complaints were rife wherever Tongamen resided, how they plundered the natives, and how, by intimidation, they forced the weaker chiefs to behave discourteously towards the whites. When Colonel Smythe visited Lakeba, he found its chief so surrounded by Tonguese, and so much under their immediate influence, that he almost repudiated the cession, and he could scarcely prevent their almost insulting him, by crowding the house in which the official meeting took place to an inconvenient degree. It is impossible to determine whether the Tonguese were emboldened by the impunity with which they had been able to show themselves so troublesome on this and other occasions, or whether the nature of the intercourse of Colonel Smythe with the Fijian chiefs was by them regarded as proof that the British Government was dissatisfied with Mr. Pritchard's checkmating them; but in October, 1860, they loudly proclaimed their intention to interfere once more in the affairs of Macuata. Chief Ritova was to be captured and sent as prisoner to Tonga, whilst the people living on his patrimonial estates of the islands of Kia and Cikobia, were to be carried over in a body to Udu, and placed under the control of Chief Bonaveidogo, whom Maafu had rewarded with the government of eastern Macuata.

 Ritova, since his loss of power, had taken up a temporary residence at Matei, in the island of Taviuni, where a party of adherents gradually gathered around him. He had repeatedly laid his case before Mr. Pritchard, showing how unjustly he had been deprived of his patrimonial estates, and asking permission to accept the offer of friendly brother-chiefs, to reinstate him by force of arms. Mr. Pritchard thought an appeal to arms unnecessary, and told Ritova that his case should be taken in hand as soon as the requisite information could be collected. The exiled chief had found a warm supporter in the late Mr. Williams, United States Consul, who called the attention of his Government to the facts, that since Ritova's removal, American whalers had been unable to obtain supplies on the northern shores of Vanua Levu, and that the *bêche-de-mer* trade of Macuata, for years carried on by enterprising American citizens, and yielding lucrative returns, had become totally extinct. Mr. Williams's able successor, Dr. Brower, took the same view of the matter. Others were not wanting who pointed out that any distribution of territories made by the Tonguese leader had become null and void by his publicly renouncing every right of interference in the affairs of Fiji.

On the 22nd of October, 1860, a meeting was held at Wai Kava (Cakaudrove), to which all the chiefs of Vanua Levu, Ritova amongst them, had been invited, in order to give Colonel Smythe an opportunity to inquire into their views respecting the cession of Fiji, and also to discuss with Mr. Pritchard the affairs of Macuata. Two of the chiefs, Tui Bua and Bete, did not appear; the former being on a journey when the message was sent, the latter pretending that the notice given was too short to enable him to attend. But Bonaveidogo, who deserted Ritova in the hour of trial and was rewarded for his treachery with the whole of eastern Macuata, had made his appearance. Bonaveidogo and Ritova had not seen each other since then, and as it was necessary, for the establishment of a durable peace, that the two should be brought face to face before the public meeting took place, Mr. Pritchard arranged an interview. Neither of them had received the slightest intimation of this arrangement, and when Ritova was conducted to a part of the house screened off by large curtains of native cloth, and suddenly found himself in the presence of a former ally and a present enemy, he was quite startled; whilst Bonaveidogo, sitting on the matted floor, evidently thought his last moment come, and involuntarily grasped his club. When the object of the interview had been explained to be a mutual adjustment of old grievances, both chiefs remained mute for some minutes. "Why did you club Bete's father?" asked Bonaveidogo, in the course of the altercations that now ensued. "Because," replied Ritova, tartly, "he had previously clubbed my father, and as a Fijian chief I was bound to resent; if I had known," he added emphatically, "that you were going to betray me, I should not have hesitated to take your life also." Words ran occasionally very high, but gradually the two disputants grew cool; they promised mutually to forget and forgive, and finally concluded a peace over a bowl of kava.

After the meeting about the cession was terminated, Mr. Pritchard declared that, having carefully gone into Ritova's case, he had made up his mind to restore him to his home on Nukubati. There should be no fighting, and every act that could give rise to provocation must be carefully avoided. This announcement caused a great sensation amongst the chiefs and landholders assembled. No Fijian chief, driven from his land, had ever been known to return without hard fighting; and here was a white man, with no armed force to back him, who promised to do in his own peaceable way what would have cost numbers of lives if done in Fijian usage. When the natives found they need no longer fear being called to account by Maafu's bullies, they openly rallied round Ritova. Tui Cakau, the ruling chief of Cakaudrove, offered his largest canoe, a recent present from Bau, for Ritova's use; and his brother Ratu Golea, chief of Somosomo, insisted upon seeing the exile safe home.

Knowing the effect produced on the native mind by acting with promptitude, the next morning was fixed for starting. At sunrise, the schooner 'Paul Jones' fired a gun by way of signal, and steered for Matei, followed by the native canoes, and having on board, besides Mr. Pritchard and myself, Ritova and three of his adherents. One of the latter was a young man, whose father was a strong supporter of Bete, Ritova's rival; and it was probably with the approbation of his parent that he joined Ritova—the Fijian knowing, as well as people nearer home did in the time of the rebellion, that it it is rather politic if, in a doubtful quarrel between two pretenders, the father fight on one side, the son on the other, when, come what may, the family property is safe, and there is always one to intercede for the captive.

Owing to the calms nearly always prevailing in the Straits of Somosomo, Matei was not reached until the second day after our departure, when Ritova went on shore to inform his people of what had passed, and order them to get ready for starting without delay for Nukubati. Great was the joy caused by this announcement, and everything was at once bustle and activity. The women were packing up the household goods; the boys and young men hastened to the forest to dig wild yams, and catch crabs for the voyage; whilst the old men busied themselves about the canoes and other matters requiring more skill and experience. Ritova's warriors were all able-bodied men with fine athletic frames, and well armed. A collision with them would have been attended with fatal consequences. They were much exasperated at the proposal of the Tonguese to dispose of their relations and friends in the manner detailed, and were quite ready to make a desperate stand against the enemy. Mr. Pritchard thought it advisable to send an official letter to Maafu, informing him that Ritova was about to be restored to his own island, and reminding him that, in accordance with the document signed, an attack on the life and property of any Fijian would not be permitted.

All being ready for starting, on the 26th of October sails were set. The schooner 'Paul Jones' had to go outside the reef encircling the eastern shores of Vanua Levu, whilst the canoes, not drawing so much water, were able to avail themselves of the advantage of going inside. Toward sunset of the following day, Naduri was reached, where Bete, the chief placed in possession of Ritova's estates by Maafu, resided. To prevent future complications it was necessary to come to some arrangement with him, and a message was dispatched to request his attendance on board. Contrary to expectation, he refused to attend, but was ready to see us on shore. As this would have been a concession implying weakness, a message was sent to the principal landholders (Mata ni vanua) that they might come to receive a communication intended for the whole community. This measure had the desired effect. Finding that the landholders were going on board, and act independently of him, Bete deemed it prudent to change his mind, and he soon after stepped on board.

Long ere this the sun had set, but the moon made every object distinctly visible. Bete was accompanied by the Tonguese teacher of his town, and his principal spokesman, who, however, hardly uttered a word during the whole interview. Having shaken hands all round, the chief was asked to sit down on deck, and all of us did the same. A Fijian chief is generally a fine man physically, considerably taller than his subjects, and possessing that commanding air which shows that he feels himself a chief. Bete, though more than the middle height, had nothing imposing in his bearing, and his face portrayed weakness and irresolution of character. Though backed by the whole influence of Maafu, he never acquired any ascendency over the people he was set to govern; they openly disobeyed his orders; and foreigners found it useless to enter into any arrangement with him about the revival of the *bêche-de-mer* trade, as he had not power sufficient to compel the necessary number of people to procure a shipload full of that valuable article. When younger, he had been guilty of murdering a white man of the name of Cunningham, who had a handsome wife from Rotuma, whom his father afterwards added to his harem. Nor had vessels going near his place been always safe: a few years ago the 'Paul Jones' and another little schooner, the 'Gladiator,' with British subjects on board, were fired into, and obliged to leave so inhospitable a neighbourhood with all possible speed. Ritova, on the other hand, is the exact contrast of Bete. He is a tall, well-made man, with intelligent features; every inch a chief. Both his mother and grandmother were the great Macuata Queens, which gave him an advantage over Bete, whose mother was a degree below them in birth. All over Fiji the rank of the mother is of importance in regulating that of her offspring, but in Macuata a still greater stress is laid upon this circumstance than elsewhere; hence, after Bete's father died, the office of Tui Macuata, or King of Macuata, vacant by his death, was offered by the landholders to Ritova as the highest chief. However, he waived his claims in favour of his son, who accordingly was duly elected, and invested with the title. After Ritova had been driven away, Maafu made Bete King of Macuata; hence there were two claimants to that dignity. In his dealings with the white men, Ritova always behaved creditably. Traders left large stocks of goods in his hand, taking no other security for their payment than his reputation for honesty, and that at a time when nearly the whole of Fiji was addicted to cannibalism, and the lives of foreigners trembled in the balance. In the complicated process of collecting and curing *bêche-de-mer*, Ritova displayed as much energy in making his people work as he did honesty in the pecuniary transactions which it involved. The benefits arising from the *bêche-de-mer* trade were felt on all hands, and when, with Ritova's removal, this lucrative traffic came to an end, even the most humble became mindful that they had not simply experienced a change of masters. What impressed me most favourably with Ritova was, that I once caught him, with his hands at his back, walking up and down in silent meditation behind his house, and on inquiry I found that such was his usual habit. Amongst Europeans this may be nothing uncommon, but amongst Fijians, or Polynesians in general, it is worth recording.

Mr. Pritchard opened proceedings by expressing regret that Bete had not visited Cakaudrove, where his opinion might have influenced the result arrived at regarding Macuata affairs. He then told him that, having refused his council, it had been settled without him that Ritova should return to Nukubati, and enjoy the undisputed rights of his patrimonial estates. Ritova was now called, and though the two chiefs had for many a long year been neighbours, separated by a few miles, they now, for the first time in their lives, shook hands with each other: interested parties on both sides had always kept up a state of enmity between them. Bete, addressed as Tui (King of) Macuata, according to a previous arrangement with Ritova, was asked to express his views on the subject; but he at once begged that Ritova might take precedence, calling him the "Vunivalu," the highest title he could apply. Ritova expressed his desire to live in peace on his lands, to devote his energies to the development of agriculture and trade; hoping, at the same time, that all old feuds might be consigned to oblivion. Bete echoed the same sentiments, and had no objection to sign a document to that effect, in which the two chiefs pledged themselves not to attack each other, or set on foot any measure or intrigue that might be attended with evil consequences to either party; to refer all matters of dispute between them to H.B.M. Consul, to disavow all allegiance or dependence on Maafu, and to suffer punishment, even to the loss of their chieftainship, in case of non-compliance with any article of the convention. A document of this nature was accordingly drawn up, ably translated by the consular interpreter, Mr. Charles Wise, signed by the two chiefs, and witnessed by Mr. Pritchard, the Tonguese teacher, the interpreter, and myself.

Early the next morning we made for Nukubati. This island, scarcely a mile in circumference, still bore ample traces of the mode of warfare carried on by the Tonguese. All the houses had been destroyed by fire, with the exception of one, the temporary residence of Maafu during the fight. The trunks of most of the cocoa-nut palms were charred by the conflagration that had consumed the town; nearly all the other fruit-trees had been cut down, and hundreds of cocoa-nut trunks felled, to make a high stockade, dividing the island into two sections, and serving as a breastwork, impenetrable to bullets. The wells had been filled up with rocks, logs, and rubbish; in fine, every damage that could possibly be conceived to change a flourishing town and a fruitful island into a wilderness, had been done. Quite recently a few settlers had collected on Nukubati, busily engaged in re-establishing the plantations and erecting houses.

Hardly had we dropped anchor when a deputation from the island, headed by the local chief, waited upon Ritova. They brought with them presents of wild yams, ready cooked, and carried on a tray of cocoa-nut leaves. The local chief, a man somewhat advanced in years, and of rather venerable aspect, came to shake hands with Ritova; whilst his followers kept at a respectful distance, and none of them ventured to stand upright as long as they were on board. This old man had been one of Ritova's most faithful friends, having shared his exile for some time. The two friends were quite overcome, and ready to cry. None of them could speak for some minutes; at last the old chief said, that he was sorry to have to come empty-handed, but they were so poor that they had nothing to give. Ritova replied, that to be able to look once more upon his dear old face was more than all the presents he could have brought; they would apply themselves manfully to rebuild their towns, and the intercourse with the white men would soon place them in possession of plenty of goods. They then went on shore, where the people were overjoyed to behold their great chief again.

The Tonguese teacher of Naduri had been invited by us to preach that day at Nukubati, for which we made him a handsome present; and all hands went on shore to attend Divine service, which, in the absence of a proper place of worship, was held in the chief's house. Instead of dwelling on the importance of the happy result that had been brought about by the arrangement just concluded, and thanking God that peace had been preserved in the land, the teacher preached a pointed sermon at Ritova, about the evils that jealousy had produced in Tonga,—Tonga is always put first by these conceited islanders,—Europe, and Fiji. Seeing several

Roman Catholics present, he dwelt on the errors of their dogmas, and abused the Virgin and the Saints in unmeasured terms. It would have been hardly possible to preach a more impracticable sermon, or exhibit worse taste or less discretion. Ritova, on pointing out the site for a church, begged the Consul to write to the headquarters of the missionaries about sending him Christian teachers; but, if possible, not a Tonguese or a man of extreme sectarian views, who, by widening the breach between Roman Catholics and Protestants, might endanger the peace, whilst a man of moderate views would have little difficulty in making the whole population of one way of thinking on religious subjects. He afterwards recurred to this topic when he saw me again, saying—though of course using different language—that the ethical part of Christianity, that which was the basis of both denominations, had a deep interest to him, but that he attached little value to mere dogmas. This was a proof to me that this man had thought much more deeply on religion than he had received credit for. When lonely pacing up and down the trodden path behind his hut, he had evidently sought to arrive at some solution respecting the conflicting views rival denominations presented to him.

One of Ritova's large canoes had come along with us, but all the others had not made their appearance the second day after our arrival. Some uneasiness being felt lest the Tonguese had captured them, heavy laden as they were with passengers, goods, and live stock, a messenger was dispatched to the island of Kia, who returned with two other canoes, having Ritova's son (Tui Macuata) on board. They had not thought it possible that affairs with Bete could be arranged amicably, and therefore had not come direct. When Ritova's son soon after stepped on shore, he could scarcely believe that he was actually treading on his native isle. "Is this really the sand of Nukubati?" he exclaimed; "really my home? Yes, it is, thanks to the Consul." His companions felt equally grateful, but gratitude in the Fijian always seeks expression in gifts, and their greatest sorrow was that they had nothing to give; even Ritova was uneasy on this point. If any brother-chief had effected his restoration, custom would have demanded that Ritova should collect all the goods he could by the twelvemonth, or later, invite his allies to a great festival, and publicly, with an appropriate speech, hand the presents over to them. The Consul explained in unmistakeable language that all he asked in return for what had been done, was the resumption of Ritova's former activity in trading with the white men, and the same friendly treatment of his customers he had invariably bestowed upon them when chief ruler of Macuata.

On the 30th of October a schooner arrived from Ovalau with dispatches, urgently calling Mr. Pritchard's attention to another part of the group. Going on shore to wish Ritova good-bye, we met deputations delivering addresses from towns which had heard of his return, and sent whales' teeth and other acceptable presents in proof of their devotion. When we returned on board, the large triangular sails of the missing canoes appeared on the horizon: all Ritova's little property was safe. We fired a salute by way of farewell, and hoisting all canvas, soon lost sight of Nukubati and its young community.*

Macuata now began to revive. Ritova eagerly set about rebuilding his town on Nukubati, and white traders again flocked to the coast, as in days of yore. This turn of affairs was far from pleasing to the Tonguese; they were indefatigable in promoting discontent and disturbance, and scarcely had Ritova's town been rebuilt than the Tonguese burned it down again. Bete, Maafu's willing tool, could not resist the temptation of playing once more the traitor. Under the pretext of making a durable peace, he coaxed Ritova over to Naduri, where he had arranged with a party of mountaineers to rush into the town and club Ritova and his family. Ritova went into the trap: fortunately his son heard of the scheme, and reported it to his father. Ritova went off in one of his canoes, professedly to drink kava, in reality to hold a council with his old men; whilst the son remained on shore to lull suspicion. Bete, in order to bring Ritova on shore, invited him to a bowl of kava; and the son, seeing the moment had arrived when all were to be massacred, told his father their imminent peril. They were all in Bete's power: what were they to do? The son urged the necessity of assuming the offensive, and killing Bete without delay; Ritova hesitated, but the young fellow went ashore, met Bete just in front of his house, charged him with the

---

* It is only up to this date that I can speak from personal experience of the events that occurred; what follows has been derived from a communication in the 'Athenæum,' from private letters, and from Commodore Seymour's and other dispatches published in the 'Fijian Blue-book.'

diabolical plot he had laid, and that had his father not followed the Consul's advice to act honestly, he would never have been in his power. " I have three balls in my musket for you, Bete;" he said, "you, who want to kill my father, his son, and all his people, in cold blood." With these words he fired, and two balls lodged in Bete's body; he died instantly. A great uproar followed; some of Ritova's friends, and they were numerous, voted for killing all Bete's followers and razing the town. Ritova, who had all the while been on board his canoes, rushed on shore, quelled the excitement by his presence, and harangued the crowd. " People of Naduri," he said, "you who deserted me, your proper chief, when the Tonguese drove him from the land of his forefathers, you may all live! Were it not for my solemn promises to the British Consul, you would all die this day with the man you followed; he has told me to spare my enemies, therefore, be pardoned; keep quiet; I will send for Christian teachers—not Tonguese—European or Fijian, and we will all endeavour to live in peace, and cultivate agriculture and trade." *

Everything was going on quietly again when Maafu dispatched his lieutenant, Wai-ni-golo, to Macuata, and troubles at once recommenced. The very excellence of this, the finest district in Fiji, makes these artful and bold Tonguese crave after it so much. Fortunately, about the middle of July, 1861, Commodore Seymour, in H.B.M.S. Pelorus, arrived at Ovalau, and extracts from his dispatch shall carry on the story.

* 'Athenæum,' No. 1791, p. 261.—Also private letters from residents in Fiji.

"Her Majesty's ship, under my command, sailed from Coromandel harbour, east coast of New Zealand, on the 8th July, and arrived at Levuka harbour, island of Ovalau, on the 15th, after a favourable passage made under sail. Having been informed by Mr. Pritchard that the trade in *bêche-de-mer* on the northwest coast of Vanua Levu was entirely stopped in consequence of a war which was being carried on there between two rival chiefs, one of whom was supported by a body of Tongans, whose usual residence is on Lakeba, one of the windward islands, I decided on endeavouring to put a stop to a state of affairs so prejudicial to British interests; and in order that my measures should be backed by the highest native authority in Fiji, I requested Mr. Pritchard to propose to Cakobau and Maafu to accompany me to the Macuata district in the 'Pelorus.' This, after a little diplomatic shuffling, they consented to do; and having received them, Mr. Pritchard, and the consular interpreter, on board, we left Levuka on the morning of the 18th, entering the great reef which encircles Vanua Levu by a pass a little to the northward of the Nadi passage, after which our course lay through a very intricate channel formed by sunken reefs and patches, of which no regular survey exists, but through which we were piloted in the most able manner by one of the English residents at Ovalau (Christopher Carr), the owner of a small *bêche-de-mer* trader. Under his direction we reached anchorage off Levuta, about twenty miles from our destination, Macuata, that evening; and the following morning, having weighed as soon as the sun was sufficiently high to enable us to distinguish the shoals, we anchored in Naduri Harbour, Macuata Bay, about 1500 yards from where some houses were visible on the beach.

"On sending on shore to ascertain the state of affairs, we found, as I had anticipated would be the case, that the combined force of the Tongans and Fijians had driven their opponents off the mainland, and that the latter had taken refuge on Kia Island, about ten miles from our anchorage. Since their expulsion their enemies had committed great havoc amongst their plantations, had destroyed nearly all the large canoes, for which this district was formerly famous, and almost daily put one or more persons to death, whose only crime was being related to the vanquished party. In these outrages the Tongans were the most prominent actors; and I may here state my opinion, that in the event of her Majesty's Government accepting the Fijis, it will be necessary, from the very first, to put a stop to the raids which the Tongans have for the last five years been in the habit of carrying into the various islands lying to the west of Lakeba.

"On the morning of the 20th I sent over to the island of Kia for Ritova, the chief of the tribe which had been driven out of Macuata, and in the afternoon he came on board in a cutter of the 'Pelorus,' followed by fifteen canoes filled with his retainers. After he had had an hour's conversation with Cakobau and Maafu, we made a preconcerted signal, on seeing which Wai-ni-golo, Maafu's lieutenant, and two Fijian chiefs, came on board; and after they and their opponents had discussed matters for an hour, I told them, through the consular interpreter, that we had no wish to injure or interfere with either the Fijians cr Tongans in any way; but that, owing to the senseless quarrels of the former, fomented by the latter, the interests of the white traders in Fiji were compromised, and that I was determined on putting a stop to a state of affairs which was equally prejudicial to their own and to British interests. I should therefore leave them to settle, by what means they could arrange, matters amongst themselves, and any advice I could give them was at their service. My observations were listened to with attention by both parties of Fijians, but were evidently unsatisfactory to the Tongan chief, who, throughout the entire business, was less manageable than either his associates or his enemies.

"The discussion, which terminated at sunset, was renewed the next day, when the following terms were agreed to by the chiefs of Fiji and Tonga present, being those which, with Mr. Pritchard's concurrence, I had decided from the first on seeing carried out :—

"*Between Ritova and Bonaveidogo, chiefs of Fiji.*

"1st. To forget all past grievances and causes of quarrel.

"2nd. To commence from this date an era of peace and friendship.

"3rd. To receive and protect the teachers of the Christian religion.

"4th. To encourage trade and commerce throughout the Macuata territories, and to protect all legitimate traders and settlers.

"5th. To dissolve all political connection, and to confine themselves to legitimate and friendly intercourse with the Tongans.

"*Between Ritova and other chiefs of Fiji and Maafu, chief of Tonga.*

"1st. That Wai-ni-golo shall, within fourteen hours, retire for ever from the Macuata territories, and shall not again appear within the line of country from Nacewa Bay on the one side, to Bua Bay on the other.

"2nd. That no Tongans shall visit the Macuata territories, or appear within the above-named limits, for twelve months from this date.

"3rd. That Tongans in the service of the Wesleyan or other missions are exempted from the above restrictions.

"4th. That if any of the above articles are infringed, Maafu agrees that Wai-ni-golo shall be sent from Fiji to his native country.

"The three last articles were inserted in the treaty at my recommendation, as I foresaw that if the Tongans were allowed to remain on the Vanua Levu, any good effect which might otherwise result from our visit would be completely done away with; and in compliance with them at dawn on the morning of the 22nd of July, the two large double canoes, in which Wai-ni-golo and his followers had come to Macuata, were launched, and by eight A.M. were under weigh, with a strong and fair wind, for Lakeba; a more picturesque scene than their departure, as they crossed the 'Pelorus's' bow, beating their drums and cheering most lustily, I have seldom witnessed. In the course of the same day Cakobau and Maafu quitted the ship, and sailed for Levuka in Cakobau's large canoe, and in the afternoon I landed at Macuata, accompanied by Ritova, and saw him and many of his people re-established in their former habitations.

"Having thus seen tranquillity re-established in Vanua Levu, I quitted Macuata on the morning of the 23rd July, having Ritova and two of his retainers on board, they being desirous of seeing the working of the engines; and on getting clear of the Mali passage we discharged them and Mr. Pritchard to the latter's schooner, after which we made sail, by noon were clear of Kia Island, and steering a course for Aneiteum." *

Commodore Seymour's visit thus proved of material benefit to Fiji, and was felt as such on all hands. " I am directed by Earl Russell to request," writes Mr. James Murray, of the Foreign Office, to Sir T. Rogers, Bart., December 31, 1861, "that you will state to the Duke of Newcastle, that his Lordship has learnt with satisfaction the steps taken by Commodore Seymour for terminating the wars which have been raging between the Tongans and Fijians."

* It will be seen how closely this statement agrees with the more condensed account in the 'Athenæum' of February 22, 1862.

CHAPTER XVI.

GENERAL REMARKS ON THE ASPECT, CLIMATE, SOIL, AND VEGETATION OF FIJI.—COLONIAL PRODUCE.—STAPLE FOOD.—EDIBLE ROOTS.—KITCHEN VEGETABLES.—EDIBLE FRUITS.—NATIONAL BEVERAGES.—KAVA.

VITI, or Fiji, is an archipelago in the South Pacific Ocean, midway between the Tongan islands and the French colony of New Caledonia, having, according to Dr. Petermann's recent calculations, a superficial area equal to that of Wales, or eight times that of the Ionian Islands. The exact number of islands and islets comprising it is merely approximately known, only a partial hydrographical survey of the whole group having as yet been made; 230 would probably be rather below than above the number. Viti Levu, Kadavu, Vanua Levu, and Taviuni, are of primary, Rabe, Koro, Gau, and Ovalau, of secondary, magnitude. Situated between latitudes 19° 47′ S. and 15° 47′ S., and longitudes 180° 8′ W. and 176° 50′ E., the climate is tropical, but the heat is moderated, in the winter season by the south-east, in the summer by the north-east trade-wind. 62° Fahr. is the lowest temperature observed in Lakeba by Mr. Williams, in Kadavu by Mr. Royce; but, though the mean temperature of the whole group may be stated to be 80° Fahr., the thermometer has been known to rise to

121° Fahr. The country is remarkably free from fever, —that curse of the Samoan group,—and the only disease Fijians and Europeans have reason to fear is dysentery, unknown, if a current belief may be relied upon, before the visits of foreigners to these shores, and hence often termed "the white man's disease" by the natives.

The time from October till April is the hottest, that extending over the other months the coolest, part of the year. It is during the former when the most rain falls, but the dry and rainy seasons do not strictly correspond with this division, nor is the difference between the wet and dry very marked. There are occasional showers during the so-called dry season in all parts of the group, and in localities like the Straits of Somosomo they may even be termed frequent. The fine weather is expected to set in about May. June, July, August, September, and October, are generally dry, and from their low temperature looked forward to by European settlers. How many inches of rain annually fall has not been ascertained; nor would a gauge kept in a single locality only give a fair approximate result of the average amount, since the difference of the meteorological conditions existing between the leeward and windward islands, the lee side and the weather side of the larger islands, are too great.*

Speaking generally, the Vitian islands may be said to

* A gauge, kept by the Rev. Mr. Whitley (probably at Levuka, B.S.), showed that ninety inches of rain had fallen in six months, and four in the night of February 12th, 1860. This statement I find in an obscure publication, the 'Primitive Methodist Juvenile Magazine,' London, 1862, vol. xi. p. 50. Not having seen it confirmed, it may possibly be incorrect, like several others in the article from which it is taken.

owe their origin to volcanic upheavings and the busy operation of corals. There are at present no active volcanos, but several of the highest mountains, for instance, Buke Levu, in Kadavu, and the summit of Taviuni, must in times gone by have been formidable craters. Hot springs are met with in different parts, earthquakes are occasionally experienced, and between Fiji and Tonga a whole island has of late years been lifted above the level of the ocean, whilst masses of pumice-stone are drifted on the southern shores of Kadavu and Viti Levu; all showing that Fiji, though not the focus of volcanic action, is not secure against plutonic disturbances and their effects. The deltas and alluvial deposits of the great rivers excepted, there is little level land. Most of the ground is undulated, all the larger islands are hilly, and the largest have peaks 4000 feet high; Voma, in Viti Levu, and Buke Levu, in Kadavu (both of whichwere ascended by me), being the most elevated. The soil consists in many parts of a dark-red or yellowish clay, or decomposed volcanic rock, which soon becomes dry, but being plentifully supplied with water, proves very productive. There is hardly a rod of land that might not be converted into pasture or be cultivated. Almost at every step one discovers that most of the land has at one time or other produced some crop. Though on the weather side dense and extensive woods exist, few of them can be regarded as virgin forests, most having re-established themselves after the plantations once occupying their site had been abandoned. Kadavu does not appear to have an acre of virgin forest beyond what is clustered around the very summit of Buke Levu. The re-establishment of the woods on ground at one time under cultivation can scarcely be adduced as a proof that the population has seriously diminished, but rather that the Fijians have for ages followed the same system of agriculture as they do at present, that of constantly selecting new spots for their crops when the old ones, which their ignorance prevents them from fertilizing by the introduction of manure, become exhausted. The displaced vegetation quickly resumes its former sway, until perhaps, after the lapse of years, it has once more to make room for cultivated plants.

The aspect of the weather side of the islands is essentially different from that of the lee side. The former teems with a dense mass of vegetation, huge trees, innumerable creepers, and epiphytical plants. Hardly ever a break occurs in the green mantle spread over hill and dale, except where effected by artificial means. Rain and moisture are plentiful, adding ever fresh vigour to, and keeping up the exuberant growth of, trees, shrubs, and herbs. Far different is the aspect of the lee side. Instead of the dense jungle, interlaced with creepers and loaded with epiphytes, a fine grassy country, here and there dotted with screw-pines, presents itself. The northern shores of Viti Levu and Vanua Levu bear this character in an eminent degree, and their very aspect is proof that rain falls in only limited quantity; the high ridge of mountains, which form, as it were, the backbone of the two largest islands, intercepting many showers, but sending down perpetual streams to fertilize the low lands of the coast. The lee side would therefore more readily recommend itself to the white settler, as it requires hardly any clearing, and would be immediately available for cattle-breeding and cotton-growing.

The coast-line of most of the islands is enriched by a dense, more or less broken, belt of cocoa-nut palms. White beaches, formed of decomposed corals, may be traced for miles; whilst good soil in many instances extends quite to the water's edge, and trees, not numbering amongst the strictly littoral vegetation, overhang the sea. Mangrove swamps are limited, and chiefly confined to the mouths of the rivers; hence the almost total freedom of the country from malignant fevers. In the windward islands, Lakeba and its dependencies, the weeping iron-wood (*Casuarina equisetifolia*, Forst.), intermingled with screw-pines (*Pandanus odoratissimus*, Linn.), abounds, and considerable tracts of country are covered with the common brake and other hard-leaved ferns: they prefer an open country, and have taken possession where little else will grow. Wherever these forms of vegetation occur on the weather side of the group, the soil may be expected to be rather poor. It would, however, be erroneous to apply the same rule to the leeward side, where they are also tolerably abundant, not because the soil is too poor to support a dense herbaceous or woody vegetation, but because the air is destitute of that excessive moisture, and the country less visited by numerous showers of rain, promoting the luxuriant growth on the weather side.

The general physiognomy of the flora is decidedly tropical; tree-ferns, branching grasses, six or seven different kind of palms, Scitamineous plants, epiphytical orchids, ferns, and pepperworts, fully accounting for this fact. Whole districts, however, possess a strictly South Australian look, owing to the presence of two phyllodineous *Acacias* (*A. laurifolia*, Willd., and *A. Richei*, A. Gray), two *Casuarinas*, several kinds of *Metrosideros*, with either scarlet or yellow blossoms, a climbing *Rubus*, *Smilax*, and *Geitonoplesium* * and *Flagellaria*, as well as the peculiar habit of various other species. There is little change in the nature of the vegetation until one reaches about 2000 feet elevation, where the plants peculiar to the coast region are replaced by mountain forms. Hollies, Myrtaceous, Melastomaceous, and Laurinaceous trees, Epacridaceous and Vacciniaceous bushes, forming the bulk; scarlet orchids, astelias, delicate ferns, mosses, and lichens, crowding their branches. None of the explored peaks have as yet disclosed any genuine alpine vegetation,—perennial herbs forming cæspitose masses and prostrate shrubs, generally bearing large and gay-coloured flowers. Should it ever be met with, there would indeed be a rich botanical harvest.

Nature has been truly bountiful in distributing her vegetable treasures to these islands; but perhaps the best proof of their extreme fertility and matchless resources is less furnished by the fact that a country with a population of at least 200,000 souls, constantly supplying provisions to foreign vessels, having an immense

* The natives term this plant Wa Dakua, from *Wa*, creeper, and *Dakua*, Kowrie pine, because its leaves closely resemble those of the Fijian *Dammara*.

number of cocoa-nuts withdrawn from consumption by a primitive and wasteful method of making oil for exportation, and cultivating, comparatively speaking, only a few acres of ground, than by the almost endless series of vegetable productions—an enumeration of which forms the subject of the succeeding pages.

Colonial produce, properly so called, such as sugar, coffee, tamarinds, and tobacco, may be expected from Fiji in considerable quantities, as soon as Europeans shall have devoted their attention to the subject; since the plants yielding them, long ago introduced, flourish so well, that a judicious outlay of capital might prove a profitable investment. The sugar-cane (*Saccharum officinarum*, Linn.), called Dovu in Fijian, grows, as it were, wild in various parts of the group, and a purple variety, attaining sixteen feet high and a corresponding thickness, is cultivated to some extent. No foreigners have as yet set up mills, nor are the natives at present acquainted with the process of making sugar; they merely chew the cane, and employ the juice for sweetening their puddings. In the greater part of the group the leaves are used for thatching the roofs of houses; it is only in Lakeba and others of the eastern islands where those of a screw-pine (*Pandanus odoratissimus*, Linn.) are preferred, whilst those of the Boreti (*Acrostichum aureum*, Linn.), a common seaside fern, are still less frequently used, though in the central islands they, in common with those of the Makita (*Parinarium laurinum*, A. Gray), supply the chief materials for covering the side walls of houses, churches, and temples. Coffee (*Coffea arabica*, Linn.) will one day rank amongst the staple products of the country; the mountain slopes of the larger islands, especially those of Viti Levu, Vanua Levu, and Kadavu, and, above all, those of the valley of Namosi, seeming well adapted for its growth. Several old coffee-trees are to be found in the Rewa district, showing the plant to be not of recent introduction. Dr. Brower, American Consul, has established a plantation on his estate at Wakaya, which gives fair promise; and Mr. Binner, of Levuka, has in his garden a number of thriving seedlings. The tamarind (*Tamarindus Indica*, Linn.) was introduced about eighteen years ago; and there is a fine tree, thirty feet high, and of corresponding dimensions, on the Somosomo estate of Captain Wilson and M. Joubert, of Sydney.

Tobacco (*Nicotiana Tabacum*, Linn.), a pink-flowering kind, is grown about towns and villages in patches, never exceeding a few rods in extent, but in sufficient quantity to keep the bulk of the population supplied. Both men and women use it for smoking only, either out of pipes or made into cigarettes with dry banana-leaves; the filthy habit of chewing or taking snuff does not seem to be practised by them, though, had they been so inclined, they might have learned it from the lower class of white settlers. Being unacquainted with the process of curing the leaf successfully, the natives greatly prefer our tobacco to their own, and are thankful for the gift of a piece, however small, but rather loth to regard it in the light of payment for goods or services rendered, preferring any other article of barter, inferior though it may be in value to the tobacco offered.

Oil and vegetable fat next claim our attention. The most valuable oil produced in Fiji is that extracted from the seeds of the Dilo (*Calophyllum inophyllum*, Linn.), the Tamanu of Eastern Polynesia, and the Cashumpa of India. It is the bitter oil, or woondel, of Indian commerce. The natives use it for polishing arms and greasing their bodies when cocoa-nut oil is not at hand. But the great reputation this oil enjoys throughout Polynesia and the East Indies rests upon its medicinal properties, as a liniment in rheumatism, pains in the joints, and bruises. The efficacy in that respect can hardly be exaggerated, and recommends it to the attention of European practitioners. The oil is kept by the natives in gourd flasks, and, there being only a limited quantity made, I was charged about sixpence per pint for it, paid in calico and cutlery. The tree yielding it is one of the most common littoral plants in the group, and its round fruits, mixed with the square-shaped ones of *Barringtonia speciosa*, the pine-cone-like ones of the sago-palm, and the flat seeds of the Walai (*Entada scandens*, Bth.), are found densely covering the sandy beaches, a play of the tides. Dilo oil never congeals in the lowest temperature of the Fijis, as cocoa-nut oil often does during the cool season. It is of a greenish tinge, and a very little of it will impart its hue to a whole cask of cocoa-nut oil. Its commercial value is only partially known in the Fijis, and was found out accidentally. Amongst the contributions in cocoa-nut oil which the natives furnish towards the support of the Wesleyan missions, some Dilo oil had been poured, which, on arriving at Sydney, was rejected by the broker who purchased the other oil, on account of its greenish tinge and strange appearance. On being shown to others, a chemist, recognizing it as the bitter oil of India, purchased it at the rate of £60 per tun; and he must have made a good profit on it, as the article fetches as much as £90 per tun. The Dilo grows to the height of sixty feet, and the stem is from three to four feet in diameter, generally thickly crowded with epiphytal orchids and ferns. The dark oblong leaves form a magnificent crown, producing a dense shade; and when, during the flowering season, they are interspersed with numerous white flowers, the aspect of the whole tree is truly noble. The exudation from the stem is, according to Bennett, the Tacamahaca resin of commerce, used by Tahitians as a scent. Carpenters and cabinet-makers value the wood on account of its beautiful grain, hardness, and red tinge. Boats and canoes are built of it, and it is named with the Vesi (*Afzelia bijuga*, A. Gray) as the best timber produced in Fiji. In order to extract the oil, the round fruit is allowed to drop and the outer fleshy covering rot on the ground. The remaining portion, consisting of a shell somewhat of the consistency of that of a hen's egg, and enclosing the kernel, is baked on hot stones, in the same way that Polynesian vegetables and meat are. The shell is then broken, and the kernel pounded between stones. If the quantity be small, the macerated mass is placed in the fibres of the Vau (*Paritium tiliaceum* and *tricuspis*), and forced by the hand to yield up its oily contents; if large, a rude level press is constructed by placing a boom horizontally between two cocoa-nut trees, and appending to them perpendicularly the fibres of the Vau. After the macerated kernels have been placed in the midst, a pole made fast to the lower end of the fibres, and two men taking hold of its end, twist the contrivance round and round till the oil, collecting into a wooden bowl standing underneath, has been extracted. Of course, the pressure thus brought to bear upon the pounded kernels is not sufficiently great to allow every particle of oil to escape, and with the proper machinery the waste would amount to little indeed.

The candle-nut (*Aleurites triloba*, Forst.), termed "Lauci," "Sikeci," and "Tuitui," in the various dialects of Fiji, contains a great deal of oil, of which, however, the natives make only a limited use for polishing, though in other parts of Polynesia lamps are fed with it, and in the Hawaiian islands the entire kernels are strung on a stick and lighted as candles. The fruit is better known as a dye, and plays an important part at the birth of a child; for no sooner is a baby born than the midwife rushes to the Lauci to gather a fruit fresh from the tree, which she places in the mouth of the interesting young stranger, with the conviction that its milky juice will clear the throat, and more effectually enable it to announce its welcome arrival. Mr. Wilson, the managing director of Price's Patent Candle Company, at Vauxhall, writes to me:—"The oil of the *Aleurites triloba* is fine and hard, worth at least as much as sesame or rape oil, in this market. It is held very lightly in its matrix, and should be pressed where grown. If the 'nuts' were brought home in their shells, the freight would be expensive; and if shelled, insects would eat them." The candle-nut tree is of middle size, common throughout Fiji, and rendered a conspicuous object by the whiteness of its leaves, produced by a fine powder easily removed. The ground underneath is always densely covered with " nuts," and large quantities might be collected.

The croton-oil plant (*Curcas purgans*, Med.), introduced from the Tongan islands, is employed for living fences in Lakeba and other parts; but the oleaceous properties of its seeds have as yet been turned to as little account as those of the castor-oil plant (*Ricinus communis*, Linn.), named " Uto ni papalagi " by the natives, and naturalized throughout the group.

The oil of the cocoa-nut palm, or Niu dina (*Cocos nucifera*, Linn.), has long been one of the articles of export; nevertheless, it is difficult to arrive at any definite result about the average annual quantity. The Wesleyan mission, in negotiating with an island trader for the transport of the oil received from the natives as contributions to its funds, were ready to guarantee that at least sixty tuns should pass through his hands. This, at the rate of £20 per tun, the average value of the oil on the spot, would give £1200 per annum—a sum tolerably well agreeing with that usually advertised on the wrapper of the 'Wesleyan Missionary Notices' as the Fijian share towards the support of the Society. Exact data for forming an opinion of the quantity shipped by the actual traders are altogether wanting. On consulting with several about this subject, they pretty nearly all agreed in fixing three hundred tuns as the utmost limit of the annual export of the whole group,=£6000 on the spot. Hitherto, there has been great waste in the making of oil, the native process being of a primitive description. To remedy this evil, Captain Wilson and M. Joubert, of Sydney, have set up proper machinery on their estate at Somosomo, after one of the partners had familiarized himself with the latest improvement in that branch of industry in Ceylon; and it is their intention to take advantage of the luxuriant manner in which Coboi, or lemon-grass (*Andropogon Schœnanthus*, Linn.), grows in Fiji, by cultivating it for the purpose of making citronella oil. Cocoa-nut oil congealing at a temperature of about 72° Fahr., and the thermometer during the cool months often falling below that degree, a proper amount of warmth will be kept up whilst the operation of pressing the pulverized kernels is going on, and thus another step be taken towards the making of the largest quantity of oil from the least number of nuts. Wilkes, upon the authority of one of the scientific men attached to his expedition, states that there were only two varieties of cocoa-nut, a green and a brown. Closer attention to the subject would have shown this to be a mistake; not only the colour, but also the average size and shape of the fruits, the height of the trees, and the insertion of the leaflets, or rather segments, offer marks of distinction between the numerous varieties with which the islands are studded. The most striking kind is the one having fruits not much larger than a turkey's egg, and bearing more than a hundred of them in each bunch. Several trees were noticed at Kadavu, about Yarabale, a narrow isthmus, where canoes are dragged across from sea to sea. The curious phenomenon of a cocoa-nut palm becoming, as it were, branched by the division of

OILS AND VEGETABLE FAT.

the trunk, has occasionally been witnessed in Fiji; and two interesting instances of it are given in Williams's 'Fiji and the Fijians,' where one of the trees is described with five branches. In Samoa Mr. W. Pritchard saw a tree with two heads, regarded with just pride by the natives who possessed it, and cut down during a war by their enemies. As in other parts of Polynesia, the trunk is made into small canoes, or supplies materials for building and fencing; stockades of it are impenetrable to bullets. The leaves are made into different kinds of mats and baskets; yam houses are occasionally thatched with them, but these roofs do not last much longer than a year. The spathe enclosing the flowers is used for torches; the fibres surrounding the nut are made into "sinnet," used for fastenings of all kinds. The young flesh is delicious eating, and the "water" contained in the nuts a refreshing drink, which, as the fruit advances, undergoes a gradual change, for all of which there are distinctive names. New-comers soon fix upon a certain stage most agreeable to their palate, and on indicating it to the natives they will readily pick it out by knocking with their fingers on the outside of either the husked or the unhusked nut, and be guided by the sound. This process requires long practice, and though I tried hard to learn at least the sound of that stage I preferred, I did not succeed in accomplishing it. The ripe nuts are grated and used for puddings, or given to fowls and pigs. Some persons have a predilection for nuts when just in the act of germinating—a taste which the Asiatic shares in eating the young palmyras, and the African in consuming the seedlings of the

Borassus? Æthiopicum, Mart. It is to be regretted that so few plantations of cocoa-nut trees are formed by white settlers. The annual value of a fruit-producing tree is never less than one dollar; and how easily might 10,000 nuts be set in the ground, and the value of an estate be permanently raised. Every part of the smaller islands and the sea-borders of the larger are suitable localities. Only Bau, Viwa, and the districts adjacent, form an exception: the trees, as soon as they have reached a certain height, become diseased; their leaves look as if dipped in boiling water, and their fruits are few in number, poor, and often drop off before they arrive at maturity; a thick layer of marl, forming the subsoil of those districts, seeming to oppose that ready drainage the cocoa-nut tree requires, and which it enjoys in so eminent a degree on the white beaches of sand and decomposed corals.

Starch is produced by four indigenous plants, viz. Roro (*Cycas circinalis*, Linn.), Yabia dina (*Tacca pinnatifida*, Forst.), Yabia sa (*Tacca sativa*, Rumph.), and Niu soria or Sogo (*Sagus Vitiensis*, Wendl.), to which of late years has been added the Cassava root of Western America (*Manihot Aipi*, Pohl), commonly termed by the Fijians "Yabia ni papalagi," *i. e.* foreign arrowroot. The Roro (*Cycas circinalis*, Linn.), a tree thirty feet high, is by no means a common plant in the islands, having been encountered only at Viti Levu and Ovalau in isolated specimens; and as the pith-like substance contained in the trunk was reserved for the sole use of the chiefs, and forbidden to the lower classes, no inducement existed on the part of those debarred from it to extend it by cultivation, as is done in the Tongan islands. The two kinds of Yabia are the arrowroot of Fiji, erroneously stated by Wilkes and others to be the *Maranta arundinacea*, Linn. They are both species of *Tacca;* their foliage springing up in great abundance in the beginning of the warm season, and their tubers ripening about June, when leaves and flowers die off. The most common is that kind termed on the Macuata coast Yabia dina (genuine arrowroot), the *Tacca pinnatifida*, Forst. It delights in light sandy soil, and is therefore most frequently encountered on the seashore; whilst the second species, known in Macuata as " Yabia sa," is almost entirely confined to the sides of hills and heavy soil. The natives prefer the first-mentioned species for the purpose of making arrowroot, though they own that there is no difference in the quality of the farinaceous substance prepared from either. In most parts of Fiji there are no distinctive names for the two kinds, both being called " Yabia;" yet the natives are perfectly well acquainted with their various characters and peculiarities of habitat. The leaf, stalks, and scape of the Yabia sa are prominently speckled, and the segments of the leaves are long and narrow, by which it is at once distinguished from its ally. The tubers, when quite ripe, are dug out of the ground and rasped on the mushroom coral (*Fungia* sp.). The fleshy mass thus produced is washed in fresh water to enable the starch to settle at the bottom of the vessel in which the operation is carried on; by pouring off the dirty water, and repeated washings, the starchy sediment may be made to assume any desired degree of whiteness. Since Fijian arrowroot has become an article of foreign demand, it has been pointed out to the natives that the impurities imparting a greyish colour to the production, caused partly by not peeling the tubers previous to rasping them, partly by not washing the sediment a sufficient number of times, must be removed in order to raise the marketable value of the article. When a satisfactory degree of whiteness has been attained, the starch is dried in the sun. For their own consumption the Fijians do not dry their arrowroot, but tie it up in bundles of leaves and bury it in the ground, when it speedily ferments, and emits a rather disagreeable odour. South Sea arrowroot fetches from threepence halfpenny to fourpence per pound in London; and, as it is invaluable when taken in cases of dysentery and diarrhœa,—the bane of the South Seas,—it is necessary to have it genuine. The Tonguese have of late years been known to adulterate it to a great extent with lime in order to increase its weight and volume, but this fraud may readily be detected by watching the arrowroot when it first comes in contact with water; if adulterated with lime, it will fizz. Care should also be taken to guard against the starch of the Cassava or Tapioco plant being passed off for Polynesian arrowroot, which, from its slightly purgative tendency and poisonous properties, is ill-adapted for bowel complaints. It is much whiter than the arrowroot made of *Tacca*, sticks to the hands like flour, and when a little water is allowed to act upon it, it assumes a pinkish colour; whilst the arrowroot made of *Tacca* has a granulated feel, does not adhere to the hand like flour, and is not changed in colour by contact with water. The Cassava root has of late years been introduced into Fiji, and grows remarkably well.

The Niu soria or Sogo (*Sagus Vitiensis*, Wendl.) is a genuine sago-palm, growing in swamps on Viti Levu, Vanua Levu, and Ovalau, and was first discovered by Mr. Pritchard and myself when on our first visit to Chief Kuruduadua. By asking the natives respecting the various palms of the islands, they described one which I was led to consider as the sago-yielding tree, and hence we made inquiries at all the places we called, but did not obtain a sight of it until we reached Taguru, on the southern coast of Viti Levu, and thence westward it was encountered in abundance. Fine groves, several miles in extent, were seen by us on the various branches and deltas of the Navua river. It was afterwards ascertained to grow on Ovalau; and Mr. Waterhouse, when accompanying Colonel Smythe, found an extensive grove on the north-eastern parts of Vanua Levu. The natives of Ovalau term this palm Niu soria, those of Viti Levu, Sogo (pronounced " Songo "); the latter name reminding one of " Sago " or " Sagu," by which some species of *Sagus* are known in other islands inhabited by the Papuan race; and rendering the discovery of this palm ethnologically as interesting as it is important commercially, by adding another raw product to the export list of the islands, and botanically, by extending the geographical range of sago-yielding palms 1500 miles further south-east than it was previously known to exist. The natives of Fiji were unacquainted with the nutritious qualities residing in the trunk, until Mr. Pritchard and myself extracted the sago from it.

The Sogo grows in swamps, and the natives occasionally take advantage of the open places among the groves to plant taro, or even clear Sogo land for that purpose. The dimensions of the finest specimens were accurately measured. The largest trees felled were from forty to fifty feet high, and their trunks, in the thickest parts, from three feet nine inches to four feet four inches in circumference. The trunk is very straight, and densely covered with aerial roots, six to twelve lines long, all having the peculiarity of being directed upwards. The crown generally consists of about sixteen living leaves in all stages of development, and there are mostly five or six dead ones still adhering to it. The pinnatifid leaves are of a dark green, seventeen feet long; whilst the leaflets, gracefully drooping at the tips, are from three and a half to four feet long, and three and a half inches broad. The petiole is covered with spines, which at its base are arranged in connected rows extending from side to side, and towards the top in horse-shoe-shaped collections. The spines are brown, and from one and a half to two and a half inches long. When the tree has attained maturity there appears a terminal panicle about twelve feet high, and divided into twenty or more branches. These branches measure eight feet in length, and are again divided into about fourteen branchlets (each averaging from fourteen to sixteen inches). The fruit, in outer appearance resembling an inverted pine-cone, is beautifully polished and of a yellowish brown, much lighter than that of *Sagus Rumphii*, Mart. This palm forms a prominent feature in the landscape, the foliage fluttering like gigantic plumes in the wind, and outbidding the cocoa-nut in gracefulness of outline and movement; the bold look of the flowers suddenly starting from the extremity of the trunk, and proclaiming, as it were by signal, that the time has arrived when nature has completed her task of laying up stores of nutritious starch, and that unless the harvest is at once gathered in, nothing will remain of the produce of years save the receptacle in which it was treasured up. Even the old dead trees, standing like so many skeletons amongst a host of young plants, present an interesting appearance, reminding one of the posts with their many arms over which the wires of electric telegraphs are carried. Mr. Pritchard and myself felled six trees, and carried two logs to Lado, where we made sago of one of them by grating and washing the yellow-white substance with which the inside was filled. The term "spongy" does not well apply to this substance; it has rather the consistency of a hard-baked loaf, and that taken from the base of the tree has a sweet and pleasant taste; towards the top it was more insipid. For the purpose of collecting sago it is of the highest importance that the tree should be cut down just at the time when the flowers begin to show themselves; if felled sooner the tree has not attained its proper development, and the quantity of farinaceous matter will not be so great as at the period indicated; if, on the other hand, the cutting down is deferred until the fruit has been formed, a considerable diminution of the quantity of sago meal will be observed; and the longer such a postponement takes place, the less chance there is of collecting a remunerative amount, as the tree, when it has borne flower and fruit, which, unlike the cocoa-nut palm, it does only *once* during the term of its existence, speedily dies and crumbles into dust. The trees are easily felled, only the outer layers of wood possessing any hardness, the central parts being as soft as bread, so that a few strokes with a good axe will bring the largest tree to the ground.*

Several kinds of spice are indigenous, or have become naturalized. Turmeric (*Curcuma longa*, Linn.), termed " Cago " by the Fijians, grows abundantly in all the lower districts. The whites use the rhizome in the preparation of curry, and the natives the powder of it as food, or more commonly to daub over the bodies of women after childbirth and those of dead friends—a custom also prevailing in the Samoan group, according to Mr. Pritchard. In the few districts that have as yet not been brought under the immediate influence of the British Consul or the missionaries, the heathen widows are painted with it before strangulation. In fact, turmeric powder is with the Fijian what rouge and Rowland's preparations are with us, a cosmetic. Promoting in their opinion health and beauty, it is put on with no sparing hand by the women, and pointed remarks are made about too great a proximity if a man be unfortunate enough to have some stains of turmeric on his body or scanty dress. The manufacture of turmeric is similar to that of arrowroot, and is generally managed by the women. The receiving pits dug in the ground are lined with herbage, so as to retain the juicy parts. The grated rhizome is afterwards placed in the body of a canoe, and

* Dr. Bennett, of Sydney, found a sago palm on Rotuma, north of Fiji, possibly identical with the Fijian, but there are no specimens.

rolled up and strained through a fine basket lined with fern leaves. It is then carried away in bamboos, and for several days exposed to the air, when the fluid is gently poured off, and a sediment, the Rerega of Fiji or turmeric of commerce, is found at the bottom. A species of ginger (*Zingiber Zerumbet*, Rosc.) also abounds in the lower districts of the group, where it is called " Beta." The rhizome, though less pungent than that of the species exported from China, has been found to make tolerably good preserves, and answers all the other purposes for which genuine ginger (*Zingiber officinale*, Linn.) is commonly employed. During our journey we often used it with turmeric, a few leaves of an aromatic Zingiberaceous plant termed "Cevuga" (*Amomum* sp.), and a few fruits of the bird's-eye pepper for making curry, which, all the ingredients being fresh, proved of excellent flavour. A species of Nutmeg (*Myristica castaneæfolia*, A. Gray), termed " Male," is found in the larger islands, forming trees sixty to eighty feet high, but yielding a very inferior kind of timber, which rapidly decays when exposed to the influence of the weather. Both its mace and nut prove a good substitute for those of the genuine nutmeg (*Myristica moschata*, Linn). The "nut" was turned to no account until the whites pointed out its valuable properties. It is about the size of a pigeon's egg; the mace (*arillus*) is of a fine pink colour, and the shape of the nut it encloses is too oblong to allow this kind of nutmeg ever to be passed off for the genuine and best sorts of the Indian Archipelago, though the Fijian produce may resemble them in every other respect. Bird's-eye pepper (*Capsicum frutescens*, Linn.)

is met with in every part of the islands, especially in places under cultivation, producing rich harvests of red pungent fruits. The Fijians call it " Boro ni papalagi" (*i. e.* foreign Boro), in contradistinction to " Boro ni Viti," or Fijian Boro (*Solanum anthropophagorum*, Seem., and *S. oleraceum,* Dun.); thus indicating that the bird's-eye pepper has been introduced by the white man, and is merely to be looked upon as naturalized, not wild.

The staple food is the same all over Polynesia, being derived, with the total exclusion of all grain and pulse, from the yam, the Taro, the banana, the plantain, the breadfruit, and the cocoa-nut; but the bulk of it is furnished in the different countries by only *one* of these plants. In the Hawaiian group the Taro takes the lead, whilst the cocoa-nut is looked upon as a delicacy, from which the women were formerly altogether cut off. In some of the smaller coral islands the inhabitants live almost entirely upon cocoa-nuts. The Samoans place the breadfruit at the head of the list. Again, the Fijians think more of the yam than of the others, though all grow in their islands in the greatest perfection and in an endless number of varieties. A striking proof of how much the yam engages their attention is furnished by the fact of its cultivation and ripening season being made the chief foundation of their calendar; and that only such of the eleven months, into which the year is divided, bear no names indicative of it, in which the crop requires no particular attention, or has been safely housed. A version of this calendar has been published by Wilkes in 'The Narrative of the United States Exploring Expedition,' and is placed in juxtaposition with one dictated to me by an intelligent Bauan chief, and the consular interpreter, Mr. Charles Wise. The names given by me, as well as their succession, do not quite agree with those given by Wilkes. This discrepancy is partly explained by Wilkes having taken down his list from the lips of Europeans imperfectly versed in Fijian, and by his adopting a loose way of spelling. The names of the months may also be different in different parts of the group. The subject, however, requires still further investigation. If, as has been averred, the Fijians invariably commenced the months with the appearance of the new moon, there would soon have been a vast difference between the lunar and the solar year. To guard against the irregularity that would thus have been introduced into the seasons, and to make the lunar year correspond with the solar, it would have been necessary either to intercalate a moon after every thirty-sixth moon, or to allow a greater period of time for one of the eleven months into which the Fijian year is divided. The latter seems to have been effected by the Vula i werewere (clearing month). Hazelwood ('Fijian and English Dictionary,' Viwa, 1850, p. 180) allows four months, May, June, July, and August, for it; but this cannot be correct, as it would derange the others. By restricting it to two or thereabouts, June and July, a proper arrangement is effected. I place the Vula i werewere first in my list instead of the month answering to January, because it is in the spring of the year (June and July), and the commencement of the agricultural operations and natural phenomena upon which the calendar is based.

Fijian Calendar.

According to Seemann.

1. *Vula i werewere* = June, July, clearing month; when the land is cleared of weeds and trees.
2. *Vula i cukicuki* = August; when the yam-fields are dug and planted.
3. *Vula i vavakadi* = September; putting reeds to yams to enable them to climb up.
4. *Vula i Balolo lailai* = October; when the balolo (*Palolo viridis*, Gray), a remarkable Annelidan animal, first makes its appearance in small numbers.
5. *Vula i Balolo levu* = November; when the balolo (*Palolo viridis*, Gray) is seen in great numbers; the 25th of November generally is the day when most of these animals are caught.
6. *Vula i nuqa lailai* = December; a fish called "nuqa" comes in in isolated numbers.
7. *Vula i nuqa levu* = January; when the nuqa fish arrives in great numbers.

According to Wilkes.

1. *Vulai were were*, weeding month.
2. *Vulai lou lou*, digging ground and planting.
3. *Vulai Kawawaka.*
4. *Bololo vava Konde.*
5. *Bololo lieb.*
6. *Numa lieb*, or *Nuga lailai.*
7. *Vulai songa sou tombe sou*, or *Nuga levu*; reed blossoms.

STAPLE FOOD. 299

8. *Vula ni sevu* = February; when offerings of the first dug yams (ai sevu) are made to the priests.
9. *Vula i Kelikeli* = March; digging up yams and storing them in sheds.

10. *Vula i gasau* = April; reeds (gasau) begin to sprout out afresh.
11. *Vula i doi* = May; the Doi (*Alphitonia zizyphoides*, A. Gray), a tree plentiful in Fiji, flowers.

8. *Vulai songa sou seselieb*, build yam-houses.

9. *Vulai Matua*, or *Endoye doye*; yams ripe. (N.B.—Vulai Endoye doye, probably is meant for Vula i doi; the Doi is a tree (*Alphitonia zizyphoides*, A. Gray), B. Seemann.)

10. *Vulai mbota mbota.*

11. *Vulai kelekele*, or *Vulai mayo mayo*; digging yams.

The yam principally cultivated is the *Dioscorea alata*, Linn., having a square climbing stem without prickles. The natives distinguish a number of varieties, all of which are known by the collective name of "Uvi." Some have large, some small roots, of either a white or more or less purplish tinge; and upon these differences, as well as their shape and time of maturity, the distinctions are founded.* At Navua, in Viti Levu, Chief Kuruduadua showed us a lot of yams six feet long and nine inches in diameter, perfectly mealy, and every part good eating; and specimens, eight feet long, and weighing one hundred pounds, are by no means rare in the group. Skilful growers maintain that in order to pro-

* These varieties are called Dannini, Keu, Kasokaso, or Kasoni, Voli, Sedre, Lokaloka, Moala, Uvi ni Gau, Lava, Namula, Rausi, Balebale, etc.

duce large and abundant roots the settings ought to be put into hard and unprepared soil. According to their notion the yam ought to meet with resistance ere it will put forth its whole strength, or, as they sometimes express themselves, it must get angry before it will exert itself. I even heard of a bet won by a woman who pursued this simple plan, and who fully made good her word, that she would produce a root large enough to feed twenty people; whilst the man who bet with her could only raise one that would not have fed one-third of that number, though he took great pains to pulverize and prepare the soil for the reception of the setting. The general signal for planting is the flowering of the Drala (*Erythrina Indica*, Linn.). As soon as its blossoms begin to appear, which happens about July and the beginning of August, all hands busy themselves about it. The land having already been cleared during the previous months, hillocks, about two feet high and four or five feet apart, are thrown up; these hillocks are known by the name of "Buke," whence the highest mountain in Kadavu, for the first time ascended on the 6th of September, 1860, by Mr. Pritchard and myself, and resembling them in shape, takes its name of Buke Levu, or large yam-hillock. There are no spades or any other iron tool for digging; all is done with staves made of mangrove-wood, and the bare hands. Pieces of old yams are set on the top of these hillocks, and within a short space of time they begin to sprout out. In less than a month they require reeds for climbing, after which little else is needed than keeping the plantations free from weeds. About February the first yams begin to ripen, and in the heathen districts offerings of them are made to the priests. In March and April the principal crop comes in, and is stored in sheds thatched with cocoa-nut leaves. As the season advances the contents of these sheds require at least a monthly overhauling; the roots exhibiting any kind of decay have to be removed to prevent their contaminating the healthy ones. Yams are eaten baked, boiled, or steamed, and the natives can consume great quantities of them. Whole cargoes have occasionally been taken with profit to New South Wales and New Zealand, and whaling and trading vessels never touch at the group without laying in a good supply.

There is another esculent root, the Kawai (*Dioscorea aculeata*, Linn.), also planted on artificial hillocks, though not so high as those of the yam. The stem of this creeper is round, and full of prickles, but it is not accommodated with reeds as that of the last-mentioned species. It ripens about June; on the 27th of that month all the leaves were dead. According to the natives it never flowers nor fruits, and I looked in vain over many a field in hopes of being able to disprove the statement. It is propagated by planting the small tubers or roots, which, like the old ones, are oblong, of a brownish colour outside, and a pure white within. When cooked, the skin peels off like the bark of the birch-tree, as Wilkes expresses it. The root is very farinaceous, and when well cooked looks like a fine mealy potato, though of superior whiteness. The taste recalls to mind that of the Aracacha of South America; there is a slight degree of sweetness about it which is very agreeable to the palate. Altogether the Kawai may be pronounced one of the finest esculent roots in the world, and I strongly recommend its cultivation in those parts of the tropics still deprived of it.

Several species of wild yam, such as the Tikau, Tivoli, and Kaile, trail in graceful festoons over shrubs and trees of nearly every wood. The Tivoli (*Dioscorea nummularia*, Lam.) has a prickly stem like that of the cultivated Kawai, and climbs very high; its roots are long, cylindrical, and as thick as a man's arm. When engaged in the forest the natives will often dig up these roots with a stick, roast, and eat them on the spot, when they taste extremely palatable. The Kaile (*Helmia bulbifera*, Kth.) somewhat resembles the Tivoli in look, and is often found entwined with it, but its stems and branches are round and unarmed, and its roots, being acrid, require to be soaked in water previous to boiling. The dish prepared from them has the appearance of mashed potatoes, and is made so thin that it can only be eaten with spoons, which are either furnished by the leathery leaves of the spoon-tree or Tatakia (*Acacia laurifolia*, Willd.), or any other substantial leaf that happens to be at hand.

The Taro, or, as the Fijian language has it, the Dalo (*Colocasia antiquorum*, var. *esculenta*, Schott), is grown on irrigated or on dry ground, perhaps more on the latter than on the former. The water is never allowed to become stagnant, but always kept in gentle motion. When planted on dry ground, generally on land just cleared, a tree or two with thick crowns are left standing in every field, which, as the natives justly conclude, attracts the moisture, and favours the growth of the crop. A considerable number of varieties are known,* some better adapted for puddings, some for bread (madrai), or simply for boiling or baking. The outer marks of distinction chiefly rest upon the different tinge observable in the leaf, stalks, and ribs. of the leaves— white, yellowish, purple. When the crop is gathered in, the tops of the tubers are cut off, and at once replanted. The young leaves may be eaten like spinach; but, like the root, they require to be well cooked in order to destroy the acridity peculiar to Aroideous plants. The Fijians prefer eating the cooked Taro when cold— a taste which few Europeans share with them; on the contrary, the latter relish them quite hot, and, if possible, roasted.

Besides the Taro, which is occasionally seen wild on the banks of rivers, there are three other indigenous Aroideous plants, the corms of which are used as articles of food: the Via mila, the Via kana, and the Daiga. The Via mila (*Alocasia Indica*, Schott), always growing in swamps, is a gigantic species, often twelve feet high; the trunk or corm of which—the edible part—is, when fully developed, as large as a man's leg: a single leaf weighing three and a half pounds. The petiole was found to be four feet long, and ten inches in circumference at the base; the blade of the leaf three feet two inches long, two feet six inches broad, and thirteen feet six inches in circumference! The plant emits a nauseous smell, amply warning, as well as the various popu-

---

* The different kinds of Dalo (Taro) are, Basaga, Bega, Dalo ni Vanua, Karakarawa, Keri, Kurilagi, Mumu, Quiawa, Sikaviloa, Sisiwa, Soki, Toakula, etc.

lar names it bears, against any incautious contact with it. Besides the name of Via mila, which signifies "acrid Via," we have that of Via gaga, or poisonous Via. What may be the meaning of Via sori, and Dranu, occasionally applied to it, I have not been able to find out. In order to remove the acrid properties, the trunk is baked, or first grated, and then treated as madrai (bread) in the manner to be explained below; yet, notwithstanding all precautions, the natives are frequently ill from eating it. The Via kau, or Via kana (*Cyrtosperma edulis*, Schott), is in every respect a similar species, also growing in swamps, not only wild, but frequently cultivated like Taro. It requires fewer preparations to render its root fit for food than that of the Via mila, and its flavour is considerd more agreeable.

The Daiga (*Amorphophallus sp.*) differs from the three preceding Aroideous plants both in habit and mode of growth. It is always found on dry ground, and appears in the spring of the year, together with arrowroot, turmeric, and ginger. Its foliage consists of a single leaf, which rises from a roundish tuber to the height of from two to four feet, having a petiole full of soft prickles, and a blade spreading out somewhat like an umbrella, and divided into numerous, deeply cut segments. The flower, or rather the spathe, is of a dull colour, not put forth until the leaf is beginning to die off, and emits an offensive carrion-like odour. In the cosmogony of the Samoans, the office of having, by means of its singular foliage, lifted up the heavens when they emerged from chaos, is assigned to this plant; and the Fijians recommend it as a safe place of refuge when the end of the world approaches, the Daiga being in their opinion a "Vasu" to heaven (Vasu kilagi). A Vasu, it should be added in explanation, is, according to widely-spread Polynesian custom, a nephew who holds the movable property of his mother's brothers at his almost absolute disposal, having the power to do whatever he pleases with it. Some Vasus even venture so far as to dispose of the very lands belonging to their maternal uncles. There are Vasus to every family, town, and kingdom. A Vasu to heaven is the climax of the whole system, cleverly employed in the charming Fijian story of the Princess Vilivilitabua. The root of the Daiga is acrid, but after being freed from that property, esteemed on account of its nutritious qualities. Being thought to assist fermentation, some of it is mixed with the leaven of bread; for the Fijians, though not growing any grain, or importing flour, prepare what they call "Madrai," or bread, from the fruits of the Ivi (*Inocarpus edulis*, Forst.), Kavika (*Eugenia Malaccensis*, Linn.), Banana, Plantain, Breadfruit, Dogo kana or mangrove, and the roots of the Taro (*Colocasia antiquorum*, Schott, var. *esculenta*, Schott), Kawai (*Dioscorea aculeata*, Linn.), Via mila (*Alocasia Indica*, Schott), Via kana, and the Daiga. A hole, having the shape of an inverted cone, is dug in the ground, and having been lined with leaves, the different materials are put in, covered with leaves, earth, and stones, to undergo fermentation, and become fused into a homogeneous mass. Two or three, ay, even nine months are allowed for that process. When taken out, the dough emits a sour foetid smell. It is then either baked on hot stones, or steamed in large earthenware pots; but the taste is such that few foreigners acquire a partiality for it, and the natives themselves infinitely prefer our bread and biscuit to their own madrai. Yet it is most fortunate that in a country where numerous kinds of fruits and edible roots, however abundant at certain seasons, are subject to such rapid decay, the natives are acquainted with a simple process, by means of which they are able to store up their provisions, and thus effectually guard against extreme want in a land of plenty.

A few other esculent roots remain still to be mentioned. Potatoes (*Solanum tuberosum*, Linn.) grown in Mr. Moore's garden at Mataisuva I found tolerably good. An attempt made by Mr. Carey, at Wairiki, to raise radishes, did not succeed. Shalots are cultivated to a considerable extent by the natives. Turnips have been produced from imported seeds. The sweet potato (*Batatas edulis*, Chois.) is an introduction probably from New Zealand, as the Fijian name (Kumara) proves identical with that given by the Maoris. It succeeds well, but does not seem to be much valued. The Masawe or Vasili Toga (*Dracæna* sp.), is a shrub with obovate leaves, cultivated, and perhaps, judging from the name Vasili Toga (= Tonga) it bears in some parts of the group, an importation from the Tongan islands. Its root is large, weighs from 10 to 14 lbs., and when baked, resembles in taste and degree of sweetness, as near as possible that of stick-liquorice. The Fijians chew it, or use it for sweetening puddings. They were ignorant of the art of extracting an intoxicating liquor from it, known to the Hawaiians. There is another species of *Dracæna* closely resembling the Masawe, and employed for making fences. It grows wild in the woods, and bears in Viti Levu the name of Vasili Kau. It is as much as fourteen feet high, and has lanceolate leaves, which, in common with those of its allies, are good fodder for sheep, goats, rabbits, and cattle. Its root is small, and thought unfit for food. The Vasili damudamu or Ti Kula (*Dracœna ferrea*, Linn.), has leaves similar in shape, but the idea of its being possibly a variety of the preceding is precluded by the fact of its having large and edible roots.

Amongst the esculent roots growing wild, and eagerly sought for just before the regular crops come in, or in times of scarcity caused by intertribal wars during the planting season, or by unfavourable weather, may be named the Yaka or Wa yaka (*Pachyrhizus angulatus*, Rich.), a Papilionaceous creeper, with trifoliated leaves and whitish flowers tinged with purple. In September and October its tubers send forth new shoots, which grow with rapidity and yield a tough fibre, invaluable for fishing-nets. The plant delights in open exposed places and a rich vegetable soil, where the roots, which generally assume a horizontal direction, often attain from six to eight feet in length and the thickness of a man's thigh. When cooked, they have a dirty white colour, and a slightly starchy but otherwise insipid flavour, much inferior, I thought, to that of wild yams. However, Mr. Charles Moore, of Sydney, ate them in New Caledonia, and is inclined to pronounce more favourably upon their taste. Living plants were brought by him to the Sydney botanic garden, where they are now growing with native vigour in the open air.

Kitchen vegetables are supplied by a number of wild and cultivated plants. The natives boil the leaves of several ferns, among them those of the *Litobrochia sinuata*, Brack., and in times of scarcity those of the Balabala (*Alsophila excelsa*, R. Br.); those of the Ota (*Angiopteris evecta*, Hoffm.), a species with gigantic foliage, are peculiarly tender, and their taste not unlike that of spinach. The common brake (*Pteris aquilina*, Linn., var. *esculenta*, Hook. fil.), though plentiful, does not seem to be used as it is by the Polynesian tribes of New Zealand. The leaves of the Boro ni yaloka in gata (*i. e.* serpent's-egg boro), our *Solanum oleraceum*, a spiny kind of herbaceous nightshade, serve as " greens " to both the natives and foreigners. The young shoots of the Vaulo of Viti Levu (*Flagellaria indica*, Linn.), known also, if I am not misinformed, by the names of Tui, Vico, Turuka, and Malava in different districts, after having been boiled, are eaten with taro and yams, but only by Fijians. Two kinds of purslane, termed "Taukuku ni vuaka" in Taviuni (*Portulaca oleracea*, Linn., et *Portulaca quadrifida*, Linn.), are common weeds which, during my stay at Somosomo, were frequently brought to table. The natives sometimes grow whole fields of the Bete or Vauvau ni Viti (*Hibiscus* [*Abelmoschus*] *Manihot*, Linn.), an erect shrub, attaining six or eight feet in height, bearing yellow flowers and lobed leaves, which, especially if not quite developed, are tender eating, relished even by Europeans. The Boro dina (*Solanum anthropophagorum*, Seem.), a straggling shrub with glabrous leaves and scarlet or yellow berries, possessing a faint aromatic smell, and resembling tomatos in shape, has also edible leaves and fruit. The Tomato (*Lycopersicum esculentum*, Mill.), as a tropical production, is quite at home. The Cajan, pigeon-pea or pea-tree (*Cajanus Indicus*, Spr.), introduced from the United States, is cultivated successfully. Its seeds, when young, make a tolerably good substitute for green peas, acceptable in a country well supplied with both wild and tame ducks. The Dralawa (*Lablab vulgaris*, Savi) grows in great abundance about Somosomo, covering whole acres of ground, and if not indigenous, has at all events become perfectly naturalized in that and various other parts of the group. It seems to bear without interruption throughout the year, its numerous white flowers being always seen wherever the plant has established itself. The beans are extremely tender, and after having been boiled in water and salt, oil and vinegar will convert them into an excellent salad. A species of *Dolichos* was noticed at Levuka, in the garden of a French settler. Indian corn (*Zea Mays*, Linn.), termed "Sila ni papalagi" (*i. e.* foreign Sila), from its resemblance in habit and foliage to the indigenous Sila (*Coix Lachryma*, L.)—our Job's tears—has as yet been raised sparingly, as the Fijians and Polynesians in general have never been accustomed to grow any grain whatever, and most of the white settlers are English, ignorant of the innumerable uses to which the Americans apply it. There is only one rather inferior kind, a small yellow-grained one, and the introduction of the larger and better sorts would be a boon easily conferred upon the islands. The settlers sadly complain that their domestic fowls (toa) become wild, and instead of keeping near the houses take up their abode in the woods, where they have to be shot when required. If more Indian corn were grown, and these birds fed with it regularly, they would probably preserve their domestic habits as thoroughly as they do in other countries. Hitherto no attempts have been made to cultivate our so-called European vegetables in the cooler regions of the mountains, where they would doubtless thrive well. None have been raised except on the coast, where the heat of the tropics is not moderated by elevation, and the unchecked influence of the sea air proves destructive to many kinds. Yet even here cabbages and turnips have been produced from foreign seeds, and parsley may be looked upon as a permanent acquisition.

Bananas and plantains—understanding by the former those *Musas* the fruit of which may be eaten raw, by the latter those which have to undergo some process of cooking before eating—are known by the collective name of "Vudi." There are about eighteen different species, or rather say kinds (for the boundary between species and variety has never been determined with accuracy in this genus)—all of which bear distinctive names.* With the exception of one, the Soaqa (*Musa Troglodytarum*, Linn.), none are found wild, and this wild one even is occasionally met with in plantations. It grows spontaneously in the depth of the forests,

---

* The following are the different kinds known to me:—Vudi ni papalagi (*Musa Chinensis*, Sweet [*Cavendishi*, Paxt.]), Soaqa (*Musa Troglodytarum*, Linn.), Balawa ni Rakiraki, Bati, Dreli, Buli, Droledrole, Gonegone, Leve ni Ika, Mudramudra, Soqo, Tumoutala, Ura, Vudi dina, Vudi Kalakala, Vudi ni Toga, Waiwai Leka, Waiwai Salusalu, Waiwai Vula, and Sei.

often in ravines, and is distinguished from all congeners by its bunches, instead of hanging down, being perfectly upright, and presenting a dense collection of orange-coloured fruits. The Polynesians, always ready to account for any deviation from a normal type, have not failed to exercise their ingenuity here. The Samoans assure us that once upon a time all the bananas and plantains had a great fight, in which the Soaqa (their Fae) came off victorious, and proudly raised its head erect; whilst the vanquished became so humiliated by the defeat sustained, that they were never able to hold up their heads again. An important addition to their stock the Fijians received in the Vudi ni papalagi (*i. e.* foreign banana), our *Musa Chinensis*, which the late John Williams, better known as the Martyr of Eromanga, brought from the Duke of Devonshire's seat at Chatsworth to the Samoan or Navigator Islands, whence again, in 1848, the Rev. George Pritchard carried it to the Tongan or Friendly Islands, as well as to the Fijis. Its introduction has put an effectual stop to those famines which previously were experienced in some of these islands. Never attaining any greater height than six feet, and being of robust growth, it is little affected by the violent winds which cause such damage amongst plantations of the taller kinds, and this advantage, coupled with its abundant yield and fine flavour, have induced the natives to propagate it to such an extent that, notwithstanding its comparatively recent introduction, the Vudi ni papalagi numbers amongst the most common bananas of the country. The fruit of the different *Musas* is variously prepared by the native cooks. Bananas split in half, and filled with grated cocoa-nut and sugar-cane, make a favourite pudding (vakalolo), which, on account of its goodness and rich sauce of cocoa-nut milk, has found its way even into the kitchen of the white settlers. Wilkes has already mentioned that the natives, instead of hanging up the fruit until it becomes mellow, bury it (occasionally, it should be added) in the ground, which causes it to appear black on the outside, and impairs the flavour. The fresh leaves are used as substitutes for plates and dishes in serving food or for making temporary clothing, the dry instead of paper for cigarettos (sulu ka). In place of the finger-glasses handed round at our tables after dinner, Fijians of rank are supplied with portions of the leafstalk of the plantain,—not a superfluous luxury when forks are dispensed with except at cannibal feasts.

The breadfruit is seen in regular forests, and in a great number of varieties, which a new-comer has some difficulty in distinguishing until he has learnt to observe that in the shape of the leaves—which are either entire, pinnatisect, or bi-pinnatisect—their size and their either bullate or even surface, the shape and size of the fruits, the time of its maturity, the absence or presence, as well as the length of the prickles on its outside, and the abortion of its ovules or their development into seeds, offer good marks of distinction. The general Fijian name for the breadfruit is "Uto," signifying "the heart," from the resemblance of the form of the fruit to that organ, whilst the varieties are distinguished by additional names. Those less frequently cultivated are, however, not known by the same names throughout the group, but bear different ones in different districts. Hence, the exact number of varieties cannot be accurately determined, until there shall be a botanic garden in Fiji, where a complete collection of breadfruits is cultivated. I have identified several names of the most prominent varieties, but hesitate about others, as I could only take the leaves with me from place to place, and often did not see the fruit, or had to carry it in my mind's eye. The principal breadfruit season is in March and April, but some kinds ripen considerably later or earlier, whilst in some districts the season itself is altogether later. It may thus be said, speaking generally, that there is ripe breadfruit, more or less abundant, throughout the year, in either one part or the other. The fruit is made into puddings or simply boiled or baked. Quantities of it are preserved underground, to make madrai or native bread. Some kinds are best adapted for puddings, some for bread, or culinary purposes of a still more simple description. Besides the fruit, the wood of the breadfruit tree is useful, but that of some kinds better adapted for canoes and buildings than others. The bark is not beaten into cloth, as in other parts of Polynesia; but the gum (drega), issuing from cuts made into the stem, is used for paying the seams of canoes.

The two most common sorts are Uto dina and Uto buco. The Uto dina, or true breadfruit, has pinnatisect leaves, the surface of which is even, and destitute of that bullate appearance which imparts to the Koqo and other varieties an almost sickly look; the fruit, bearing abortive ovules, is nearly round, smooth on the outside, and supported on stalks four to five inches long, which from the very first are bent downwards. It is this variety which most botanists consider as the type of the species, and the adjective " dina," true or genuine, given by the Fijians, may be cited as a proof of the correctness of this surmise. But if we have to look for an original stock from which all other sorts have sprung, we ought not to select one which, like the Uto dina, has invariably abortive ovules, and can therefore not produce seeds from which new varieties can be raised. The Uto sore, Uto vaka sorena, or Uto maliva, as it is termed in different districts, has not that deficiency, but does yield ripe seeds in abundance, and has, therefore, greater claims to be regarded as the type from which all the other varieties may have been raised. The name of Uto dina (true or genuine breadfruit) may perhaps have been applied on account of its goodness, which, I believe, is undisputed. The Uto buco also has pinnatisect leaves with an even surface as opposed to the bullate one of other sorts, and an obovate obtuse fruit of larger size than that of the Uto dina, and quite free from any prickles on the outside when fully ripe.*

---

* In order to obtain a clearer insight into the varieties, it will be best to subjoin a synopsis of all the breadfruits cultivated in Fiji:—

I. LEAVES ENTIRE OR QUITE ENTIRE.

1. *Uto lolo* bears this name in the Straits of Somosomo, and is called *Uto cokocoko* in the Rewa district; perhaps, also, identical with the *Uto dogodogo* and *Uto draucoko* mentioned in the Fijian dictionary. It looks different from all others, the leaves, especially when the tree gets older, being quite entire; in young plants they are sometimes obscurely lobed. The fruit is without seeds.

II. LEAVES PINNATISECT.

2. *Uto dina.*—Known by that name, and that name only, throughout Fiji. Leaves with an even surface; fruit without seeds, nearly spherical,

Other edible fruits, some of delicious flavour, are met with throughout the group, either perfectly wild or in a state of cultivation. Most of them have been in Fiji from time immemorial, and only a few, such as the pine-

with a smooth surface, and supported on stalks, four or five inches long, nodding from the first.

3. *Uto buco.*—Known by that name throughout the group. Leaves with an even surface. Fruit ovate obtuse, larger than that of most sorts, destitute of seeds, and with a smooth surface when ripe.

4. *Uto koqo.*—Known by this name throughout the group, but in some dialects called *Oqo* and *Qoqo*. Leaves bullate; fruit without seeds, and as large as that of *Uto dina*, smooth on surface.

5. *Uto votovoto.*—Known under this name throughout the group. Leaves with an even surface; fruit oblong without seeds, and covered with prickles three-quarters of an inch long.

6. *Uto varaqa.*—Known by this name in Rewa and Bau; *Uto varaka* in some dialects. Leaves larger than those of any other kind; fruit roundish, of middle size, without seeds, and with a rough surface.

7. *Uto bokasi.*—Known by that name in Rewa and Ovalau. Leaves with even surface; fruit obovate, with a smooth surface, without seeds, erect when young, nodding when ripe, and arriving at maturity early in the season.

8. *Uto sore.*—Known by that name in Rewa, by that of *Uto vaka sorena* in Ovalau, *Uto asalea* in the Straits of Somosomo, and *Uto maliva* at Nukubalaon. *Uto sasaloa* may also prove a synonym. "*Sore*" or "*Sorena*," signifies a seed; hence *Uto sore*, or *Uto vaka sorena*, is the seed-bearing breadfruit; the only kind in which the ovules develope into seeds, rendering it probable that this kind is the parent of all the others. Leaves with even surface.

9. *Uto rokouta.*—Known by that name at Namara, near Bau. Leaves bullate, giving the tree a sickly look.

10. *Uto balekana.*—Known by that name in the Straits of Somosomo and at Ovalau. Leaves with even surface; fruit small but of superior quality, according to the natives.

11. *Uto qio.*—Known by that name in Ovalau. Fruit almost as large as that of *Uto buco*. "Qio" is the name for shark, and was probably given to this fruit from the surface its resembling in roughness that of the fish.

12. *Uto vonu.*—Known at Somosomo. Leaves . . . ; fruit largish.

III. Leaves bi-pinnatifid.

13. *Uto kalasai.*—Known by that name in Rewa, and by that of *Uto*

apple, the papaw, the custard-apple, and the Chinese banana, have been introduced of late years. The most prominent place among the native fruits undoubtedly belongs to the Wi (*Evia dulcis*, Comm., = *Spondias dulcis*, Forst.). The tree appears to be self-sown, and is met with in abundance about towns and villages. It is often sixty feet high; the bark is smooth and whitish, the leaves pinnate, glabrous, and of a dark green, forming a fine contrast with the yellow oval-shaped fruits with which the tree is heavily laden. The fruit has a fine apple-like smell, and a most agreeable acid flavour, rendering it highly suitable for pies; indeed, the Wi is the only Fijian fruit which recommends itself for that purpose. At Rewa I weighed and measured several highly developed ones, and found the largest to be exactly one foot in circumference, and one pound two ounces in weight. The natives are as fond of Wis as the white settlers, and quite content to make their dinner of Taro and Wis. The Dawa (*Nephelium pinnatum*, Chamb., = *Pometia pinnata*, Forst.) is more plentiful than the Wi; entire forests of it are frequently

sawesawe in the Straits of Somosomo. The leaves, especially when the plant is young, are distinctly bi-pinnatifid, in which respect this kind differs from all others; fruit, according to natives, rather oblong and covered with prickles.

Of the following I know nothing, save the names, partly taken from Hazelwood's Dictionary, partly from a list of breadfruits known at Ovalau, and kindly communicated by Mr. Binner, of Levuka. Most of them will doubtless prove synonyms of those enumerated above :—Draucoko (= Cococoko?), Bucotabua, Utoga (= Koqo), Waisea, Utoloa (= Uto lolo?), Matavesi, Dregadrega (N.B. Drega is the name of the gum issuing from the stem), "Buco uvi." The "Bucudo" of Wilkes's Narrative, and is probably identical with Buco, though he mentions the latter name spelt "Umbuda;" but what can be meant by his "Botta-bot"?

encountered, and there appear to be several varieties. It is sixty feet high, and shares with most Fijian fruit-trees the peculiarity of yielding a useful timber. The leaves are pinnate, the leaflets serrate, and when first opening, display a brilliant red tinge, which at a distance looks as if the tree were in bloom. The flowers, arranged in terminal panicles, are whitish and of diminutive size. The fruit, ripening in January and February, has rather a glutinous honey-like taste, and attains about the size of a pomegranate. The Fijians deem the Dawa peculiar to their islands. It certainly does not occur to the eastward in a wild state, as the Tonguese are said to have obtained it from Fiji; but it seems to be quite common in all the groups lying westwards, the New Hebrides, New Caledonia, and others. A native of Were assured me it was plentiful in his island, and Dr. Bennett, of Sydney, found it cultivated under the name of "Thav," at Rotuma, a little island to the north of Fiji, as recorded in his 'Gatherings of a Naturalist.' I succeeded in carrying living plants to the botanic garden at Sydney, where they were left in charge of Mr. Moore, and whence they may perhaps find their way to the new colony of Queensland, and prove acceptable additions to the fruits of that country.

The Kavika or Malay-apple (*Eugenia Malaccensis*, Linn.) abounds in all the forests. As in the Hawaiian and other Polynesian islands, there are two varieties; the purple (Kavika damudamu) and the white (Kavika vulavula). When the tree, which attains about forty feet in height, is in flower, the ground underneath is densely covered with petals and stamens, looking, especially if the two varieties grow together, like a fine Turkey carpet. I have often seen the natives gathering handfuls of them to strew on their heads. In their idea, there is scarcely a finer tree than the Kavika; and when in their fairy tales the imagination runs riot, and describes all that is lovely and beautiful, the Kavika is rarely omitted. The Hawaiians, as I have stated elsewhere ('Narrative of the Voyage of H.M.S. Herald,' vol. ii. p. 83), thought this tree worthy of supplying materials for their idols; and thus, like the Fijians, recorded their veneration for it. A botanist, himself more than half a tree-worshipper, can fully sympathize with them. The fine oblong leaves, their smooth shining surface, the deep purple or pure white flowers, and afterwards the large quince-shaped fruits, with their apple-like smell and delicate flavour, are well calculated to justify much of the praise Polynesians bestow upon the tree. The Ivi, or Tahitian chestnut, as it has been called by voyagers (*Inocarpus edulis*, Forst.), is one of the common trees, and when fully grown has a most venerable aspect. I still see in my mind's eye a fine group on the banks of a rivulet between Wairiki and Somosomo, diffusing a dense shade. Sixty, often eighty feet high, the Ivi bears a thick crown of oblong leathery leaves, small white flowers emitting a delicious perfume, and kidney-shaped fruits, which contain a kernel resembling chestnuts in taste. The kernel is either baked or boiled, and eaten without further preparation, or grated on the mushroom coral (*Fungia*), and made into puddings or bread (madrai). The stem is most singular. When young, it is fluted like a Grecian column; when old, it has regular buttresses of projecting wood. Ferns, orchids, and wax-flowers frequently take up their abode on the soft spongy bark. The roots of old trees appear above the ground somewhat like those of the bald cypress of North America (*Taxodium distichum*, Rich.). Thousands of seedlings are continually springing up around the old plants, and nothing, save the dense shade of their parents, and the close proximity in which they grow to each other, exercise a check upon their engrossing all the adjacent ground. If the fruit of the Ivi is compared with the chestnut, that of the Tavola (*Terminalia Catappa*, Linn.) may be likened to the almond, both in shape and whiteness, though not in taste—the Tavola having none of the flavour imparted by the presence of the essential oil of almonds; hence the name of "Fijian almonds," given by the white settlers, must be received *cum grano salis*. The natives are extremely fond of the Tavola as a tree, and frequently plant it around their houses and public buildings. The branches, arranged in whorls, somewhat like those of pines, though perhaps not quite so regular, have a horizontal tendency, upon which the natives improve by placing weights upon them. The large obovate leaves are deciduous, and before falling off assume a variety of tints, —brown, red, yellow, and scarlet, such as one is wont to behold in a North American forest before the approach of winter. The flowers are white and small; the wood hard and applicable to a variety of purposes. A close ally of the Tavola is the Tivi (*Terminalia Moluccana*, Lam.), a timber-tree, always growing on the sea-beach, and bearing seeds sometimes eaten by children.

Like its congener, it changes the colour of its foliage, but the tints are neither so rich as those of the former, nor is the general habit of the tree so striking. The Oleti or papaw-tree (*Carica Papaya*, Linn.), has been introduced in the early part of this century, and has spread with such rapidity that there is hardly a part of the group in which it is not to be found; neither the natives nor the white settlers (who sometimes will persist in calling it mamey-apple, a very different fruit) seem to care much for it. Only a few seem to be aware that saponaceous properties reside in the leaves, which, in the absence of soap, may be, and in tropical America are, turned to advantage; that both the leaves and the fruit act in an hitherto unexplained way upon the animal fibre, and make the toughest meat tender, if either boiled with portions of them, or even wrapped up in the leaves;* that the fruit is very good eating, either raw or boiled; and that the seeds, distinguished by a mustard-like pungency, are an efficacious vermifuge for children. The Guayava (*Psidium Guayava*, Raddi) is another fruit of recent introduction, that has spread rapidly over the country, and is eaten either raw or made into sweetmeats. One of the custard-apples (*Anona squamosa*, Linn.) has not made such progress. I met a few trees on the Somosomo estate of Captain Wilson and M. Joubert, of Sydney, and a few at Levuka, in the garden of a French settler. The loquat (*Eriobotrya Japonica*, Lindl.) is of recent introduction, and seems to

---

* I heard a wag telling a story of an old bachelor, who, sitting for a while under this tree with a young lady, became so te..der-hearted as to pop the question.

promise fair results. A number of healthy-looking plants grow in the garden of Mr. Binner, of Levuka, where the grape-vine (*Vitis vinifera*, Linn.) and various other useful plants, recently brought to the islands, are also to be met with. The different species of *Citrus*, shaddocks, oranges, lemons, and Seville oranges, are known collectively as " Moli," and distinguished from each other by additional names. The shaddock (*Citrus Decumana*, Linn.) or Moli kana (*i. e.* edible Moli), is extremely common, and thickly lines the banks of rivers; as, for instance, that of Namosi in Viti Levu, where, during our stay in August, 1860, the stillness of night was frequently broken by the heavy splash of the falling fruits. There is a variety with white, another with pink, flesh, both of which are much liked by the natives. The Moli kurukuru (*Citrus vulgaris*, Risso) is equally common, but the Fijians do not make use of it as an article of diet. The Moli kara, or lemon (*Citrus medica*, Risso), has been brought from Tahiti, about 1823, by Mr. Vanderford, and is almost exclusively confined to the neighbourhood of present or former habitations of white settlers. The Moli ni Tahaiti (*Citrus Aurantium*, Linn.) is the common variety of the orange, also derived, as the native name indicates, from the Society Islands, whence it was introduced simultaneously with the lemon, by Mr. Vanderford. Like the other species of *Citrus* just mentioned, it succeeds well, and small cargoes of it have occasionally been shipped to New Zealand. The pomegranate (*Punica Granatum*, Linn.) is a recent acquisition. The pine-apple (*Ananassa sativa*, Lindl.), vernacularly termed " Balawa ni papalaqi," or foreign screw-pine, thrives well, especially near the sea. There is, besides the common variety, a proliferous one, having many different sprouts emerging from the top of the fruit. The water-melon (*Citrullus vulgaris*, Schrad.) is as plentiful as the Vaqo, or bottle-gourds (*Lagenaria vulgaris*, Ser.), which supply the natives with vessels for their oil. Melons (*Cucumis melo*, Linn.), cucumbers (*Cucumis sativa*, Linn.), and pumpkins or squashes (*Cucurbita Pepo*, Linn.), have also found their way to the islands, and, in common with indigenous Cucurbitaceous plants, are collectively called " Timo."

There is besides a number of fruits eaten and even esteemed by the natives, but most insipid to a European palate. Foremost amongst them stands the Tarawau (*Dracontomelon sylvestre*, Blume), which is also connected with native superstitions. The Tarawau does not seem to be regarded as a sacred tree in the light of those mentioned above (p. 87), it not being worshipped; but it is held to be the business of the dead to plant it, and believed to grow not only in this world, but also in Naicobocobo, the Fijian nether-world, or perhaps, more correctly, the general starting-place for it. Hence arose the expression, " Sa la'ki tei tarawau ki Naicobocobo," literally, " He has gone to plant Tarawaus at Naicobocobo ;" *i. e.* he is dead. It is difficult to guess why these trees should have been deemed worthy of such distinction; they grow to the height of sixty feet, have flattish branches, pinnated leaves, insignificant whitish flowers, and a tough insipid fruit, only palatable to the natives; moreover, they are regarded as the emblem of the truth-speaking man, not having, as so many others, a number of false or sterile flowers. There is also the Loselose (*Ficus* sp.), the Kurá (*Morinda citrifolia*, Linn.), the Balawa (*Pandanus odoratissimus*, Linn.), the Wa gadrogadro (*Rubus tiliaceus*, Smith) having a fruit resembling the raspberry in appearance, and being occasionally used by white settlers for pies; the Bokoi (*Eugenia Richii*, A. Gray), with a fruit somewhat like the Kavika (*Eugenia Malaccensis*, Linn.), but inferior in flavour; the Sea (*Eugenia* sp.), still more insipid, if possible, than the last-mentioned; and the Nawanawa (*Cordia subcordata*, Lam.), producing an edible kernel—a tree twenty-feet high, often mistaken for the *Cordia Sebestena*, Linn., of the West Indies, which it closely resembles in habit, but its orange flowers are neither so brilliant nor so numerous. Nor must the Vutu kana, or Vutu kata (*Barringtonia excelsa*, Blume), be forgotten. It is a tree forty feet high, cultivated about Bau, and distinguished from the other *Barringtonias* of Fiji by its egg-shaped, not angular, fruit, eaten either raw or cooked. Another species of *Barringtonia*, closely resembling the foregoing, is the Vutu dina, which is also edible; but whilst the fruit of the Vutu kana (*i. e.* edible Vutu) has a soft outside, that of the Vutu dina has a hard one, requiring the application of a knife before the edible portion can be got at. Finally must be mentioned the Somisomi, Sosomi, Tomitomi or Tumitumi, as the different dialets have it,—the *Ximenia elliptica*, Forst. It is a sea-side shrub, having simple leaves, and a spherical fruit containing a kernel like a cherry-stone, and emitting, especially when green, a most powerful smell of essential oil of almonds. The fruit when quite ripe is orange-coloured, and has, though a tart, not a disagreeable flavour. The natives share a partiality for it with the wild pigeons, which flock to it in numbers. The wood of the shrub is very hard, and used for making those peculiar pillows (Kali) of the country, which the Fijians doubtless invented in order to prevent the derangement of their enormously large heads of hair, curled and dressed as they are with infinite care.

The national beverage is the Kava, or, as the Fijians term it, "Yaqona," prepared from the root of the *Piper methysticum*, Forst., or, as its modern name is, *Macropiper methysticum*, Miq., a species of pepper, of which there are six varieties, distinguished by the height of the entire plant, the length and thickness of the joints, and the more or less purplish or greenish tinge of the stem and leaves. The best Yaqona, for this name applies to the plant as well as to the beverage extracted from it, grows from 500 to 1000 feet above the sea-level, and in the islands of Kadavu and Viti Levu. The plant is cultivated throughout the group in small patches, and isolated specimens are frequently noticed around public and private houses. It is propagated by offshoots. The highest shrubs are about six feet, and their stem from an inch to an inch and a half in diameter; the leaves are cordate, and either green or more or less tinged with purple. The root and extreme base of the stem are the parts of which the drink is prepared; they are preferred fresh, but are nearly as good when dry. After the roots have been dug up, they are placed in an airy spot, generally on a stage over the fireplace. In order to prepare the beverage, it is necessary to reduce the roots to minute particles, which, according to regular Polynesian usage, is done by chewing—a task in Fiji devolving upon lads who have sound teeth, and occupy a certain social rank towards the man for whom they perform the office. In other Polynesian islands it is done by young women. When a sufficient quantity has been chewed, the masticated mass is placed in a bowl made of the wood of the Vesi (*Afzelia bijuga*, A. Gray), and having four legs and a piece of rope attached to it, which, when the bowl is brought in, is thrown towards the greatest man present, and guides those who happen to arrive in ignorance of his rank in observing the ceremonies required from them. Some Fijians make it a point to chew as great a quantity as possible in one mouthful; and there is a man of this sort at Verata, famous all over the group, who is able within three hours' time to chew a single mouthful sufficient to intoxicate fifty persons. Fortunately, Kava, unlike distilled spirits, does not render people quarrelsome; and Fijians, on extolling the virtues of their national beverage, often make this observation. On public occasions, or at convivial meetings, when the chewed root is placed in the bowl, and water is poured on, the whole assembly begin to chant appropriate songs, accompanied by the beating of little sticks on a bamboo or log of wood, and this is kept up until the dregs of the root have been strained through the fibres of the Vau (different species of *Paritium*), or in the absence of them, through fern leaves. When the beverage is ready, the chant is discontinued, and the priest or any head man present pronounces a toast or prayer over it, after which the first cup—a cocoa-nut shell—is handed to the person of highest rank in the company. The Kava is taken out of the bowl by means in the strainer, which is dipped into the fluid, and then squeezed. Although both bowl and cup are always carefully dried and cleaned after having been used, a crust invariably forms at the inside, giving them the appearance as if they had been enamelled. This crust, after a lapse of three or four months, is carefully scraped off, and makes the strongest of all Yaqona. The beverage has the look of coffee with plenty of milk in it, and an aromatic slightly pungent taste, which, when once acquired, must, like all acquired tastes, be perfectly irresistible. Drunk in moderation, it has probably no bad effect, and acts upon the system somewhat like betel-nut; but taken in excess, it generates all sorts of skin-diseases, and weakens the eyesight. Nearly all the lower class of whites in the Fiji are Kava drinkers, some regular drunkards; and it is generally accepted as a proof of a man belonging to the more respectable portion of society if he refrains from touching this filthy preparation. Most of these whites prefer it prepared in true Polynesian fashion; only a few have the root rasped on a grater—a process said to impair the flavour considerably. Roots of Yaqona are presented to visitors as tokens of goodwill, and to the temples as offerings. I have also seen the leaves of the plant hung up in the temples, together with the little twigs of the *Waltheria Americana.* As we in Europe, when engaging soldiers or servants, hand a small coin in proof that the bargain has been accepted, so the Fijians, when effecting a bargain or sale give or take a small deposit, which is called the " Yaqona," and either consists of a piece of Kava-root, or any other article that may prove acceptable. Drinking Kava being peculiar to all Polynesian tribes, Thomson ('Story of New Zealand:' London, 1859: vol. i. p. 193) expresses surprise that the Maoris of New Zealand should have forgotten the art of extracting it, "seeing that the plant (*Piper methysticum*, Forst.) grows abundantly in the country." But the *Piper* found wild in New Zealand is not, as Dr. Thomson supposes, the *Piper methysticum*, Forst. (the true Kava plant), but the *Piper excelsum* of the same author (*Macropiper excelsum*, Miq.). Hence it can form no surprise that a genuine Polynesian people should have forgotten the art alluded to during the long lapse of time intervening between their departure from Samoa and their discovery by Europeans. They have, however, preserved the name of "Kava," which they have transferred to their indigenous pepper (Kawa-kawa), and also to a beverage (Kawa) made of the fruits of the *Coriaria myrtifolia*, Linn., by them termed Tupa-Kihi, Tutu, or Puhou. Kawa-kawa, according to Colenso's statement in J. D. Hooker's 'Flora of New Zealand,' signifies "piquant." Thomson attempts to trace Kawa, Kava, or Ava, as the various Polynesian dialects have it, to the Sanscrit "Kasya," which seems to be a general term for intoxicating beverages.*

* The medicinal properties of the Kava-plant have of late claimed some attention. In the French translation of Golding Bird's work on Calculous Affections, Dr. O'Rorke has inserted, amongst others, the following remarks:—
"The Kava-plant is the most powerful sudorific in existence, and its stimulant qualities render it applicable in those cases in which colchicum is prescribed. . . . The intoxication it produces is not like that caused by spirituous liquors, but rather induces a placid tranquillity, accompanied by

Besides their favourite yaqona, the Fijians drink the natural liquor of young cocoa-nuts; but they had absolutely no other beverage save water. They were ignorant of extracting an intoxicating drink from the saccharine roots of their Masawe (*Dracæna terminalis*, Linn.), so much employed for that purpose by the Hawaiians and other Polynesians. They were even strangers to infusions and decoctions made of aromatic leaves, and the so-called Fijian tea, vernacularly termed Matadra, was never used by the natives. The European settlers, who first employed it as a substitute for Chinese tea, by drying and then *boiling* the leaves, brought the custom from the eastern groups of Polynesia. The Matadra (*Missiessya corymbulosa*, Wedd.) is a straggling shrub, belonging to the nettle-tribe, having slender branches, and generally growing as underwood. It attains from six to eight feet in height, has leaves somewhat resembling those of the elm, but white underneath, and minute flowers and fruits arranged in corymbs.

incoherent dreams. Kava is as powerful in its therapeutic action as lignum vitæ or guaiacum, sarsaparilla, etc., and the islanders use it as a specific against the diseases brought over to them by foreign vessels. On the other hand, this drug, used to excess as an intoxicating agent, over excites the skin by its sudorific effects, and eventually even occasions elephantiasis. . . . The chemical constituents, according to Gobley, are as follows:—carbon, 62·03; hydrogen, 6·10; nitrogen, 1·12; oxygen, 30·75. It contains 26 per cent. of cellulose, 49 per cent. of starch, one of methysticine, a crystallizable principle, two of an acrid resin called Kawine, and about 7 per cent. of gum, iron, and magnesia, and a few substances of minor importance." In a paper, which M. Cuzent laid before the Academy of Sciences at Paris, in 1861, the chemical composition of the Kavahine (thus it is spelt in the report at hand), the active crystallizable principle of the Kava, identical, it would seem, with what Gobley terms "Methysticine," is thus given: no nitrogen, 66 per cent. of carbon, 6 of hydrogen, and 28 of oxygen.

Some people have drunk a decoction of its leaves without perceiving it to be different from Chinese tea.

There is another negative fact of singular ethnological importance connected with this subject. Neither the Fijians nor the Polynesians in general were acquainted with the art of extracting toddy from the unexpanded flowers of the cocoa-nut palm. It is only in quite recent times that Europeans have instructed them in it. This, in a great measure, seems to strengthen the position of those who maintain that the Polynesians did not come from the Malayan or any other district of Asia; that they would never have migrated contrary to the direction of the prevailing trade-winds; and that the identity of certain Malayan and Polynesian words, thought to be an overpowering argument in favour of that exodus, cuts both ways, and may be made to prove either that these words came from purely Malayan to Polynesian districts, or from a genuine Polynesian to a Malayan; and exactly the same dilemma is encountered in dealing with the geographical distribution of Polynesian plants and animals. Passionately fond as are the Polynesians of intoxicating drinks, they would never have discontinued making toddy, if they had ever known the way to make it, especially as a tree yielding it, the cocoa-nut palm, is common throughout Polynesia. In order to reconcile this fact with the hypothesis that the Polynesians are of Malayan origin, it might be assumed that they left the cradle of their race before the extraction of toddy from the cocoa-nut tree, or even the cocoa-nut tree itself, was known there. Tradition, historical evidence, and observed facts, all agree in showing the progression of the cocoa-nut tree from west to east. Numerous cocoa-nuts are annually drifted on the eastern shores of New Holland, where they often germinate and grow, until the seedlings are killed by the low temperature of the winter months. Ceylon, now covered with immense forests of cocoa-nut palms, has a distinct tradition that at one time the tree was unknown there, and there is even a statue not far from Galle, recording the event of its becoming known there ; whilst the oldest chronicles of the island, known by the name of the Marawansa, and the historical value of which is now fully admitted, are absolutely silent on everything relating to the cocoa-nut, while they never fail to record every accession to the plantations of other fruit-trees made by the native princes. This seems to prove that the cocoa-nut was not always known, and that it would have much sooner found its way there than it did if it had been indigenous to India Proper; whilst the fact that all other species comprising the genus *Cocos* are strictly confined to the interior of tropical America, and only this one species (*C. nucifera*, Linn.), a sea-side plant, unaffected by drifting on sea-water, is spread over Polynesia and the Old World generally, offers another important consideration. But even if the introduction of the cocoa-nut tree to Asia took place *after* the assumed departure of the Polynesian tribes, the latter must have been well acquainted with the art of making toddy, as there is a number of palms in Asia, about the true native country of which there is no doubt whatever, yielding toddy—a beverage of so ancient a date that even the oldest language of that continent has a name for it,—toddy being only a corruption of the Sanskrit word " tade." Had, therefore, the Polynesians once known the process by which they might have obtained, not only a strong liquor, but also sugar, vinegar, and yeast, they would have operated as readily upon the cocoa-nut tree in the South Sea, as the people of Southern Asia did when the cocoa-nut tree came to their shores. Taking, probably, its departure from Western America, the cocoa-nut was drifted by prevailing winds to Polynesia, where its toddy-yielding properties were not suspected; thence it drifted on towards Asia, and there was perceived to be as capable of yielding a favourite beverage as the Palmyra, the wild date-tree, the *Arenga saccharifera*, and the various species of the singular *Caryota* palms had done from time immemorial.

CHAPTER XVII.

VEGETABLE POISONS.—MEDICINAL PLANTS.—SCENTS AND PERFUMES.—MATERIALS FOR CLOTHING.—MATS AND BASKETS.—FIBRES USED FOR CORDAGE.—TIMBER.—PALMS.—ORNAMENTAL PLANTS.—MISCELLANEOUS.

VEGETABLE poisons are extracted by certain natives who make a profound secret of their art, and it would require an intimacy of years before any reliable information on this point could be elicited. I was ready to make presents of hatchets, knives, and other valued articles, to get some insight into their toxicology; but Mr. Pritchard begged me to abstain : the natives would take alarm at my inquiry, and if perchance any great man should be taken ill or die during my visit, it would at once be said that I, availing myself of the knowledge acquired, had administered a fatal dose—a most undesirable charge in the present state of political ferment. The Fijians have both slow and acute poisons, and when a man is gradually sinking (often, no doubt, from a very different cause), it is readily believed that " he has had a dose." He will then seek the advice of some skilful native physician, if possible one at Bau, the capital, to administer the necessary antidotes, and restore him to health. However, very often there is no time to interpose between the fatal dose and its consequences, the effect being almost instantaneous. When, in October, 1860, I revisited Cakaudrove, a poisoner had just been strangled by orders of the ruling chief; he having been detected in putting a certain drug into a cigarette, which proved fatal to the smoker. The poisoner, on finding himself condemned to die, not only pleaded guilty to this crime, but also confessed to having been instrumental in bringing about the death of no less than three hundred people, all victims to his infamous art.

There being no chance of gaining any direct information about the more subtle poisons from the lips of the natives themselves, an examination of all plants possessing narcotic properties would supply the deficiency, if it were not for an anomaly, as yet insufficiently explained, that certain species shunned as poisonous in one country, are eaten with impunity in another. There are mushrooms which in England are absolutely noxious, and on the Continent wholesome food. In Fiji, the leaves of the Boro yaloka ni gato (*Solanum oleraceum*, Dun.), a spiny species, closely allied to *Solanum nigrum*, Linn., and those of the Boro dina (*Solanum anthropophagorum*, Seem.) as well as the fruit of the latter and that of the Bora Sou or Sousou (*Solanum repandum*, Forst.), are eaten; the latter in soups or with yam. I was in some measure prepared for this, having seen quantities of the first-named species, as well as another nightshade (*Solanum nigrum*, Linn.), exposed for sale in the market of Port Louis, Mauritius, and learnt on inquiry that they were common pot-herbs, eaten both by the white and coloured population, as intimated by Bojer in his Hortus Mauritanus. *Strychnos colubrina*, Linn., is met with in

Viti Levu, but I have not been able to learn whether the natives are aware of its containing *strychnine*. A kind of Upas tree (*Antiaris Bennettii*, Seem.), commonly termed "Mavu ni Toga," probably because it has been introduced from the Tonga islands, was formerly planted about heathen temples, and is even now to be found in towns and villages. It is a middle-sized tree, with a thick crown of dark foliage, oblong glossy leaves, and a fleshy fruit of the size of an apricot, covered with a velvety skin of a most beautiful crimson colour. A gum exuding from the stem and branches is used for arrows. The exact nature of its poisonous properties has not yet been ascertained. That they are not equal to those ascribed to the true Upas tree of Java (*Antiaris toxicaria*, Leschen.) is proved by the manner in which the natives handle it; but it is impossible to say whether one of the reasons for its cultivation near temples, and its probable introduction from Tonga, may not be found in its yielding a poison of which the heathen priests may have occasionally made use. Sir E. Home gathered it in Wallis Island, and Dr. Bennett, of Sydney, found it cultivated in Tucopia for making bark-cloth.

Amongst the trees most dreaded by the natives on account of their noxious qualities, the Kau Karo, literally itch-wood, occupies a prominent place, and seems to act somewhat like *Rhus venenata* or *Semecarpus Anacardium*. Mr. Pritchard and myself first heard of its existence during our visit to the southern shores of Viti Levu, in July, 1860, and on the banks of a river were fortunate enough to obtain specimens of the tree, proving it to be the *Oncocarpus Vitiensis*, A. Gray =

Rhus atrum, Forst., an Anacardiaceous plant. The tree, when fully developed, is about sixty feet high, bearing large oblong leaves and a very curious corky fruit, somewhat resembling the seed of the walnut. On handling the specimens a drop of the juice fell on the hand of one of our party, and instantly produced a pain equal to that caused by contact with a redhot poker. Mr. E. A. Egerström, a Swedish gentleman, residing on the island of Naigani, had been still more unfortunate in his accidental contact with the Kau Karo; and on visiting his hospitable roof on the 2nd July, 1860, he was just recovering from the effects of the accident. Having desired a native carpenter to procure him a spar suitable for a flag-staff, one was brought of Kau Karo, about forty-two feet long, and twenty-two inches in girth at the foot, having a white wood and a green bark, not unlike that of the Vau dina (*Paritium tiliaceum*, Juss.) and light-coloured when peeled off. Ignorant of the poisonous properties of the tree, Mr. Egerström himself peeled off the bark, and found the sap beneath it very plentiful. " In the evening,"—I quote Mr. Egerström's own words, in a letter to the British Consul,—" I was troubled with considerable itching about my legs, and every part of my body which had come in contact with the spar, especially about the abdomen and lower parts, having sat across the tree when barking it. All the parts affected became red and inflamed, breaking out in innumerable pustules, which emitted a yellowish matter with a nauseous smell. The itching was exceedingly painful and irritating, and my arms having been bare when operating upon the tree, also became inflamed and broke out as already described. The neighbouring natives, who came to watch my proceedings, now warned me, too late, not to touch the tree, as it was a poisonous one, and advised my keeping quiet and not to touch or scratch the parts inflamed. This advice, however, I could not follow, the irritation for several days being excessive. I employed no remedy, but bathed daily, as usual, in fresh water, although advised to the contrary, but did not get rid of the injurious effect of the itch-wood for nearly two months."

Another tree, the contact with which is avoided by the Fijians, is the Sinu gaga (*Excœcaria Agallocha*, Linn.) or poison Sinu, called so in contradistinction to the Sinu damu (*Leucosmia Burnettiana*, Bth.) and the Sinu mataivi (*Wikstrœmia Indica*, C. A. Meyer), both of which, like the Sinu gaga, are littoral plants. The Sinu gaga is found in mangrove swamps or on dry ground, just above high-water mark. It is sixty feet high, has a glossy foliage, oblong leaves, and minute green flowers arranged in catkins. It is difficult to exterminate, for unless the stumps are taken up, innumerable young shoots spring up the moment the main stem is felled. When the tree is wounded abundance of white milky juice flows, which causes a burning effect on coming in contact with the skin. Some natives, however, can handle this poisonous juice with perfect impunity (*era sinu dranu*), analogous to what I witnessed in the Manzanillo or Manchineel tree of tropical America, the sap of which caused me the greatest agony after it had accidentally entered my eyes, and never raised even as much as a blister on being allowed to dry on the hands of a travelling companion. The smoke of the burning wood affects the eyes with intolerable pain, exactly as does that of the manchineel tree, of which I gave an instance in the 'Narrative of the Voyage of H.M.S. Herald,' vol. i. p. 141,—one of our boat's crew becoming blind for several days after lighting a fire with manchineel wood. None, save those who have been sufferers from the effect of these poisons, can form any adequate conception of the agonies endured, and the courage displayed, by a Fijian who voluntarily submits himself to being cured of leprosy by the smoke of the Sinu gaga wood. The Rev. W. Moore, of Rewa, was well acquainted with a young man of the name of Wiliami Lawaleou, who underwent the process of being smoked. Mr. Moore gave me the full particulars of this remarkable case when I was his guest in 1860, and he has also published a full account of it in 'The Wesleyan Missionary Notices,' Sydney, 1859, p. 157. After stating that he knew Wiliami as a fine healthy young fellow, Mr. Moore was surprised to find him one day so much altered by the effects of leprosy. Some time after he again met him full of health, and on inquiry learnt the treatment adopted to bring about this change. Taken to a small empty house, the leper is stripped of every article of clothing, his body rubbed all over with green leaves, and then buried in them. A small fire is then kindled, and a few pieces of the Sinu gaga laid on it. As soon as the thick black smoke begins to ascend the leper is bound hand and foot, a rope fastened to his heels, by means of which he is drawn up over the fire, so that his head is some fifteen inches from the ground, in the midst of the poisonous smoke. The door is then closed and his friends retire a little distance, whilst the poor sufferer is left to cry and shout and plead from the midst of the suffocating stream; but he is often allowed to remain for hours, and finally faints away. When he is thought sufficiently smoked the fire is removed, the slime scraped from the body, and deep gashes cut into the skin until the blood flows freely. The leper is now taken down and laid on his mats to await the result. In some cases death—in many, life and health. Wiliami had undergone this fearful process. He had taken some of the youths of the place, and on his way to the smoking-house told them his pitiable condition, his shame as an outcast, and his willingness to suffer anything to obtain a cure, and much would depend on their firmness. They were not to be moved by his groans and cries, and for the love they bore him he begged them to do the operation well, and threatened to punish them if they performed it only half. Imagine the scene! They proceed to the lonely house. Wiliami's companions, as much afraid of overdoing as underdoing their sad task, leave the poor leper drawn up by his heels in the midst of a thick black smoke; they retire to some distance, and presently are horrified by his piteous cries and groans. Some weep, some run home, others rush into the smoking-house to take him down; but, with Spartan-like endurance, he commands them not to terminate his suffering until the process is complete. At last they take him down—he is faint and exhausted—the operation has been successful. Wiliami is no longer a leper, but again walks God's earth a healthy man.

VEGETABLE POISONS.

The materials employed by the natives for poisoning, or rather stupefying, fish, a custom as prevalent all over Polynesia as it is amongst the Indians of America, are the square fruit of the Vutu rakaraka (*Barringtonia speciosa*, Linn.) and the stem and leaves of the Duva gaga (*Derris uliginosa*, Benth.), both plants growing in abundance on the sea-beach, just above high-water mark. As soon as these materials,—pounded to render them more efficacious,—are thrown into the water, or drawn through it by means of a line or creeper to which they have been attached, the fish turn on their back and appear on the surface. They are perfectly stupefied, and are thus easily taken; but they soon recover their lost activity, and are believed not to die from the effects of the treatment they have received.

The nettles,—those mosquitoes of the vegetable kingdom, irritating but never killing as they do,—are collectively termed "Salato"—a name also including those animals familiarly known as sea-nettles. There are two kinds. The Saloto ni coro is an annual weed (*Fleurya spicata*, Gaud., var. *interrupta*, Wedd.), which abounds about towns and villages (hence the specific appellation of "ni coro"); and although the virulence of its sting is not to be compared with that of our European nettles, the natives so carefully avoid all contact with it, and ran away in such fright when I gathered specimens of it for the herbarium, that one is tempted to fancy their skins more keenly affected by it than ours. Still greater is their dread of an Urticaceous tree (*Laportea*, sp.), forty to fifty feet high, which they simply term "Salato" (nettle), and which, when touching the skin, produces a burning pain similar to that ascribed to the sap of Malawaci (*Trophis anthropophagorum*, Seem.). Milne (Hook. Jour. and Kew Misc. ix. p. 110) states, that "if you should be so unfortunate as to sting yourself, you will feel the effects for some months. I am suffering at this moment," the writer continues, "from an accident which occurred a month ago. There is no eruption; but it is most painful when exposed to the influence of water."

The medicinal plants employed by the natives are as difficult, perhaps more difficult, to find out than the poisonous ones used for illegal purposes. Those who profess to be acquainted with their properties—often women, and answering to our herbalists—cannot be tempted by any presents to disclose secrets which to them prove a lucrative source of income for life. It is only the virtues of plants generally known that a casual inquirer has any chance of learning. The high estimation in which the oil of the Dilo (*Calophyllum inophyllum*, Linn.) is held by the whole population, as an efficaceous remedy for rheumatism and other pains, has been mentioned in another place. The leaves of the Kura (*Morinda citrifolia*, Linn.), a middle-sized tree, with shining leaves and white flowers, not unlike those of the coffee-shrub, are heated by passing them over flame, and their juice squeezed into ulcers, whilst the leaves themselves are put on the wound as a kind of bandage. The bark of the Danidani (*Panax fruticosum*, Linn.), a shrub about eight feet high, and cultivated about the native houses on account of its deeply-cut, ornamental foliage, is scraped off, and its juice taken as a remedy for macake, the thrush—ulcerated tongue and throat. The properties of the Sarsaparilla (*Smilax* sp.), as a means of purifying the blood, are well known. The creeper is found throughout the group, especially on land that has at one time been cleared, and might be gathered in quantities if there were any demand for it. In the London market it would at present be unsaleable. It belongs to that section of sarsaparillas distinguished by pharmacologists as the " non-mealy," the most valued representative of which is the Jamaica sort. Moreover, it has no " beard," or little rootlets The natives of Ovalau, Viti Levu, and Vanua Levu, name it Kadragi and Wa rusi ; those of Kadavu, " Na kau wa," literally, " the woody creeper." I met with it years ago in the Hawaiian group ; it is said to be also common in the Samoan and Tongan groups, and prepared sarsaparilla occasionally imported to the two last mentioned has found no market, the indigenous being preferred to the foreign production. Curious to add, in Fiji it is not, as with us, the rhizome that is used, but the leaves, which are chewed, put in water, and strained through fibre, like the Yaqona or Kava (*Piper methysticum*, Forst.), before being taken. Strong purgative properties reside in the Vasa or Rewa (*Cerbera lactaria*, Ham.), a sea-side tree, twenty-five feet high, with soft wood, smooth shining leaves, and white scented flowers, used for necklaces by the natives. The aromatic leaves of the Laca (*Plectranthus Forsteri*, Benth.), a weed abounding in cultivated places, and having purple bracts supporting pale blue flowers, cure, it is said, " bad eyes " and headaches on being brought in contact with the affected parts. It is also recommended for coughs and colds, in common with an Acanthaceous herb inhabiting swamps (*Adenosma triflora*, Nees), which shares its aromatic properties. The people of Somosomo declare that the leaves of the Vulokaka (*Vitex trifoliata*, Linn.), with which their beach is thickly lined, when reduced to a pulp by chewing, are employed by them for stuffing hollow teeth. The leaves and bark of another sea-side shrub, the Sinu mataiavi (*Wikstrœmia Indica*, C. A. Meyer), are employed for coughs, the bark alone for sores.

Through a native connected with the Wesleyan mission, I succeeded in purchasing a knowledge of the drugs employed about Bau for procuring abortion. It appears there are five plants which furnish them, two Malvaceæ, a Büttneriacea, a Convolvulacea, and a Liliacea —namely, the Kalakalauaisoni (*Hibiscus diversifolius*, Jacq.), a spiny shrub, growing in swamps; the Wakiwaki (*Hibiscus* [*Abelmoschus*] *moschatus*, Moench), closely resembling the latter, and bearing large yellow flowers like it, but being destitute of spines, and invariably preferring dry ground; the Siti (*Grewia prunifolia*, A. Gray), a small tree, abounding in the groups, and producing a fruit eaten by the Fijian bat; the Wavuti (*Pharbitis insularis*, Chois.), a blue-flowering seaside creeper, and the Ti Kula, Te Kula, or Va sili damudamu (*Dracæna ferrea*, Linn., var.). Of the Kalakalauaisoni, Wakiwaki, and the Wa buti, the juice of the leaves,—of the Ti kula, that of the heart of the leaves and surface of the trunk, are used. The Ti kula is held to be the most efficacious, and only administered when the other drugs have failed to produce their murderous effects.

Perfumes for scenting cocoa-nut oil, which the natives profusely apply to their hair and naked body, are supplied by the wood of the Yasi (*Santalum Yasi*, Seem.), the bark of the Macou (*Cinnamomum* sp.), the flowers of the Uci (*Evodia hortensis*, Forst.), the Makosoi (*Uvaria odorata*, Lam.), the Balawa (*Pandanus odoratissimus*, Linn.) and the Bua (*Fagræa Berteriana*, A. Gray), and the fruit of the Makita (*Parinarium laurinum*, A. Gray), and the Leba (*Eugenia* [*Jambosa*] *neurocalyx*, A. Gray).

The Yasi or sandal-wood (*Santalum Yasi*, Seem.) is confined to the south-western parts of Vanua Levu, and formerly abounded near Bua or Sandal-wood Bay. The high estimation in which it was held by the Tonguese early induced them to undertake regular trading voyages to Fiji, long previous to those attempted by ourselves. Mariner, who was a resident in Tonga from the year 1806 to 1810, affords us a tolerable insight into them (J. Martin's Account of the Natives of the Tonga Islands: London, 1817: p. 319, 333), in narrating the adventures of Cow Mooala, a Tonguese chief, who had been about fourteen years from home, and had originally set out on a sandal-wood expedition. Attempts had been made, he assures us, to extend the range of the wood by cultivation, both in Fiji and Tonga; but the tree, though successfully transplanted, yielded a produce with little or no scent, absolutely useless for the purposes for which it was required. The demand continuing, and the article becoming scarcer every day, prices went up. At one time the Fijians would give a considerable quantity for a few nails. "But now," Mariner continues, "they demand axes and chisels, and those, too, of the best quality, for they have gradually become judges of such things: whales' teeth are also given in exchange for it. The chiefs of the Fiji islands very seldom oil themselves, and consequently require very little of this wood, the principal use of it being to scent the oil. The natives of the Tonga islands, however, who require a considerable quantity of it for the above purpose, complain heavily of its scarcity; and what renders the matter still worse for others is, that the Fiji people, demanding a greater number of axes and chisels for a given quantity of wood, these implements are growing very scarce at the Tonga islands, and plentiful at Fiji. Before the Tonga people acquired iron implements, they usually gave whales' teeth, gnatoo (bark cloth) mats for sails and platt; but whales' teeth are exceedingly scarce, and the other articles are too bulky for ready exportation. The *sting* of the fish called sting-ray was also occasionally given; but these *stings*, which they use for the points of spears, are by no means plentiful. This fish is found in the greatest quantity at an island called Ovoa, which lies about midway between Vavau and Samoa. Another article of exchange is a peculiar species of shell, which they find only at Vavau, and is also scarce." It does not seem that Europeans engaged in the sandal-wood trade until towards the close of the eighteenth century, when it was taken up by Manila vessels for shipment to China. However, so great was the demand for this article, both in the Chinese and Polynesian markets, that about the year 1816 there was scarcely enough left for home consumption—several thousand tons having probably been exported, worth in China from £20 to £30 a ton. In 1840, the United States Exploring Expedition with difficulty obtained a few specimens for the herbarium. To save the tree from utter destruction in the islands, the Rev. Mr. Williams planted one in the garden of the mission station, at Bua, which, when I visited the place, in 1860, was in full vigour and bloom. When sandal-wood was still plentiful, a butcher's knife was usually exchanged for ten sticks of three feet long. At present, fancy prices are readily given for the little that now and then turns up. In 1859, Tui Levuka, chief of Ovalau, had nearly half a ton of it in his possession, but that seems to have been the largest quantity of late years brought together; a year later Mr. Hennings, a German, trading in Fiji, could only succeed in obtaining a few pieces. On visiting Bua, in October, 1860, a log, six feet long and two or three inches in diameter, was presented to me, and thought quite a valuable gift by my native attendants. The Yasi has very much the appearance of a Myrtaceous plant, and the Fijians, who possess a quick eye for discerning natural affinities, class it with several species of *Eugenia*, which they respectively distinguish as Yasi ni wai, Yasi dravu, etc. The leaves are opposite and lanceolate, and the flowers very minute, and on first opening they are white, but gradually change to pink, and ultimately to a brownish purple. The fruit is in shape, size, and colour like that of the black currant.

The wood is of a light-brown, and highly charged with aromatic oil, especially in the central portion of the stem and branches, developed in the highest degree in the oldest trees and near the root. It is grated on the mushroom coral (*Fungia*) and mixed with cocoa-nut oil by the Fijians, as well as by all the Polynesian tribes who are fortunate enough to obtain possession of it. In China, the larger pieces were used for ornamental work, and the sawdust and other remnants made into joss-sticks, burned before idols and images.

The bark of the Macou, as it is termed in the Bau dialect, " Mou " in that of Kadavu, and " Maiu " in that of Namosi, is a kind of Cassia bark, which may prove of commercial importance, and is used by the Fijians for scenting cocoa-nut oil. The tree yielding it—a species of *Cinnamomum*—is about thirty feet high, four to five inches in diameter, and is met with above an elevation of 1500 feet, in dense virgin forests. I met it on Buke Levu, island of Kadavu, and on Voma peak, Viti Levu; and Mr. Pritchard received fine specimens from the island of Gau, where they had been collected by W. Berwick, a coloured man, residing there. The bark has a fine aromatic smell and flavour, a light-brown colour, is thicker than that of the cinnamon of commerce, and resembles some of the laurineous barks, such as the Sintoc and Culilawang, brought from the Moluccas. In Namosi it is used as a sudorific. Unfortunately, I did not see the tree in flower, and hence am unable to determine whether the " buds " are equal to the best " Cassia buds " of commerce. The resemblance of the Fijian names to that of " Massoy," given to a fine quality of Cassia bark, from New Guinea, deserves investigation.

The flowers of the Uci or Sacasaca (*Evodia hortensis*, Forst.) diffuse, like those of most *Diosmaceæ*, an overpowering, rather sickly odour, highly esteemed by the natives, but only appreciated by those Europeans who can enjoy patchouly, musk, and scents of a similar category. The perfume emitted by the flowers of the Makosoi (*Uvaria odorata*, Lam.) and of the Balawa (*Pandanus odoratissimus*, Linn.) commands a greater number of European admirers, whilst that of the Bua (*Fagræa Berteriana*, A. Gray) may be said to be universally inhaled with delight. The Bua blossoms in September and October, and one of the months of the Fijian calendar is occasionally called the Vulai Bua, or Bua month. The flowers, or rather corollas, are gathered after they have dropped on the ground, and brought home in baskets. They are tubular, white, and fleshy, and are either strung into necklaces, which retain their delicious and powerful perfume long after they are dry, or they are placed while still fresh in cocoa-nut oil, in order to impart scent to it. Sandal-wood and Bua flowers are often put into the same vessel of oil. The abundance of the tree (which yields a hard, white wood) at Sandal-wood Bay may have given rise to its native name " Bua,"—a form of " Pua," by which the plant is known in the Society Islands.

Another perfume largely employed in scenting oil is furnished by the Makita (*Parinarium laurinum*, A. Gray), a tree about fifty feet high, supplying tough spars for canoes, and having oblong leathery leaves, formerly used exclusively in thatching heathen temples, but now also for common dwelling-houses. The flowers are small and white, slightly tinged with purple, and the fruit has a rough, woody outside, of a light-brown colour, containing a large kernel, which possesses a scent much esteemed by the Fijians, but in which we detect nothing remarkable either as regards strength or beauty. The fruit of the Leba (*Eugenia* [*Jambosa*] *neurocalyx*, A. Gray), a middle-sized Myrtaceous tree, with large flowers, considering the natural order to which it belongs, has much more to recommend it to the notice of Europeans. It ripens about September, and its odour gravitates between that of the apple and the melon. It is roundish, strongly ribbed, often three inches long and eight inches in circumference, of a dark purple, and contains five large seeds, of an angular shape, and a beautiful crimson colour. The natives wear a whole fruit, or part of it, suspended around their necks, and also use it for scenting cocoa-nut oil.

Materials for the scanty clothing worn by the Fijians are readily supplied by a variety of plants, foremost amongst which stands the Malo or Paper Mulberry (*Broussonetia papyrifera*, Vent.), a middle-sized tree, with rough trilobed leaves, cultivated all over Fiji. On the coast, the native cloth (Tapa*) and plaitings are gradually displaced by cheap cotton prints introduced by foreign traders,—a fathom of which is considered enough for the entire dress of a man. In the inland heathen districts the

---

* Tapa = Kapa of some dialects, I take to mean originally "covering;" Atap, the name for *thatch* in the Indian Archipelago, doubtless belongs to the same set of words.

boys are allowed to run naked until they have attained the age of puberty, and publicly assumed what may be termed their *toga virilis*—a narrow strip of native cloth (Malo) passing between the legs, and fastened either to a waistband of string or to a girdle formed by one of the ends of the cloth itself. The length of the Tapa hanging down in front denotes the rank of the wearer; the lower classes not having it longer than is absolutely necessary for the purposes of securing it to the waistband, whilst the chiefs let it dangle on the ground, and when incommoded by it in walking, playfully swing it over their shoulder. In the christianized districts of the coast, a piece of Tapa, at least two yards long and one yard broad, is worn around the loins, and distinguished persons envelope their body in pieces many yards long, and allow long trains to drag after them on the ground. A fine kind of Tapa (Sala) is worn in the shape of a turban by those who still adhere to the old custom of letting their hair grow long. From a laudable desire to promote cleanliness the missionaries have pronounced against long hair and the use of the Sala, but in doing so they deprived the natives of a capital protection against the sun; the immense mass of hair curled and frizzled to make it stand off many inches, and covered by a piece of snow-white Tapa, must have kept the head cool. Now most of the Christian natives move about without any covering for their head, and with their hair cut short, which, in a tropical climate, cannot improve their intellect. The abolition of the old custom might have proved more beneficial if immediately followed by the institution of some kind of head-dress. The manufacture of native cloth is entirely left to women of places not inhabited by great chiefs, probably because the noise caused by the beating out of the cloth is disliked by courtly ears. The rhythm of Tapa-beating imparts therefore as thoroughly a country air to a place in Fiji as that of threshing corn does to our European villages. The Masi tree is propagated by cuttings, and grown about two or three feet apart, in plantations resembling nurseries. For the purposes of making cloth it is not allowed to become higher than about twelve feet, and about one inch in diameter. The bark, taken off in as long strips as possible, is steeped in water, scraped with a conch shell, and then macerated. In this state it is placed on a log of wood, and beaten with a mallet (Ike), three sides of which have longitudinal grooves, and the fourth a plain surface. Two strips of Tapa are always beaten into one with the view of strengthening the fibres—an operation increasing the width of the cloth at the expense of its length. It is easy to join pieces together, the sap of the fibres being slightly glutinous; and in order to make the junction as perfect and durable as possible, a paste is prepared of arrowroot, or a glue of the viscid berries of the Tou (*Cordia Sprengelii*, De Cand.). I have seen pieces of native cloth, intended for mosquito curtains and screens, which were nearly one hundred feet long and thirty feet broad. Most of the cloth worn is pure white, being bleached in the sun as we bleach linen; but printed Tapa is also, though not so frequently, seen, whilst that used for curtains is always coloured. Their mode of printing is by means of raised forms of little strips of bamboo, on which the colour is placed, and the tops pressed; indeed, the fundamental principle is the same as that of our printing books, the little strips of bamboo standing in the place of our types. The chief dye employed is the juice of the Lauci (*Aleurites triloba*, Forst.), and the pattern, though rudely executed, often displays much taste. It is stated that in times when the Malo plantations have failed to produce a sufficient quantity of raw material, recourse is had to the Baka (*Ficus* sp.); but this is only a makeshift, whilst the bark of the Breadfruit-tree seems never to be resorted to as in other parts of Polynesia.

When the men have no native cloth of any sort, they make a dress by splitting a cocoa-nut or plantain leaf in halves, and tying one of these parts around their waist. There is an old monkish tradition that our first parents, when adopting dress in the garden of Eden, availed themselves of the leaf of the plantain, hence called *Musa paradisiaca*; and it must be owned that a Fijian, having assumed this dress, presents a most primitive appearance, the more striking because his movements are entirely free from any approach to indecency, which a European who has never lived amongst races going naked would naturally fancy associated with so scanty a garb. It is, perhaps, the most simple form of an article of dress much worn in Fiji, and called "*Liku*," consisting of a number of fringes simply attached to a waistband. The length of these fringes is subject to certain rules of custom. Men can wear them very long; but women, particularly young unmarried ones, must not have them longer than two or three inches. Liku is made of many different plants, and might be classified into temporary and permanent. Amongst the temporary Likus ought to be placed those made of plantain and cocoa-nut leaves, or those made of a climbing plant, the Vono (*Alyxia bracteolosa*, Rich., A. Gray), the stem of which is partially broken to give it greater flexibility, and also to bring out an agreeable smell peculiar to the Vono, on account of which it is also worn as garlands around the head. Amongst the permanent Likus is one termed " Sausauwai," the long black fringes of which, playing on the white Tapa, or on the fine limbs of the natives, has a most graceful appearance. Both on account of the scarcity of the materials of which it is composed, and its being unaffected by water, especially when greased with cocoa-nut oil, the Sausauwai is highly valued by fishermen, and all people living on the coast of Fiji; they will give twenty fathoms of white Tapa, and the Tonguese and Samoans as much as £1 sterling, for a single one of these elegant articles of dress. The fringes of which it is composed are of the thickness of a common wire, rather flexible, and occasionally ornamented with small beads. Placed under the microscope, the vegetable origin of these fringes becomes at once evident, and they are found to be composed of glossy black joints, of unequal length. None, save a few natives, had ever seen the plant producing them, and it was the general belief of all the foreign residents in Fiji that they were the roots of a certain tree, until Mr. Pritchard and myself made the subject a point of special inquiry during our first visit to Navua. A few words from Chief Kuruduadua, and two large knives held out by us as a reward, induced two young men to procure a quantity of this singular production sufficient for scientific examination; proving it to be, not the root of a tree, as had been believed, but the entire body of a species of *Rhizomorpha*. The plant is vernacularly termed "Wa loa," literally, black creeper, from *wa*, creeper, and *loa*, black—a name occasionally applied to the Liku made of it also. The Wa loa is confined to the south-western parts of Viti Levu, where it grows in swamps on decaying wood fallen to the ground; the threads of which it consists are several feet long, leafless, not much branched, and they are furnished here and there with little shield-like expansions, acting as suckers, by means of which the plant is attached to the dead wood upon which it grows. The threads, having been beaten between stones in order to free them from impurities adhering, are buried for two or three days in muddy places, and are then ready for plaiting them to the waistband.

The Liku worn by the women, always speaking of those who have not as yet adopted foreign calico, are principally made of the fibres of the different species of Vau, the Vau dina (*Paritium tiliaceum*, Juss.), the Vau dra (*Paritium tricuspis*, Guill.), and the Vau damudamu (*Paritium purpurascens*, Seem.). The bark of these trees is stripped off, steeped in water to render it soft and pliable, and allow the fibres to separate. The fibres are either permitted to retain their original whiteness, or they are dyed yellow, red, or black. The yellow colour is imparted with turmeric, the black with mud and the leaves of the Tavola (*Terminalia Catappa*, Linn.), and the red with the bark of the Kura (*Morinda citrifolia*, Linn.), and that of the Tiri (*Guttiferæ?*). The Liku worn by the common women consists of one row of fibres, all of the same colour; whilst those worn by ladies of rank are often composed of two or three rows or layers—flounces, I suppose, would be the proper term —every one of which exhibits a different colour.

Mats, with which the floors of houses and sleeping-places are thickly covered, are made of two kinds of screw-pines: the coarsest, of the leaves of the Balawa (*Pandanus odoratissimus*, Linn.); the finest, of those of the Voivoi (*Pandanus caricosus*, Rumph.). The Balawa, or Vadra, as it is termed in some districts, is a tree twenty-five feet high, indicative of poor soil, growing in exposed positions, and being one of the first plants appearing on newly-formed islands. Its singular habit has often been dwelt upon. The smooth white branches, with their dense heads of foliage, not inaptly compared to the arms of a huge candelabrum; the strong aerial roots, covered with minute spines, and serving as so many props; the curious corkscrew-like arrangement of the leaves, the leathery, sword-shaped leaves themselves, and their spiny edges; the long spikes of male, and the shorter branches of female flowers, their delicious perfume strongly recalling to mind that of the vegetable ivory of South America; finally, the bright orange-coloured drupes, formed into large heads of fruit, to say nothing of their insipid taste, appreciated only by natives, are all so essentially different from what a European traveller is accustomed to in his own country, that his attention is involuntarily arrested, and he hardly ever fails to record it. The Voivoi or Kiekie (*Pandanus caricosus*, Rumph.) is a stemless species, with leaves ten to twelve feet long, which delights in swampy localities of the forests, and is occasionally cultivated to meet the demand. Fans, baskets, and the finest mats —even those on which newly-born babes, naked as they are for more than a twelvemonth, are carried—are made of its bleached leaves. Occasionally neat patterns are worked in, by introducing portions of the material dyed black, whilst the borders of highly-finished mats are tastefully ornamented with the bright-red feathers of the Kula,—a parroquet (*Coriphilus solitarius*, Latham) not found in the groups eastward of Fiji, and therefore highly esteemed by the inhabitants of those islands. The bleached leaves are also employed for decorating the body, being tied by the men over their head-dress (sala), around their breast, upper part of the arms, wrists, and above the calves. The custom is not restricted to any particular class, but freely practised by all, serfs, commoners, and chiefs, when they go to war, or wish to look smart. The bright-coloured leaves of the Ti kula (*Dracæna ferrea*, Linn., var.), and a number of flowers, ferns, and leaves, are used by both sexes as wreaths, garlands, necklaces, and similar ways, evidently showing their great love for flowers and graceful foliage. A certain kind of mats, worn as articles of clothing, are called " Kuta," from a species of sedge (*Elæocharis articulata*, Nees ab Esenb.), supplying materials for them, growing in swamps to the height of six feet or more, and going either by that name or by that of Ya. Baskets are also made of the leaves of the cocoa-nut palm, and the stem of the *Flagellaria Indica*, Linn., split up in narrow strips; those of the former are the most easy to make, but they do not last long, whilst those of the latter are the neatest and last the longest.

Fibre used for cordage is derived from three species of Vau (*Paritium tiliaceum, P. tricuspis*, et *P. purpurascens*), the cocoa-nut palm; the Yaka or Wayaka (*Pachyrhizus angulatus*, Rich.), the Kalakalauaisoni (*Hibiscus diversifolius*, Jacq.), and the Sinu Mataiavi (*Wikstrœmia Indica*, Meyer). Plaiting cocoa-nut fibre into "sinnet," afterwards to be made into rope, or simply used for binding material, and as such a good article of exchange in the group, is a favourite occupation of the men, even of high chiefs, when sitting in bures and discussing politics or other topics of the day. According to Mr. Pritchard, none of the Polynesians produce so great a quantity of this article as the Fijians, though the Tonguese excel them in colouring it. I have seen—he continues in the memorandum from which I quote—a ball of "sinnet" six feet high, and four feet in diameter. Some heathen temples, Bure ni Kalou, used to be entirely composed of such plaiting, and their completion must have been a task extending over a considerable period, since a model of them, four feet high, ordered for the Museum of Economic Botany at Kew, could not be finished in less time than six weeks, and at a cost of £5. The fibre of the Yaka or Wayaka (*Pachyrhizus angulatus* Rich.=*Dolichus bulbosus*, Linn.) is principally sought for fishing-nets, the floats of which are the square fruits of the Vutu rakaraka (*Barringtonia speciosa*, Linn.). The Sinu Mataiavi (*Wikstrœmia Indica*,

Meyer), a sea-side shrub, perhaps identical with the Sinu ni vanua, serves the same purpose, its bark, like that of other *Thymeleæ*, containing a readily-available fibre—a fact also known, according to Mr. Pritchard, in the Samoan islands, where the plant is termed "Mati." Only a limited use is made of the fibre of the Kalakalauaisoni (*Hibiscus* [*Abelmoschus*] *diversifolius*, Jacq.), a plant abounding in swamps all over Fiji.

Timber of excellent quality, both for house and ship-building purposes, abounds on the large islands, and a trade in it has already sprung up with the Australian colonies. The timber-trees belong principally to the natural orders *Coniferæ*, *Casuarineæ*, *Guttiferæ*, *Myrtaceæ*, and *Leguminosæ*. The most valuable woods are those produced by the Dakua, Vesi, Dilo, and Vaivai, and a list of nearly one hundred useful kinds might be drawn up.

The Dakua or Fijian Kowrie-pine (*Dammara Vitiensis*, Seem.) is a noble addition to a genus of Conifers, of which several species are known, scattered over New Zealand, Southern Queensland, New Caledonia, Aneitum, the Moluccas, Java, and Borneo. Dakuas have been found in Vanua Levu, Viti Levu, Ovalau, and Kadava; but European sawyers have already made such sad havoc amongst them, that it is only in the two former islands where they are still abundant. Wilkes alludes to a fine one near Levuka, Ovalau, which measured five feet in diameter, or 15 feet in circumference. Those which I saw at Korovono, Vanua Levu, displayed greater dimensions, the largest stem being, at four feet above the base, eighteen feet; and another, also four feet above the base, sixteen feet in circumference. Milne (Hook. Jour. Bot. and Kew Misc. ix. p. 113) gives from eighteen to twenty-seven feet circumference as the maximum, but he does not state at what height above the base his measurement was taken. Some of the trees at Korovono were from 80 to 100 feet high, and up to a height of 60 feet free from branches. The bark was whitish on the outer, red on the inner, surface, peeling off like that of Australian gum-trees. Old specimens did not have regular whorls of branches, as is the case with most Conifers. The wood of the Korovono tree was white, but there is said to be also a red-wooded kind, which may perhaps prove distinct from this plant. Dakua is used for masts, booms, and spars, for flooring houses, and for all those purposes for which deal is usually employed by us. Spars, from sixty to eighty feet long, and two to three feet thick, were seen at Taguru, Viti Levu. The Dakua is not gregarious, but found always isolated in forests of a mixed composition. Like other Kowrie-pines, the Fijian exudes a gum, or rather resin, called "Makadre." Lumps weighing 50 lbs. have occasionally been found under old rotten stumps; and a good deal might be collected in districts whence these trees have disappeared, if the natives could be made acquainted with the peculiar way in which the New Zealanders sound the ground for their kowrie-gum. There has never been any foreign trade in this article, because the Europeans in Fiji, ignorant of its average market-value, rejected the offer of the natives to collect it. Captain Dunn, an American, is said to have taken away half a ton of it, but it has not transpired whether he was able to dispose of it to advantage. New

Zealand kowrie-gum has for years past fetched at public sales in London from 14s. to 16s. the cwt. In consequence, however, of the rebellion in New Zealand, it gradually advanced in 1860 to from 25s. to 28s.; in the spring of 1861 it was quoted at from 18s. to 20s., and it will no doubt ultimately be sold again at its former prices. The Fijians principally use the gum for glazing pots (vakamakadretaka),—the substance being put on while the vessels are yet very hot,—and for burning. The older it gets the better it burns. At first it is of a light whitish colour, but becomes more and more that of amber, as well as transparent with age. The natives, fearing demons, ghosts, and other creations of their wild fancy, are always anxious to be housed before sunset, and when compelled to venture out in the dark or when benighted, set up loud yells to drive away evil spirits, and light a torch made either of the resin of the Dakua (bound round with rushes), the stem of the Wavuwavu (*Erigeron albidum*, A. Gray), the trunk of the bamboo, or the flower-stalks of the cocoa-nut palm. In the smaller islands and certain coast-districts of Vanua Levu and Viti Levu, lamps fed with cocoa-nut oil are common; but in the interior of the principal islands, where that oil is an imported article difficult to obtain, the resin of the Dakua is burnt, either in the form of pastiles about two inches long, or in ribbon-like strips surrounded by slips of wood, so as to constitute a kind of candle. When burnt in the first-mentioned way, the resin is protected by crocks from running about and igniting the Pandanus matting or other inflammable materials of the houses. A dye obtained from the smoke of the burning resin is used for the hair and for painting native cloth black, or mixed with a certain red earth to make a brown pigment. Amongst the lower classes it is employed for tatooing women instead of the juice of the Lauci fruit (*Aleurites triloba*, Forst.), resorted to by ladies of rank: the skin being punctured with thorns of the shaddock tree.

Besides the *Dammara Vitiensis*, Seem., there are five other cone-bearing trees, all of which yield valuable timber, viz. the Kau solo, the Gagali, the Kuasi, the Kau tabua, and the Leweninini. The Kau solo represents a new genus peculiar to Fiji, and growing abundantly in the southern parts of Viti Levu, where it attains from sixty to eighty feet in height and nine feet in girth. It has the appearance of the Yew,—dark, lanceolate leaves, about an inch long, and solitary nuts at the ends of the branches. The Gagali (*Podocarpus polystachya*, R. Br.) is common on the banks of rivers. It is never seen higher than thirty or forty feet, and on the Navua I noticed that during the season when the river overflows its banks, the trees must often be under water, as dead twigs, leaves, and herbage, carried down by the tide, were lodged in their crowns. The wood is peculiarly elastic, and would probably do well for keels of boats and schooners. The Kuasi (*Podocarpus elata*, R. Br.) is confined to the summits of mountains, and forms the chief vegetation of Voma peak, Viti Levu. Its wood is used for outriggers of canoes. Another cone-bearing tree is the graceful Kau tabua (*Podocarpus cupressina*, R. Br.), common in the mountains of the Indian Archipelago, and in Aneitum. Milne found it in Viti Levu. Its native name is derived from the wood (Kau), resembling in its yellowish tinge a well-oiled whale's tooth (tabua), formerly esteemed the most precious article in the group. The tree is from fifty to eighty feet high, with spreading pendulous branches, presenting a beautiful appearance. The Leweninini (*Dacrydium elatum*, Wall.) is found in mixed forests from the sea-shore to the highest peaks. The branches are very delicate, and the youngest hang down in graceful fringes, clad with needle-shaped leaves of about half an inch in length. The slightest breeze—and there is scarcely ever a calm in Fiji—causes the branchlets and foliage to tremble (ninini), somewhat like our aspen; hence the natives of Ovalau have given it the name of "Leweninini." When coming from Somosomo to Levuka, the crew on board the 'Paul Jones' gave me an account of a moving plant, which they assured me grew in the mountains of Ovalau, and which excited my curiosity in an eminent degree. No sooner had I landed than two boys were dispatched for specimens of the Leweninini; but instead of bringing this *Dacrydium*, they brought a club-moss, common in the tropics (*Lycopodium cernuum*, Linn.), and which I found was termed Leweninini sa, on account of a certain resemblance to it. Macdonald (Jour. Geog. Soc. Lond. xxvii. p. 247) fancied this *Dacrydium* identical with the New Zealand *Dacrydium cupressinum*, Sol.; but this is a mistake. He also expresses his belief that the wood called Dakua salusalu is the produce of this tree, and in this he is supported by Mr. Storck, who, being now a permanent resident in Fiji, had ample opportunity to go into the question. My inquiries respecting the last-mentioned point have not been attended with success.

Nearly every native consulted pointed out a different tree as the source of that timber. Mr. Pritchard also took some pains about it, as the subject was brought before him in his consular capacity. A resident in Ovalau had made a contract with a man for a supply of Dakua salusalu. When the timber was delivered, cut on Vanua Levu, it was found to be that of the common Dakua (*Dammara*), quite unlike the wood going by the name of Dakua salusalu in Ovalau. Payment being refused, the Consul's interference was invoked. There being no scientific work to which an appeal could be made, Mr. Pritchard solved the difficulty by deciding that, although the wood tendered might bear or bore the name of Dakua salusalu in Vanua Levu, it was not the one recognised by that name in Ovalau; and whereas the contract had been entered into in the latter island, only such wood as was called "Dakua salusalu" *there* need be paid for.

The Nokonoko (*Casuarina equisetifolia*, Forst.) produces a wood much used for clubs and all purposes in which hardness and heaviness is an object. It is most frequent in the eastern parts of the group, its prevalence indicating a poor soil. Its sombre aspect, and the wailing sound caused by the playing of the breezes in the branches, forcibly appeal to the poetical sentiment; hence the Nokonoko is planted in masses about tombs, and a fine grove of that kind is seen at Lakeba, surrounding the burial-place of a departed chief. The young branches are drooping, imparting to the tree a peculiarly graceful look, and forming a beautiful contrast to the erect and rigid growth of its congener, the

Velao (*Casuarina nodiflora*, Forst.), which is occasionally met with in its company, and also yields a useful timber Whilst the Nokonoko assumes a more or less pyramidal form, is scarcely ever higher than forty feet, and has a greyish hue, the Velao is often sixty feet and even more in height and three feet in diameter, and has a green mossy-looking crown, which, by its flatness on the top, reminds one of the stone-pine so characteristic of the Italian landscape. The Velao almost invariably grows in good soil, generally in mixed forests; whilst the Nokonoko shuns, as it were, a close contact with other kinds of trees, and it scarcely ever associates with any save the Balawa or Screw-pine (*Pandanus odoratissimus*, Linn.).

The Dilo (*Calophyllum inophyllum*, Linn.), a sea-side tree, grows to a large size, and its wood is used for canoes and boats. Several of the little coasting vessels, cruising about Fiji, are almost exclusively built of it and the Vaivai (*Serianthes Vitiensis*, Gray); their masts being supplied by the Dakua (*Dammara Vitiensis*, Seem.). Dilo wood has, besides, a beautiful grain and takes a fine polish. Allied to the Dilo is the Damanu (*Calophyllum Burmanni*, Wight), a large inland forest tree, also furnishing materials for boats, canoes, masts, and all kinds of carpentry. The Tivi (*Terminalia Moluccana*, Lam.), a littoral tree, and its congener, the Tavola (*Terminalia Catappa*, Linn.), add their share to the Fijian woods. That of the Tavola is made into drums called "Lali," the beating of which is resorted to when distinguished guests arrive, on festive occasions, or to call the Christians to Divine service; and it is a curious coincidence, but certainly nothing more save a coincidence, that the ancient Egyptian term for rejoicing was "lali," as in the Arabian song of '*Doos ya-lel-lee.*' These drums are beaten with two short and thick pieces of wood, and the sound produced can be heard within a circle of several miles. Great praise is bestowed on the Mulomulo (*Thespesia populnea*, Corr.), a tree common on the sea-beaches of the Eastern hemisphere, on account of the almost indestructible nature of its wood whilst under water. When fully developed it is about fifty feet high, and the stem from one to two feet in diameter, bearing heart-shaped leaves and flowers somewhat resembling those of the hollyhock, but changing their colour as the day advances,—a peculiarity they share in common with those of several other Malvaceous plants. Its thick foliage renders it suitable for avenues, and I have seen it planted for the sake of its shade both in Ceylon and the Hawaiian islands. The centre of old stems generally decays in the way our European elms do, and the wood towards that part presents a deep claret colour. The Mamakara (*Kleinhovia hospita*, Linn.) and the Marasa (*Storckiella Vitiensis*, Seem., so called in honour of my able assistant in the botanical exploration of Fiji, Mr. Jacob Storck) should not be omitted in a list of timbers. The Mamakara is from forty to fifty feet high, and rather a social tree, indicating its presence during the flowering season by its numerous and large panicles of pink blossoms. The Marasa, discovered on the southern side of Ovalau by Mr. Storck, is a noble object, attaining eighty feet or more in height, nine feet in girth, having a remarkably straight stem, a dense, dark-green foliage, pinnate leaves, flowers of a bright yellow colour, arrayed in terminal panicles, at first sight easily mistaken for those of a *Cassia*, and a curious cultriform fruit (*legumen*). A hard and durable timber is produced by the Sagali (*Lumnitzera coccinea*, Wight et Arn.), a tree with blackish wood, glossy foliage, and bright scarlet flowers, abounding in maritime swamps, as well as by another inmate of the same localities, the Dogo or mangrove (*Rhizophora mucronata*, Lam.). The sap of the latter has a blood-red colour, much employed by the natives, amongst whom it is almost as fashionable to dye their hair red as it was amongst the ladies of ancient Rome, after their roving husbands and brothers had become acquainted with the fair locks of the Teutonic race. When first put on, the sap is allowed to run freely over face and neck, producing an effect much like that a crown of thorns is represented as doing. On Nukubati, off the Macuata coast of Vanua Levu, I saw it employed by potters for painting their crockery. Just after the pots had been baked and were still quite hot, a mixture, consisting of this fluid and the sap of the Wakiwaki (*Hibiscus* [*Abelmoschus*] *moschatus*, Linn.), was used for that purpose, the colour of the paint remaining almost unchanged after the vessels had become cool and dry. The aerial roots of the Dogo being very elastic, offer good materials for bows, of which the Fijians avail themselves; whilst the fruit is made into bread (madrai) in times of scarcity.

The Vuga (*Metrosideros collina*, A. Gray), a tree with glossy foliage and scarlet flowers, yields a hard wood of good grain; and several other Myrtaceous plants, among them the Yasi dravu (*Eugenia rubescens,* A. Gray), are esteemed for their durable timber. A sea-side tree of middle size, the Tatakia (*Acacia* [§ *Phyllodineæ*] *laurifolia,* Willd.), has a hard wood, useful for axe-handles and smaller pieces of carpentry. The Qumu (*Acacia Richii,* A. Gray), another phyllodineous species, also yields a hard wood, even more useful, as the tree is larger than the last-mentioned, and supplies the paint with which the heathen natives blacken their faces, when they dress for war or wish to look particularly smart, hence "Qumu" paint. The Vaivai (*Serianthes Vitiensis,* A. Gray), often seen in company with the Qumu, produces one of the most valued of all Fijian woods; but the Vesi (*Afzelia bijuga,* A. Gray), which in outward appearance is not unlike our beech (*Fagus sylvatica,* Linn.), having the white smooth bark, the colour, and somewhat the shape of the leaves of that familiar forest-tree, is held in the highest estimation. It is used for canoes, pillows, kava-bowls, clubs, and a variety of other purposes, and seems almost indestructible. One of the most common tree-ferns, the Balabala (*Alsophila excelsa,* R. Br.), is much used for building purposes by the natives. Its trunks make excellent posts, lasting an incredibly long time, and possessing moreover the advantage of being almost fire-proof. After a house has been burnt down, these posts are almost the only trace that remains. It is also customary to make the ridge pole of houses and temples of this tree-fern, and to surround it with the Wa-Kalou (holy creeper), a species of that curious genus of climbing ferns (*Lygodictyon*), partially no doubt from some superstitious notions, but partially also to keep out the wet. The trunks of the Balabala, cut into ornamental forms, are frequently observed around tombs, temples, churches, and bures, presenting a pretty effect. The little sticks which the chiefs carry, stuck under their turban, and with which they scratch their heads, are also made of Balabala. The young leaves are eaten in times of scarcity, while the soft scales covering the footstalks, or more correctly speaking the stipes, of the fronds, are used for stuffing pillows and cushions by the white settlers, in preference to feathers, because they do not become so heated, and are a real luxury in a sultry tropical night. The Balabala is common all over the group, especially on the weather-side, and its trunk attains the height of about twenty-five feet, and eight or ten inches in thickness. The fronds form a magnificent crown of gigantic dimensions, rendering the plant a noble feature in the landscape.

Palms play an important part in the domestic economy of the natives. The Fijians are the only people who in their barbarous state had a collective term for the great natural order of palms, applying that of "Niu" to all those inhabiting their islands, and adding specific names to distinguish the one from the other; viz.:—

Niu dina = *Cocos nucifera*, Linn.
Niu sawa = *Kentia exorrhiza*, Wendl.
Niu niu = Cagicake = *Ptychosperma filiferum*, Wendl.
Niu soria = Sogo = *Sagus Vitiensis*, Wendl.
Niu masei = Sakiki = Viu = *Pritchardia pacifica*, Seem. et Wendl.
Niu Balaka = *Ptychosperma Seemanni*, Wendl.

The word "Niu" is common to most Polynesian languages, often taking the form of "Nia" and "Niau;" the New Zealand "Nikau," by which the Maoris designate their indigenous palm (*Areca sapida*, Sol.), does belong, and perhaps even "Nipa," the Philippine name of *Nipa fruticans*, may belong, to the same group of words. We further trace the Fijian "Niu," or with the article "a" (a niu) before it, in the Anao, Anowe, Anau, and Nu, by which names a sugar-yielding palm, the *Arenga saccharifera*, is known in different parts of the Indian Archipelago. The existence of a collective term for "palms" never having been pointed out, the passage in John xii. 13, "Took leaves of the palm-trees," is rendered both in the Viwa and the London edition of the Fijian Bible, "Era sa kauta na drau ni balabala,"— literally, "Took leaves of the tree-fern," for balabala is a tree-fern (*Alsophila excelsa*, R. Br.). "Niu" is the term that ought to have been used, there being two kinds of real palms in Syria, but no tree-ferns.

Only one of all the palms as yet discovered in Fiji is a fan-palm, the rest having pinnatifid leaves. This is the Niu Masei, Sakiki or Viu, a new genus of *Coryphinæ* (*Pritchardia pacifica*, Seem. et Wendl.), differing from all described ones in several important characters. The blades of the leaves are made into fans, "Iri masei" or "ai Viu," which are only allowed to be used by the chiefs, as those of the Talipot (*Corypha umbraculifera*, Linn.) formerly were in Ceylon. The common people have to content themselves with fans made of *Pandanus caricosus*. Hence, though there is not a village of importance without the Sakiki, or, as it is termed in the

Somosomo dialect, which suppresses the letter *k*, Saii, there are never more than one or two solitary specimens to be met with in any place, the demand for the leaves being so limited, that they prove sufficient to supply it. The fans are from two to three feet across, and have a border made of a flexible wood. They serve as a protection both from the sun and rain; in the latter instance the fan is laid almost horizontally on the head, the water being allowed to run down behind the back of the bearer. From this the Fijian language has borrowed its name for " umbrella," a contrivance introduced by Europeans, terming it " ai viu," that being one of the names by which fans are known. The leaves are never employed as thatch, though their texture would seem to recommend them for that purpose; the trunk, however, is occasionally used for ridge-beams. The palm seldom attains more than thirty feet in height. Its trunk is smooth, straight, and unarmed, and from ten to twelve inches in diameter at the base. The crown has a globular shape, and is composed of about twenty leaves, the petioles of which are unarmed and three feet four inches long, and densely covered at the base with a mass of brown fibres. The blade of the leaves is rounded at the base, fan-shaped, four feet seven inches long, three feet three inches broad, and when young, as is the petiole, densely covered with whitish-brown down, which, however, as the leaf advances in age, gradually disappears. From the axil of every leaf flowers are put forward, enveloped in several very fibrous flaccid spathes, which rapidly decay, and have quite a ragged appearance even before the flowers are open.

The inflorescence never breaks out *below* the crown, as it does in the Niu sawa (*Kentia? exorrhiza*, Wendl.). The spadix is three feet long, stiff and very straight, bearing numerous minute hermaphrodite flowers, of a brownish-yellow colour. The fruit is perfectly round, about half an inch in diameter; and, when quite matured, it has exactly the colour of a black-heart cherry, the outside having a slight astringent taste. The seeds germinate freely, and out of a handful thrown carelessly into a Wardian case in Fiji, more than thirty had begun to grow when they reached New South Wales, where they were taken care of in the Botanic Gardens, and will duly be distributed amongst the various establishments forming collections of rare and beautiful palms—for such this species certainly is.

The Niu sawa (*Kentia? exorrhiza*, H. Wendl.) is a pinnatifid palm of considerable beauty, of which there is a characteristic sketch, representing the vegetation of the Rewa river, in 'The Narrative of the United States Exploring Expedition.' This palm is found all over Fiji, ascending mountains to the height of two thousand feet. Mr. Charles Moore, of Sydney, met with it in New Caledonia; and there is reason to believe that it is also found in the Tongan group, where, as in Fiji, it is known by the name of "Niu sawa," I am told; "sawa," signifying "red" in Tonguese (and having no meaning in Fijian), being doubtless given on account of the fruit, which merges from bright orange into red. This palm is remarkably straight, and often more than sixty feet high. The trunk is unarmed, smooth, and of a whitish colour; it is a couple of feet above the base, from two to three feet in circumference. When the tree gets old, numerous aerial roots, all covered with spines, begin to appear, forcibly reminding one of the *Iriartea exorrhiza* in tropical America. The leaves are from ten to twelve feet long, pinnatifid, and the segments four feet long and two inches broad. Before expanding they are perfectly erect, looking like a pole inserted into the heart of the foliage; their petiole and midrib and veins are in that stage densely covered with a very short brown tomentum, which more or less disappears as the foliage advances in age. The flowers appear *below* the crown of the leaves, growing out of the old wood; they are enveloped in thick coriaceous boat-shaped spathes, which, unlike those of the Sakiki (*Pritchardia pacifica*, Seem. et Wendl.), are not subject to rapid decay. The spadix, on which the minute monœcious green flowers are inserted, is much branched, and the branches are "yarring," forming large bunches, which, when loaded with ripe fruit, are rather weighty. As many as eight of these bunches are often seen on a tree at one time in various stages of development. The fruit is ovate, acuminate, and about the size of a walnut. At first green, it gradually changes into bright orange, and ultimately merges into red at the base. The kernel has a slight astringent taste, and is eaten by the natives, especially by the youngsters. The wood is used for spars. Fine specimens of the tree, brought by Mr. Moore from New Caledonia, and by me from Fiji, are cultivated at the Sydney Botanic Garden.

The Niu Niu, or as it is more commonly termed, Cagicake (*Ptychosperma filiferum*, Wendl.), is found in the depth of the forest, where it shows its feathery crown above the surrounding trees, forming what St. Pierre poetically called "a forest above a forest," and what the Fijians less skilfully wished to express by the name of Cagicake, literally "above the wind." Before I had seen the fruit the natives described it to me as being exactly the same shape and colour as that of the Niu sawa, but only very much smaller in size; and in this they were pretty correct. Whilst the fruit of the Niu sawa is as large as a walnut, that of the Cagicake is about the size of a coffee berry. The trunk is smooth, unarmed, and about eight inches in diameter, furnishing capital material for rafters, which the natives declare are so durable that they last for ever. The leaves are pinnatifid, ten to twelve feet long, and the lowermost segments being narrower, and at least three or four times as long as the uppermost, hang down in long fringes. When in the dusk of the evening I first encountered this singlar palm on the Macuata coast of Vanua Levu, it was this peculiarity that first attracted my attention, otherwise I should have taken it to be a Niu sawa. It was pitch-dark before the tree was felled and dragged out of the thick jungle in which it grew, when passing my fingers over the surface of the segments, I felt a thick marginal and elevated vein, which at once assured me that an undoubtedly new addition had been made to my collection. The disproportionate length of the lower segments, and the thick marginal vein pointed out, though they had been first discovered in the absence of regular daylight, are amongst the most striking peculiarities, and ought to be seized upon by those giving a popular description of this palm; the upper segments are four feet long and three inches broad. The spadix, like that of the Niu sawa, is much branched, and may be said to be a miniature imitation of it. The palm is found both in Vanua Levu and Ovalau, and doubtless also in Viti Levu, for a palm which grows in the interior of the latter islands, and is termed about Namosi " Tankua," must, from the description given to me by natives, be identical with the Cagicake. According so the superstitious notion of the inland tribes of Viti Levu, the diminutive fruit of the Tankua and those of the Boia (*Heliconia?* sp.), a plantain-like species, is the chief food of the Veli, spirits half fairy, half gnome, with a fair complexion and diminutive body. The Tankua is their cocoa-nut, the Boia their plantain, and the Yaqoyaqona (*Macropiper puberulum*, Benth.), their kava plant, none of which mortals can destroy or injure without exposing themselves to the danger of being severely punished by those dwellers in the forests, the Veli.

The Balaka (*Ptychosperma Seemanni*, Wendl.) is a diminutive palm, growing as underwood in dense forests. It was met with both in Vanua Levu, on the southern side, and on the mountains of Taviuni. The trunk is remarkably straight, ringed, and about an inch in diameter when fully developed. On account of its strength and straightness it is used for spears by the natives, and would make good walking-sticks. The leaves are pinnatisect, about four feet long; and the segments are erosodentate at the point, like those of *Caryota* and *Wallichia*. The flowers appear below the few leaves, forming the crown of this, the smallest of all Fijian palms.

In Wilkes's 'Narrative of the United States Exploring Expedition,' mention is made of a *Caryota*, as growing in Fiji, and being used for rafters in building. "Its straight stem, with its durable, hard, and tough qualities, render it well adapted for this purpose." No one has subsequently met with a true *Caryota*, one of the most remarkable genera: and I fancy that the botanists of Wilkes's expedition may have mistaken the erosodentate leaves of a timber-yielding palm, probably *Ptychosperma Vitiensis*, Wendl., abounding in some parts of Viti Levu, for those of a *Caryota*. It is about forty feet high, has a smooth trunk, pinnatifid leaves, and was seen by me at Nukubalavu. I have not been able to learn its native name. Two other species, the sago and the cocoa-nut palm, already treated of above, and three discovered by the United States Exploring Expedition, augment the list of Fijian palms to ten.

Ornamental plants are highly appreciated by both natives and white settlers, especially those having either variegated leaves or gay-coloured flowers, since the Fijian flora shares with that of most islands the peculiarity of possessing only a limited number of species displaying gay tints. Those most frequently seen about the native houses are what gardeners call "leaf plants," including the Danidani (*Panax fruticosum*, Linn.), with its deeply-cut foliage, several beautiful varieties of the *Dracæna ferrea*, some of which have been introduced from various Polynesian islands, the *Croton pictum*, the indigenous *Acalypha virgata*, Forst., termed Kalabuci damu, the foliage of which changes from dark-green to brown, yellow and scarlet, and two kinds of ornamental grass (*Panicum*), the one having purple, the other variegated leaves. The couch-grass is also spreading fast through the islands, and there is a fine lawn of it in front of the king's house at Bau, blending well with the number of fine shrubs and trees which, at Mrs. Collis's instigation, were planted around the royal residence. Of the Kauti, Senitoa, Senicicobia, or Shoe-black plant (*Hibiscus Rosa-sinensis*, Linn.), a single pink and purple as well as a double variety are cultivated. When the *Cassia obtusifolia* and *Cassia occidentalis* were first brought to Fiji, the natives took them under their special protection, and disseminated them freely, being highly pleased with their leaves "going to sleep" at night, whence the names of Mocemoce and Kaumoce, *i.e.* sleeping plants. But they became weary of their pets when it was found that they speedily proved two most troublesome weeds, which, in common with the *Datura Stramonium*, *Euphorbia pilulifera*, *Plantago major*, *Erigeron albidum*, and other foreign intruders, caused them a great deal of additional labour.

Most of the white settlers have little gardens in which all flowers derived from warm countries are grown with great success. The pride of Barbadoes (*Poinciana pulcherrima*, Linn.), both the red and yellow variety, may be seen in perfection; the same may be said of the white trumpet-flower (*Brugmannsia candida*, Pers.), the balsam (*Impatiens Balsamina*, Linn.), the *Quamoclit vulgaris*, Chois., the scented Acacia (*Acacia Farnesiana*, Willd.), the blue *Clitoria Ternatea*, Linn., the *Gomphrena globosa*, Linn., *Vinca rosea*, Linn., *Calendula officinalis*, and the well-known Marvel of Peru (*Mirabilis*

Jalapa, Linn.). Prince's feathers (*Amarantus cruentus*, Linn.), and its congener, Driti damudamu (*Amarantus tricolor*, Linn.), have become perfectly naturalized in some districts. Attempts to grow the flowers of colder regions have not been so successful. Carnations are kept alive with difficulty; roses, though growing and blooming freely, possess little or no scent, and are chiefly valued from the pleasing associations connected with them; dahlias were introduced in 1860, by Dr. Brower, but I have not yet learnt the fate that attended them; a species of honeysuckle (*Lonicera*), noticed on the mission premises at Viwa and Bau, concludes the limited list of foreign garden plants cultivated in Fiji, a list, for any additions to which the inhabitants would feel very grateful.

The natives do not content themselves with merely looking at or smelling plants, but profusely decorate their persons with them: elegant-formed leaves, passion flowers, the bright-red leaves of the dracænas, or the bleached ones of the stemless screw-pine, are made to grace their heads or turbans. Great aptitude is displayed in making necklaces (taube or salusalu), the materials for which are principally furnished by monopetalous, white, and odoriferous flowers, strung upon a piece of string. I noticed those of the Bua (*Fagræa Berteriana*, A. Gray), Buabua (*Guettarda speciosa*, Linn.), Vasa or Rewa (*Cerbera lactaria*, Ham.), and Sinu dina (*Leucosmia Burnettiana*, Bth. = *Dais disperma*, Forst.). The flowers of the Sinu dina, or as it is also termed Sinu damudamu, are capitate, and the necklaces made of them are called " sinucodo," a term also applying to a chain. The shrub is about fourteen feet high, has fine dark-green shining foliage, odoriferous flowers, which on opening are pure white, but gradually change to cream-colour, and bright-red drupes, about as large as a hazel-nut.

Numerous plants serve for miscellaneous purposes. The flat round seeds of the Walai (*Entada scandens*, Bth. = *Mimosa scandens*, Linn.), called "ai Cibi," or "ai Lavo," have suggested to the Fijians a comparison with our coins, and supplied a word for money (ai Lavo), of which their language was formely destitute, because that article was entirely unknown to them, all commercial exchange being carried on by barter. The Walai or Wataqiri is a creeper, always associated with mangroves and other maritime vegetation. Its stem, when young used in place of ropes for fastenings, occasionally attains a foot in diameter, and forms bold festoons, whilst its pods arrest attention by their gigantic dimensions, measuring as they do several feet in length. The greyish bony involucre of the Sila, or Job's tears (*Coix Lacryma*, Linn.), a grass growing in swamps and having the aspect of Indian-corn, as well as the seeds of the Diridamu, Quiridamu, or Leredamu (*Abrus precatorius*, Linn.), which resemble those of the Drala (*Erythrina Indica*, Linn.) in having a bright red colour and a black spot, are affixed with breadfruit gum to the outside of certain oracle boxes, of which Wilkes has given fair illustrations in his 'Narrative of the U.S. Exploring Expedition.' These boxes have a more or less pyramidal shape, and are kept in the temples, as the supposed abode of the spirit consulted through the priests. Toys, consisting of cocoa-nut shells, and covered with these materials, are occasionally seen in the hands of native children, and they have rather a pretty effect. The bamboo, vernacularly termed "Bitu," is represented by two species, a large and a small one, both of which are rather local in their geographical range. The trunk of the larger is in general use for vessels to contain water, some of which are six feet long. It requires a certain knack, with some difficulty acquired by foreigners, to pour the water out of the small hole on one side of the upper end without spilling some of the contents. The natives drink out of these vessels by pouring the water in their mouth without allowing their lips to touch them: sipping the fluid as we do would be considered an act of impropriety. Bamboo split up in narrow strips makes capital torches, which do not require, as has been stated, to be dipped in cocoa-nut oil in order to make them give a clear and bright light. Fishing rafts, pillows for sleeping, instruments for beating time to national songs, pan-flutes, fences for gardens and courtyards,—all are constructed of these giant grasses. At Nagadi, in Viti Levu, I visited a heathen temple surrounded by a bamboo fence, some of the sticks used being the young shoots entire, with unexpanded leaves, and looking like so many fishing rods. The priest in charge of this building exhibited a bundle of bamboos, which on being struck on the ground with the opening downwards produced a peculiarly loud and hollow sound. Two single bamboos of different lengths are beaten contemporaneously with this large bundle in religious ceremonies. An amusing sight is presented by a grove of bamboos on fire. When returning from Namosi, I passed several places where, to clear the land, fire had been set to these groves. As soon as the flame fairly embraced the canes a loud explosion succeeded, the general effect of which being that of a well sustained skirmish between two hostile parties of sharp-shooters. In Ecuador I once saw a sugar-cane plantation on fire, but the noise of the bamboo by far exceeded that caused by the former. The leaves of the Qangawa, a species of pepper (*Piper Siriboa*, Linn.), climbing and rooting like our ivy, and, if report may be trusted, those of the Vusolevu (*Colubrina Asiatica*, Brongn.) are used for washing the hair, to clean it and destroy the vermin. The Moli kurukuru (*Citrus vulgaris*, Risso) serves the same purpose, a remark also applying to the vine called Wa roturotu (*Vitis saponaria*, Seem.), the stem of which, especially the thicker part, is cut in pieces from a foot to eighteen inches long, cooked on hot stones, and when thus rendered quite soft produces in water a rich lather almost equal to that of soap. The fruits of the Vago, or bottle-gourd (*Lagenaria vulgaris*, Ser.), are readily converted into flasks for holding oil and other fluids, by allowing their pulp to undergo decomposition. The juice of the Vetao or Uvitai (*Calysaccion obovale*, Miq.), a useful timber-tree, yields a dye, at present only employed by the natives for changing their black hair into red; but when it is remembered that its congener, the *Calysaccion longifolium*, Wight (= *C. Chinense*, Wlprs.), furnishes the buds known as the Nag-kassar of Indian commerce, it is not unlikely that the Vetao or Uvitai may yet be turned to better uses.

This enumeration by no means exhausts the catalogue of the useful products in which a Flora of about a thousand different species, such as the Fijian is, abounds. Enough, however, has been stated to show how bountiful nature has been in supplying these islands with edible roots and fruits, with drugs, spices, fibres, timber, dyes, vegetable fats, and other articles of commercial importance. The long list of cultivated plants shows that the natives are not ill prepared for entering on agricultural operations on a large scale, whilst the fact that the varieties of the different products grown are almost endless, furnishes a striking proof of their succeeding to perfection. The numerous plants introduced from every direction of the compass, and their successful naturalization, may justly be regarded as indicative of the climate being of that happy medium which, in a similar way, enables the English gardener to assemble in his domain a far greater collection of species than his continental rival.

CHAPTER XVIII.

REMARKS ON THE FAUNA OF FIJI.—MAMMALS.—BIRDS.—FISHES.—REPTILES.—MOLLUSKS.—CRUSTACEA.—INSECTS.—LOWER ANIMALS.

No attempt has as yet been made to draw up a list of the animals of Fiji, and all the materials for it are scattered through various periodicals and other publications. There are very few mammals in the group; indeed, except the rat (Kalavo), four Cetaceous animals, and five species of bats, collectively termed Baka, we have none belonging to this fauna. One of these bats or flying foxes has been named *Notopteris Macdonaldii*, in honour of its discoverer. Three of them are tailless, two have tails. There are two kinds of porpoises and two of whale in the adjacent seas and amongst the islands, but, though whales' teeth are highly valued, and were so still more formerly, the Fijians have never taken to whaling in any form, and always seem to have purchased their stock from foreign traders. The dog (Koli), the pig (Vuaka), the duck, and the fowl (Toa) were the only domestic animals known to the natives. Dogs were not eaten and suckled by the women, as was and is the case in other Polynesian islands; indeed, the custom of eating dogs seems to have been restricted in the Pacific to the islands and countries north of the line, and was apparently brought from the Sandwich Islands to Tahiti. The white settlers have introduced cattle, horses, goats, sheep, rabbits, and cats, all of which seem to thrive well. The horses are as yet few in number, and they are not much valued, as most inhabited places can be reached by water, and there are as yet no roads in the large islands. The terror of the natives at first seeing a horse and a man on its back seems to have been quite equal to that recorded of the ancient American nations; they ran away in wild dismay, or climbed trees and rocks to get out of the reach of the monster. Cattle succeed well; and I saw some very fine young bullocks on Kadavu, the property of Mr. Boyce. Fijians not fencing in their plantations, they have rather a dislike to cattle, and in some instances they have killed them, as their crops have frequently suffered from their devastation. They are very fond of beef, and as there was no native name for it, they have compounded one, calling it "Bulla-ma-kau," because it is derived from a bull and a cow. Goats have become very numerous, and most of the white settlers have flocks of them for the sake of their milk; but I am not aware that any of the natives have as yet reared any. Sheep were first introduced, if I am rightly informed, by Dr. Brower, the present American Consul, and several extensive sheep-runs have lately been bought on the northern shores of Viti Levu and Vanua Levu by British subjects from Australia. It was formerly supposed that the climate of Fiji was too warm for sheep, but that does not seem to be the case. Some specimens of Fijian wool were sent to the London Exhibition of 1862. "We find sheep answer well," writes a friend to me; "the wool grows rapidly, the sheep fatten well, and the ewes breed rapidly, frequently having three at a birth, so that we can by-and-by export wool as well as cotton. In one of the boxes sent to the Exhibition there is some wool of a sheep five months old, born on Wakaya, and the property of Dr. Brower." Cats are now quite common, and the natives have taken to them in order to kill the mice and rats which European vessels have introduced.

Birds are much more numerous than mammals. I have a list of forty-six different species, among them parroquets, owls, bitterns, teal, hawks, ducks, pigeons, etc. The feathers of some of them are collected for ornamental purposes, and the high value set upon the Kula (*Coriphilus solitarius*, Latham) has already been noticed. Ducks and pigeons, excellent eating, are very abundant, the former about the rivers, the latter in the woods. The fowls (Toa*) which the natives had were very small, and could scarcely be termed domesticated, indeed they have become perfectly wild in many districts. Europeans have introduced better kinds, and also turkeys, but I do not remember seeing any geese. I fancy that the domestic ducks must have come to the islands early in this century from some Spanish ships.

* Toa is the Fijian form of the word "Moa," applied throughout Polynesia to domestic fowls, and by the Maoris to the most gigantic extinct birds (*Dinornis* sp. plur.) disentombed in New Zealand. The Polynesian term for birds that fly about freely in the air is Manu or Manumanu, and the fact that the New Zealanders did not choose one of these, but the one implying domesticity and want of free locomotion in the air, would seem a proof that the New Zealand Moas were actually seen alive by the Maories, about their premises, as stated in their traditions, and have only become extinct in comparatively recent times.

My list contains a hundred and twenty-one species of fish. Some of them are excellent eating; indeed a great part of the native food is derived from this source. They are secured by nets, spears, fish fences, or stupefaction, by the different plants enumerated above (p. 339). The night is a favourite time for fishing on the reefs, and large parties are made up, chiefly women, who, torch in hand, traverse the reefs laid bare by the ebb-tide, and gather what they can. Such a fishing party is a pretty sight; and when suddenly disturbed from my sleep by shouts and merry laughter, I have often watched the long lines of torches moving along in the depth of night on the shores of Ovalau. The fences made in the sea are constructed with great care, and so that the fish will enter them in large bodies and have little chance of escaping. There were generally some about Lado, and baskets full of their produce were daily sent to us as presents. The fences were not allowed to remain for more than a few days in the same place, as the natives maintained that the fish become aware of their existence and would not enter them. Besides the edible fish, there are a number of different sharks about the group, and one hears of frequent accidents caused by them. The natives, being excellent swimmers, do not mind being capsized in their canoes, but are in great dread of the sharks. The latter are called collectively "Qio," and nine salt-water and several fresh-water species are enumerated. One day we encountered a very large one on the reef, where he had been left in a shallow pool by the receding tide. Our boat being near, an axe was fetched to kill him, but no sooner did he catch sight of the weapon than he made off in great haste, moving along over many hundred yards of dry reef like a serpent, without our being able to stop him. There is a curious tradition about a species of sole called "Davilai." Mr. Davilai used to be the leader of the songs amongst the fishes, and one day, when all his band were together and he was requested to commence the strain, he obstinately refused to comply. Enraged at such behaviour, the other fishes trod him under foot till he became flat; and hence, when a person refuses to pitch a song, the proverb is, "Oh, here is Mr. Davilai." There is also a most beautiful fish, about as large as a gold fish and of the finest ultra-marine colour; it is very frequent about the coral-beds, and a finer sight can scarcely be imagined than this creature playing in the crystal water over what looks like so much mosaic-work.

Reptiles are comparatively few in species. There are about ten different kinds of snakes, but none of them larger than about six feet. A good many inhabit trees, and often drop down; some are eaten. Snakes are collectively termed "Gata," and every species has a distinctive name. A large frog, Boto or Dreli (*Platymantis Vitianus*), is common about the swamps. There are three kinds of turtle, collectively known as "Vonu." The green turtle is called "Vonu dina," and that which yields the shell—the tortoise—"Vonu taku." But there is besides one which the natives term "Tovonu," said to be from six to ten feet long; however, I never have seen it; those which the chiefs often have in their turtle-ponds are the two first-mentioned kinds. The lizard tribe is represented by a chameleon and four other species. The largest is *Chloroscartes fasciatus*, Günth., with a body two feet long, and of a beautiful green colour, somewhat like that of the German tree frogs; indeed, the *Chloroscartes* inhabits trees, and I had one alive for some time. Crocodiles are not indigenous, but about the beginning of this century a large one made its appearance in Fiji, probably having been drifted thither from the East Indies. The natives, as related by Mariner ('Tonga,' vol. i. p. 334), fancied it had come from Bulu,—from heaven, —and they had some difficulty in catching it, not, however, before it caused some mischief.

There is a great variety of both salt, fresh-water, and land shells, probably several hundred species, and a number of them are quite peculiar to Fiji. The collective name for shells is " Qa ni Vilivili," Vilivili being the animal, Qa the shell. The most famous Fijian shell is the orange cowry (*Cyprœa aurantium*, Martyn), which is found in no other part of the world, though some works state it to have been found in Tahiti—an error originating in Mr. Cuming having purchased a single specimen in that island. There are several other cowries also used, as the orange cowry is, for necklaces and ornaments by the natives. Canoes, houses, temples, and churches are frequently decorated with the Buliqaqau (*Ovulum ovum*, Sowb.), not the *Cyprœa ovula*, as stated in some works. Several other species of shells are also used for ornamental purposes; the Sóbú or Sovui is on that account much valued. Armlets (Qatos) are made of the Sici, Taluvi, Tebe, Tebetabe, or Toru (*Trochus Niloticus*, Linn.). A pearl-oyster shell, Civa or Cove of the natives, is ground, and serves for ornament. Some fine pearls have occasionally been found, but actual pearl fishery has as yet not commenced on a large scale; and the Fijians in some of the islands act on the idea, that in order to preserve these treasures they must be boiled. The Davui (*Triton variegatus*, Lamk.) is made into horns and trumpets, invariably found in all larger canoes. Ai Kaki or Ai Koi, a species of *Dolium*, is used for scraping, as is also another univalve, the Tuasa or Ai Walui. Several kinds of oysters are eaten, and a fresh-water *Cyrena* is made into soup.

Crustaceous animals are well represented. Shrimps, prawns, crayfish, lobsters, and crabs, are plentiful and esteemed as food by the natives. In some of the smaller islands, for instance Qelebevu and Vatuvara, a very large kind of land crab, called " Ugavule " (probably *Birgos latro*, and the same of which C. Darwin speaks in his 'Journal of a Naturalist'), is common. Being fierce and strong, it is taken with some difficulty when on the ground, and throws earth and stones into the face of its pursuers. It climbs the highest cocoa-nut trees, and not only pierces the nuts, but removes the husk from the old nuts and breaks them, in order to get at the flesh. When up a tree, the natives take a bundle of grass and bind it round the body of the tree, about halfway up. The Ugavule comes down backwards, and when it gets to the grass it fancies the bottom has been reached, and, relinquishing its hold on the tree, falls twenty or thirty feet, and thus stunned is easily captured.

The insect tribe is very numerous, both in species and individuals. Mosquitoes (Namu) are very troublesome in some parts, as has already been related; and equally irritating are the flies (Lago), which keep one's hands constantly employed, and in order to have a meal in peace a boy must be kept continually employed in driving them away. Fleas, to finish the catalogue of irritants, are not so plentiful as I have found them in Spanish America or Southern Europe, nor are foreigners much troubled by the vermin so abundant in the large heads of hair worn by the heathen natives. Cockroaches are swarming in most houses, canoes, and vessels, and often disturb one during the night, not only by running over one's body but also by attacking it in right earnest. Some very fine beetles and butterflies are met with; and at dusk the woods begin to swarm with myriads of fireflies. Highly curious are what are popularly termed leaf- and stick-insects, species of *Mantis;* the wings of some of them can scarcely be distinguished from real leaves. Some large kinds of spider, amongst them a stinging one, have to be noticed. Centipedes, nearly a foot long, were frequently encountered by us in the woods, and scorpions are more frequent than one could wish.

There is a goodly display of the lower evertebrate animals, amongst them a long series of sea-slugs, seacumbers, and *bêche-de-mer*, annelidans, starfish, and medusas.

It would well repay a zoologist who has some funds at his command—without them he must not go to this expensive place—to spend a couple of years in investigating the Fauna of Fiji. Judging from what has been collected, mostly in great haste, a number of new genera and species may be expected from a thorough zoological examination of the group.

CHAPTER XIX.

FIJIAN RELIGION.—DEGEI, THE SUPREME GOD.—INFERIOR DEITIES.—WORSHIP OF ANCESTORS.—IDOLIZED OBJECTS.—TEMPLES.—CREATION AND ULTIMATE DESTRUCTION OF THE WORLD.—A GREAT FLOOD.—IMMORTALITY OF THE SOUL.—CONCEPTION OF FUTURE ABODE.—PROPS OF SUPERSTITION.

THE supreme god in Fiji is Degei (pronounced Ndengei), known in the other groups of Polynesia as Tanga-roa, or Taa-roa; Tanga being his proper name, "roa" an adjective, signifying 'the far removed,' perhaps also 'the most high.' To him is attributed the creation and government of the world; and no images of him are made, nor of any of the minor gods, collectively termed "Kalou." His sway is universally acknowledged in Fiji, and no attempts are ever made to elevate any local gods above him. For this reason I think that in teaching our Christian religion it would have been advisable to select the name of Degei for the Supreme Being rather than that of "Kalou," which seems to be used not only collectively for all gods, but also for anything superlative, good or bad. When the natives saw us doing anything inspiring them with admiration or surprise, they would say, "Ah, you are Kalous," which, of course, could not be translated, 'You are gods,' but 'You are clever fellows!—men of genius!' etc. As no images were ever made of Degei, nor indeed any other god, it would have been very easy to strip the conception of him of any heathen superstitions. Degei, like Jupiter, had a bird, and is supposed to be enshrined in a serpent,—the world-wide symbol of eternity,—lying coiled up in a cave of Na Vatu, a mountain on the Rakiraki coast of Viti Levu, indicating his turning about by occasional shocks of earthquakes. (Compare p. 223). Some traditions represent him with the head and part of the body of a serpent, the rest of his form being stone, emblematic of everlasting and unchangeable duration; in fact, Degei seems to be the personification of eternal existence.

Besides Degei, there is a host of inferior gods, but their rank is not easily ascertained, as each district contends for the superiority of the deity it has adopted and specially worships. Tokairabe and Tui Lakeba Radinadina seem to stand next to Degei; they are his sons, and act as mediators in the transmission of prayers to their father. Rokomoutu is a son of Degei's sister, and insisted upon being born from her elbow. Some of the gods find employment in Bulu, some on earth, and the latter are the tutelary deities of whole tribes or individuals; thus Rokova and Rokola are invoked by the carpenters, Roko Voua and Vosavakadra by the fishermen, whilst every chief has a god in whom he puts his special trust.

One of the most universally known gods is Ratu mai Bulu; he is the Ceres of Fiji, and comes once a year from Bulu to cause the various fruit-trees to blossom and yield fruit. During his stay it is forbidden to do most kinds of work, to go to war, sail about, plant, build houses, beat the drums, or make much noise, lest he should take offence and depart with his work unfinished. In December the priests bathe Ratu mai Bulu, and then announce his departure from earth by a great shout, which is quickly carried from village to village, from town to town.

One of the most universal beliefs of all mankind is, doubtless, that in the aid or protection departed ancestors are able to afford. All nations participate in it more or less, and even Christianity has not been able to uproot an idea which poetry and art have rivalled to perpetuate. What educated man could be so cruel as to wish to prove to an orphan child, left alone in the wide world, that, according to strict orthodoxy, the spirit of its mother could not possibly watch over it, because the lost one would quietly slumber in her grave till the great day of judgment? The Chinese, Japanese, South African tribes, and Polynesians, do not clothe their ideas in so poetical a garb, or banish admiration for the mighty deeds of their ancestors from the region of religious sentiment. They supplicate their formidable shades when misfortune befalls them, or fear of the future takes possession of their minds. With the Fijians, as soon as beloved parents expire, they take their place amongst the family gods. Bures, or temples, are erected to their memory, and offerings deposited either on their graves or on rudely constructed altars—mere stages, in the form of tables, the legs of which are driven in the ground, and the top of which is covered with pieces of native cloth. The construction of these altars is identical with that observed by Turner in Tanna, and only differs in its inferior finish from the altars formerly erected in Tahiti and the adjacent islands. The offerings, consisting of the choicest articles of food, are left exposed to wind and weather, and firmly believed by the mass of Fijians to be consumed by the spirits of departed friends and relations; but, if not eaten by animals, they are often stolen by the more enlightened class of their countrymen, and even some of the foreigners do not disdain occasionally to help themselves freely to them. However, it is not only on tombs or on altars that offerings are made; often, when the natives eat or drink anything, they throw portions of it away, stating them to be for their departed ancestors. I remember ordering a young chief to empty a bowl containing kava, which he did, muttering to himself, " There, father, is some kava for you. Protect me from illness or breaking any of my limbs whilst in the mountains."

Besides their regular gods and deified spirits, the Fijians have idolized objects, such as sacred stones, trees, and groves, of which I have already spoken (p. 87); and in addition to these, certain birds, fishes, and some men, are supposed to have deities closely connected with or residing in them. He who worships the god inhabiting a certain fish or bird, must of course refrain from harming or eating them.

All Fijian temples—at least those about the coast— have a pyramidal form, and are often erected on terraced mounds, in this respect reminding us of the ancient Central American structures We meet the same terraced mounds also in Eastern Polynesia, with which Fiji and all other groups of the South Sea share the principal features of religious belief.

FIJIAN TEMPLE (BURE KALOU).

There is in most of them a shrine, where the god is supposed to descend when holding communication with the priests, and there is also a long piece of native cloth hung at one end of the building, and from the very ceiling, which is also connected with the arrival and departure of the god invoked. The revelations, however, are made by means of the spirit of the god entering the body of the priest, who, having become possessed, begins to tremble most violently, and in this excited state utters disjointed sentences—supposed to be the revelations which the god wishes to make by the mouth of his servant. It is the oracle at Delphi over again. Mankind will be deceived, whether by a Fijian priest, a Grecian Pythia, or an American spirit-rapper.

The conceptions which the Fijians have of the origin of their islands is, that they were made and peopled by Degei. This god, when walking along the beaches, wore long trains of native cloth, like those worn by great chiefs at the present day; and whenever he allowed them to drag the ground, the beach, becoming free from vegetation, showed the white sand; whenever he took them up, and cast them over his shoulder, the trees and shrubs remained undisturbed.* What Humboldt pointed out as one of the characteristics of all religions is not wanting in that of Fiji. There is a tradition of a flood. Degei was roused every morning by the cooing of a monstrous bird, called "Turukawa," who performed his duty well until two youths, grandsons of the god, accidentally killed it with bow and arrow, and, in order to conceal their deed, buried it. Degei, accustomed to being roused at sunrise by his favourite bird, was greatly annoyed on finding it had disappeared, and he at once dispatched his messenger, Uto, all over the island in

* Williams ('Fiji and the Fijians,' p. 250) makes Roko Mouta, another god, take this walk.

search of it; but all endeavours to discover any traces of the lost one proved unsuccessful. The messenger declared that it could nowhere be found. Degei had a fresh search instituted, which led to the discovery of the body of the dead bird, and that of the deed which had deprived him of life. The two youths, fearing Degei's anger, fled to the mountains and there took refuge with a powerful tribe of carpenters, who willingly agreed to build a fence strong enough to keep Degei and his messengers at bay. They little knew the power they had attempted to balk. Degei, finding the taking of the fence by storm impossible, caused violent rains to fall, and the waters rose to such a height that at last they reached the place where the two youths and their abettors had fortified themselves. To save themselves from drowning they jumped into large bowls that happened to be at hand, and in these they were scattered in various directions. When the waters subsided, some landed at Suva, some at Navua and Bega; and it is from them that the present race of carpenters and canoe-builders claim to be descended.*

* The late Rev. J. Hunt has published a version of this story, which he himself terms as being between an imitation and a translation of the original. I quote a few verses. It begins with one of the boys trying his arrow:—

"'I'll try, I mean no harm, I'll only try,'
Pointing his arrow as he fix'd his eye:
His brother strikes his hand, the arrow flies,
And prostrate at their feet old *Turukawa* lies.

"Stretch'd on the fatal ground, upon his back,
They see the deadly arrow's fatal track;
His entrails all turn out, his flowing blood
Stains the white sand, and dyes the ocean flood.

"'This is no common bird,' one faintly said,
'His glaring eyes retain their crimson red;

Those who make a philosophical digest of such myths as these, will at once perceive the points of resemblance it exhibits with the Mosaic narrative:—The anger of the supreme god has been roused by certain transgres-

> His sacred legs, with many a cowry bound,
> Crash'd as the monster fell upon the ground.
>
> " ' My brother, can it be ? is this the bird
> Whose office long has been to wake the god
> Whose serpent form lies coil'd in yonder cave,
> Boasting the dreaded power to kill or save ? '
>
> " They strip him of his coat, by Nature given,
> And, lo, his feathers rise in clouds to heaven,
> Fly o'er the mountains on the gentle breeze,
> Cover the mystic grove of sacred trees.
>
> " A grave, at once convenient and secure,
> They find beneath the threshold of the door;
> They bury him with vows of self-defence,
> Should Degei's anger visit their offence.
>
> " The god lies sleeping, nor has power to wake;
> He turns himself, and rocks and mountains quake;
> When gloomy night has laid aside his pall,
> He lists intent for *Turukawa's* call.
>
> " Three suns have risen, but no call he hears;
> His heart now beats with boding god-like fears;
> The god, exhausted with suspense so sore,
> Sends Uto his dominions to explore.
>
> " 'Go search my favourite bird, my precious store;
> Oh, shall I never hear his cooing more?
> If distance weary, or the sun shall burn,
> Refreshing draughts shall wait thy glad return.
>
> " 'Go search 'mong tow'ring heights, 'mong vales beneath,
> 'Mong gloomy caverns, and the cloud-capp'd cliffs;
> There dwell the murderers, so report declares;
> Vengeance shall now absorb our god-like cares.' "

The result was, that Degei made war on the two youths, but without effect; he then caused a flood of water, with which they were drifted to the Rewa district.—The mystic grove of sacred trees referred to in verse 5, are the *Balawas* (screw-pines) at the top of Degei's mountain, which

sions, as a punishment for which a flood rises; and it is only by embarking—not in ordinary vessels—that certain people save their lives, afterwards to become the progenitors of a powerful race. But there is one essential difference. Whilst Noah and his family were saved *Deo volente*, the Fijian transgressors effected their escape notwithstanding Degei was resolved upon their destruction. Williams adds, that in all, eight persons were saved, and that two tribes of people became extinct, one of them distinguished by a tail like that of a dog.*

As the Fijians believe in the creation, so they believe in the ultimate destruction, of the world. This appears incidentally from their tradition of the *Daiga*, a species of *Amorphophallus*, the foliage of which consists of a single leaf, supported on a stalk two to four feet long, and spreading out somewhat like an umbrella. In the cosmogony of the Samoans, the office of having, by means of its single foliage, pushed up the heavens when they emerged from chaos, is assigned to this plant, and the Fijians recommend it as a safe place of refuge when the end of the world approaches, the *Daiga* being a "vasu" to heaven (Vasu ki lagi: see p. 304).

The immortality of the soul, and a life hereafter, is

are sacred. The spirits of the dead are said to throw a whale's tooth at these trees, that their wives may be strangled. When a shock of an earthquake is felt, Degei is turning himself. This, and a few other little things, are not in the original.

* The existence of savage tribes of people with a tail, somewhere in Africa, has as a popular belief been frequently alluded to in the newspapers. Dr. Kieser, the President of the Imperial Academy of Germany, has made numerous inquiries about them; and when Heuglin set out in search of Edward Vogel, his attention was particularly directed to this singular topic.

one of the canons of Fijian belief. It is from this conviction that, on the death of a man, be he chief or commoner, all his wives are strangled, so that he may not have to go alone on his journey or arrive at the future abode of bliss without anybody near and dear to him. Only in the christianized districts has this cruel custom been abolished. The Tonguese restricted the possession of a soul to chiefs and gentry, but the Fijians go further, allowing it not only to all mankind, but to animals, plants, and even houses, canoes, and all mechanical contrivances. The ultimate destination of the soul is Bulu, identical with the Tonguese Bolotu, and the general starting-place (Cibicibi) is supposed to be at Naicobocobo (= Naithombothombo), the extreme western or lee side of Vanua Levu, to which pilgrimages are occasionally made. It is not a little singular that the Fijians agree with the Tahitians, Samoans, Tonguese, and Maoris, in fixing this starting-place invariably on that side of their respective countries. The ancient Egyptians, it will be remembered, coincided with them in supposing their souls to depart westward.* But I must not accumulate coincidences. Those theory-spinners who are always on the look-out for traces of the lost tribes, and similar losses that give them uneasiness, might propound an hypothesis purporting to account for the westward movement common to the souls of the ancient Egyptians and the modern Polynesians, and, taking a hint from the incidental observation that Fijian temples have somewhat the shape of

* In Tahiti this place is called Fareaitu, in Samoa Fafa; the Maoris start from Cape Maria Van Diemen.

pyramids, and that "lali" in Egyptian means 'to rejoice,' and that "lali" in Fijian is the name of a drum-beater when people do rejoice, advance conclusions of a startling description.

About five miles east of Naicobocobo there is a solitary barren hill on the top of which grows a sacred screw-pine, which the soul of a married man must hit with the spirit of the whale's tooth,—remember, in Fiji all things have souls!—if he wishes to make sure of his wives being strangled to follow him to his future abode. A similar screw-pine stood on the east end of Vanua Levu, and was cut down by Chief Mara (p. 229); and I may further add that an identical belief attaches to some on the top of Degei's mountain: so that superstition seems to have placed these trees very conveniently within the reach of all who desired to avail themselves of their power.

It is by no means clear where Bulu, the ultimate abode of bliss, is situated, and whether it is, as in the Tonguese mythology, a distant island; but the fact that it cannot be reached except in a canoe shows that it is separated from this world by water, across which the souls have to be ferried by the Charon of Fiji. Before embarking they have to do battle with Samuyalo, the killer of souls, informed of their approach by the cries of a parroquet; should they conquer, they are allowed to pass on towards the judgment-seat of Degei, but if they should be wounded or defeated, they have to wander amongst the mountains. Again, if to any questions they should return untrue answers, Samuyalo gives the lie direct and fells them to the ground. Bachelors have a still greater difficulty to encounter, and stand scarcely any chance whatever of getting to Bulu. First they have to meet the spirit of a great woman, and, having eluded her fatal grasp, face a still more powerful foe. Naganaga, a bitter hater of all unmarried men, is on the look-out for them, and if he catches them, dashes them to pieces on a large black stone.

Some of the traditions speak of Bulu as Lagi (= Langi), the sky, the heavens; others again as being under the water: all however assert that in this future abode there are several districts. The names of Lagi tua dua, Lagi tua rua, and Lagi tua tolu, the first, the second, and the third heavens, are given to them by one set of traditions, and that of Murimuria and Burotu by the others. Murimuria seems to be a district of inferior happiness, where punishments and rewards are awarded. Burotu is the Fijian Elysium, where all that the natives most desire, value, and enjoy, is abundant. The manly nature of the Fijian is nowhere better displayed than in the conception of his future abode. He does not expect to exist there in indolent ease, reclining on soft couches, and sipping nectar handed by lovely houris, but hopes to resume all the out-door exercises to which he has been habituated during his stay on earth. Food will be plentiful, it is true, but there will be lots of canoes, plenty of sailing, fishing, and sporting —plenty of action. In fact, he hopes to lead very much the same life as he does here, and his admiration for fine, well developed people will be gratified; for, if accounts may be trusted, all will be larger than they were on earth. There does not seem to be any separation between the abodes of the good and the wicked, nothing that corresponds to our heaven and hell, no fire and brimstone. Punishment is evidently inflicted upon evil-doers in the same locality where the good enjoy their fair rewards. Women, not tatooed, are chased by their own sex, allowed no repose, scraped up with shells and made into bread for the gods. Men who have not slain any enemy are compelled to beat dirt with their club,—the most degrading punishment the native mind can conceive,—because they used their club to so little purpose. Others are laid flat on their faces and converted into taro-beds.

In order to uphold the whole fabric of heathen superstition, the priests had recourse to the same means which all religions have had in dealing with doubting minds. Punishment was sure to overtake the sceptic, let his station in life be what it might. What could be more terrible than that which was inflicted upon Koroika? He, a chief high in rank at Bau, made bold to doubt the existence of the god Ratu mai Bulu; and, as the god was then enshrined in a serpent of a neighbouring cave, he determined to put the question to the test. Embarking in a canoe with a cargo of fish, he steered for the very spot where the god was reported to be. On arriving, a serpent issued from the cave; and the chief asked, "Please, good Sir, are you the god Ratu mai Bulu?" "No, I am not," was the reply; "I am his son." The chief made him a present of fish, and requested an interview with his father. Presently another serpent appeared, but that proved to be the grandson, and the same present and request was made to him as had been made to the son. At length there issued a serpent, so large, so noble and commanding, as to leave little doubt in the mind of the chief that the god himself was now before him. Fish was presented to him; and just as the god was retiring with it, Koroika hit him with an arrow, and then retreated in all possible haste. But the voice of the god followed him, exclaiming, "Nought but serpents!— nought but serpents!" Arrived at home, and scarcely recovered from his state of agitation, he ordered dinner to be brought. The cover was removed from the pot, when, oh! horror, it was full of serpents! The chief seized a jug of water, saying, "At any rate, I will drink;" but, instead of the limpid fluid, he poured out crawling serpents. Unable to eat or drink, he sought comfort in sleep. He unrolled his mat, and was in the act of lying down upon it, when innumerable serpents appeared. Mad with excitement, he rushes out of doors, and passing a temple, hears, to his dismay, a priest revealing that the god has been wounded by the hand of a citizen, and that punishment will overtake the city. There is now no escape but to make a suitable atonement for the terrible offence committed. He returns home, collects all the valuables he can lay his hands on, presents them to the god, is pardoned, and his name handed down to unborn generations as a sceptic, and a fit example of the danger to which all men of his disposition expose themselves.*

A different but equally severe punishment awaited unbelievers in Bulu. One day, two young men paint and oil themselves, and put on a new piece of native cloth (just as the dead are prepared for the grave), and

* Compare Waterhouse, 'Vah-ta-ah,' p. 46.

approach Naicobocobo. One calls, " Please, Sir, we want a canoe to take us to Bulu." An invisible hand places a canoe, built of the timber of the breadfruit tree, within their reach. " Oh, Sir," said the spokesman, "we are not slaves; we want to go to Bulu like chiefs." The canoe is withdrawn, and its place supplied with one built of ironwood. No sooner is it near them, than the sceptics throw their spears at it, and exclaim, with a derisive laugh, " Oh, we are not going to die just yet." A voice was heard, " Young men, unbelievers, you have called for two canoes : they have not returned empty ; both have conveyed your own relatives. There is death in the houses of both of you." Thoroughly alarmed, they hurry home. The sounds of wailing are heard as they near their town. Both their mothers are dead.

But I must conclude, for fear that I may be served as Dr. Brower, the American Consul in Fiji, served a man residing on his estate at Wakaya, who nightly would persist in attracting all the boys of the neighbourhood by telling stories, and inflaming their youthful imagination to such an extent, that not one of them would stir abroad for fear of meeting some of the mighty personages to whom he had been introduced. Dr. Brower, not liking the whole troop to sleep on his premises, hit upon the expedient of requesting the story-teller to accompany every one of those he had frightened to his respective home, and, as the youthful listeners live in every direction of the compass, it takes him a good time to comply with the request; still, it does not prevent him from again and again indulging in his old weakness of telling fairy and ghost stories.

CHAPTER XX.

HISTORICAL REMARKS ON FIJI.—DISCOVERY OF THE ISLANDS.—SANDAL-WOOD TRADERS.—EARLY WHITE SETTLERS.—MISSIONARIES.—FOREIGNERS AT PRESENT RESIDING IN THE GROUP.—MY DEPARTURE FROM FIJI IN THE 'STAGHOUND.'—TERRIFIC STORM OFF LORD HOWE'S ISLAND.—ARRIVAL IN SYDNEY.—RETURN TO ENGLAND.—CONCLUSION.

BEFORE bidding farewell to the islands, I must say a few words about their history as connected with the white race. In the year 1643, Abel Jansen Tasman, when exploring the South Seas, discovered, between longitudes 19° 50′ E. and 180° 8′ W., a group of islands which he named " Prince William's Island," and which the inhabitants collectively term "Viti," and the Tonguese, who cannot pronounce the *v*, as well as other nations who have not this excuse, erroneously designate as " Fiji," spelt in a variety of ways. Although nearly two centuries have elapsed since the event, this archipelago of more than two hundred islands was only nominally known until visited by D'Urville and Wilkes; Captain Cook, who merely sighted Vatoa or Turtle Island, Captain Bligh, who twice passed through parts of this group, and Captain Wilson, of the 'Duff,' whose vessel was nearly lost on the reef off Taviuni, having scarcely added any save secondhand information to our stock of knowledge.

HISTORICAL REMARKS. 405

Towards the close of the eighteenth and the beginning of the present century, Viti began to be visited by vessels from the East Indies in search of sandalwood and *bêche-de-mer*, or Trepang, for the Chinese market. At that time the aborigines were regarded as ferocious savages, and great caution was exercised by the traders in dealing with them. The vessels were well armed, and none of the crew ventured on shore until chiefs of high rank had been sent on board as hostages, only to be given up after all business transactions had been concluded, and the loaded vessels were far enough at sea to be safe from surprise or any sudden attack. Some of these vessels were wrecked, on board of others mutinies occurred, and the crew took up its residence on shore; again, between some of the traders differences arose, which induced the natives to attack the foreign vessels, and kill the whole or portion of their crew. These were the materials which probably formed the first white immigration. In 1860, there was at Cakaudrove an old Manila man, named Jetro, who had been a boy on board a sandalwood ship, and who gave me a detailed account of the murder of the captain by the crew, the goods being given up to the king of Bau because no one was able to navigate the ship, which had to be abandoned, and it being thought best to purchase the goodwill of a powerful chief in order that the mutineers might have a protector. Jetro could give no clue to the date of this event, except that it took place shortly after Charles Savage had died, which would make it about the year 1814.

Charles Savage is said to have been a Swede by birth.

T. Williams* thought him to have been one of a number of convicts who in 1804 effected their escape from New South Wales; but, according to more authentic information,† he was an honest sailor belonging to the American brig 'Eliza,' wrecked in Fiji in 1808, and of which Dillon was mate. He seems to have possessed some redeeming qualities, was acknowledged as a head-man by the companions of his own race, and acquired great ascendency at Bau, the capital of the group. Up to this time the natives seem to have solely depended upon clubs, spears, and slings, for success in intertribal wars. The foreigners who had now come amongst them taught them the use of fire-arms, rendering the teachers highly welcome allies to the states then struggling for supremacy in the group. Bau and Rewa received them with open arms, and in return for their alliance gratified all their whims and demands, of whatever nature they might happen to be. From the ascendency thus acquired, it would have seemed that the absolute government of the whole Fijis lay within their grasp, if their ambition, rising beyond a life of indolence, had prompted them to consolidate and improve the power thus won; however, this was far from being the case. There is good proof that Savage at least made a fair attempt to take advantage of these favourable circumstances. Firmly establishing himself at Bau, in the very heart of the most powerful Fijian state, he exacted all the honours paid to exalted chiefs, and, knowing that no man can attain

---

* 'Fiji and the Fijians,' p. 3.

† Dillon, 'Discovery of the Fate of De la Pérouse,' vol. i.; Captain I. Erskine, 'Western Pacific,' p. 197.

position in Polynesia who is not a polygamist, he demanded a number of wives, amongst them some of the highest ladies of the realm. Thus far his native friends seem to have been willing to allow his carefully concealed plan to succeed. Every additional step in advance was rendered impossible; the natives were fully aware that if any of his sons whom a great chief, as Savage was considered to be, had by the daughters of powerful kings and leaders, should ever attain manhood, they would be in a position to exercise an unmitigated despotism, and set on foot a centralizing influence, which the centrifugal tendency of the Fijian mind has ever as strongly resisted as the Teutonic. According to Fijian polity, the sons of great queens, such as Savage had for his wives, would, in virtue of their right as "*Vasus*," or nephews, hold the territory and property of their uncles at their absolute disposal, which, combined with their position as sons of a great chief, would have given them an immense preponderance. It was therefore deemed politic to allow none of Savage's children to be other than still-born; he might have wives of the highest rank, but there must be no offspring. On this point the natives seem to have been inflexible, though Savage seemed to have strained every nerve to frustrate their cruel determination. The stand which the natives made, became the rock on which the hopes of the white men to establish their permanent sway in Fiji were wrecked. Savage died in March, 1814, near Vanua Levu, where he carried on a war with the natives in order to procure a cargo of sandalwood for an English trading vessel, the 'Hunter,' of Calcutta. Together with portions of the crew, he was put to death and eaten, whilst his bones were converted into sail-needles, and distributed amongst the people as a remembrance of victory.*

However, it was not only from shipwrecked mariners and runaway seamen, that the early white population was recruited. In 1804, a number of convicts escaped from New South Wales, in all about twenty-six, who took up their abode in Fiji, who however died out rather rapidly, either in the intertribal wars, in desperate fights amongst themselves, or in consequence of the irregular life led in a tropical climate. In 1824 only two, in 1840 only one of them, an Irishman of the name of Connor, survived, who occupied the same position towards the king of Rewa as Savage had done towards that of Bau. Connor does not seem to have been of such a deep, plodding nature as his comrade, or to have troubled his head much about the affairs of the future. Even when, after the loss of his royal patron, misfortune overtook him, he appears to have preserved all the humour for which his nation is proverbial, and was fully aware that the natives would never let him starve as long as he could while away an idle hour by the narration of a telling tale—upon which he depended towards the close of his days, quite as much, or perhaps even more, for a livelihood, than upon the rearing of fowls and pigs.

On the whole, the natives seem to have treated the first white men that came to live among them with hospitality and kindness. This is exactly what, from the nature of their country, might have been predicted. A sanguinary custom may have demanded that bodies slain

* Dillon, 'Discovery of the Fate of De la Pérouse.'

in battle should be baked and eaten, but the Fijian never displayed that determined hostility towards foreigners which is common to all natives in their barbarous state, and found vent even in civilized countries in a system of protective laws, which modern science still struggles to clear away. In some of the smaller islands of Polynesia, where food is scarce, and famine a common occurrence, every addition to the population is regarded rather as a calamity than as a matter of rejoicing, and the shores are jealously guarded against an infliction by which the whole community must suffer. It is therefore emphatically islands of this nature which our tract charts still mark as the most dangerous for landing. Viti, on the contrary, is so fertile, that food, as a general rule, is abundant at all seasons; and its inhabitants being well fed, and taking plenty of out-door exercise, do not seriously differ from other nations who enjoy the same advantages. A man who has every day a good dinner is a differently-disposed being from him who has to go very often without his daily meals; and the same process continued for generations must produce very opposite results in their respective characters. If any of the early white settlers met with a violent end, it was generally the foreigner, not the native, that furnished its primary cause. Taking undue advantage of the easy terms on which they lived with the chiefs, the white men often applied insulting epithets or used foul language to their hosts and protectors, provoking that contempt which familiarity, with a certain class of minds, invariably engenders. It was generally language of this kind, or demands which the chiefs deemed it below their dignity to comply with, which led to fatal consequences.

Some of the old convict gang were still alive when a few of a more respectable class of white traders and missionaries took up their abode in the group, principally at Lakeba, Levuka, and Rewa. Of the traders we know little except the incidental notices here and there preserved; but of the doings of the missionaries ample records have been placed before the world in their own publications. When the latter commenced their labours the political state of Fiji was little understood, and we can therefore not wonder that they should have made a serious mistake in the very outset. They began their work of christianization at Lakeba, one of the windward islands. Now Lakeba is dependent on Cakaudrove, and the chiefs of the latter state were naturally jealous to see vassals assume a greater importance than themselves, and they opposed the spread of the new doctrine with all means in their power. When, after a time, missionaries established themselves at Somosomo, then the capital of Cakaudrove, at Viwa and Rewa, they struggled against similar disadvantages. These three states were more or less dependent on Bau, and Bau, irritated at seeing its subordinates in possession of all the good things that an active intercourse with the Christian teachers threw in their way, tried to crush the new doctrine by its mighty influence. There can be no doubt that many atrocities were committed in the native capital, merely to prove how little Bau was influenced by the religious change going on in other parts of the group. It appears that at an early date Cakobau had invited the missionaries to come to Bau, but that they did not put sufficient confidence in him. The doubt thus cast upon his honour, together with the constant irritation of seeing parts of the group under the suzerainty of Bau daring to desert heathenism when still upheld by the leading state, and a daily diminishing political influence, turned King Cakobau into a deadly foe to Christianity. Had the missionaries taken the bull by horns, and endeavoured to obtain a footing at Bau before they took up their residence in any other part of the group, their labours would have been easy in comparison to what they have been, and the whole group would have renounced heathenism long ere this.* It was all up-hill work, yet results have been attained, to which no right-minded man can refuse admiration. According to the latest returns, the attendance on Christian worship in 1861 was 67,489, and there were 31,566 in the day-schools. For the supervision of this great work the Society had only eleven European missionaries and two schoolmasters, assisted by a large class of native agents, who are themselves the fruits of mission toil, and some of whom, once degraded and cannibal heathens, are becoming valuable and accredited ministers of the Gospel.

The white settlers at present in the group may amount to about two thousand souls, the greater number of whom have arrived within the last few years and

* Cakobau "was offended with Mr. Cross, because he would not trust himself at Bau on his first visit, but turned aside and opened a mission at Rewa. The proud spirit of the chief was hurt at being placed second." (Calvert, 'Fiji and the Fijians,' vol. ii. p. 234.) Additional passages might be cited from missionary writings to prove the view I have taken of Bau's hostility.

principally taken up their residence in Levuka and the Rewa districts. They are traders, agriculturists, and sheep farmers. Several have turned their attention to cotton growing. Most of them live in native-built houses, and only a few, including the consuls and missionaries, have weather-boarded houses. They belong to all nations; I have seen English, Americans, Germans, French, Poles, and Russians, but the greater number are British subjects. Nearly all have acquired more or less land from the natives, and several have bought extensive tracts. Small islands are in great request, and generally paid for at a much higher rate than pieces on the larger islands, which require fencing in, and are apt to give rise to disputes about boundaries. All the land sold is registered at the British Consulate, and Mr. Pritchard, before he did so, was always very careful to have the sellers acknowledge before him, and in the presence of a number of their townsmen, that they were satisfied with the bargain and had obtained the price stipulated. The land originally belongs either to individuals or to whole families, and the title confirmed by the ruling chiefs is supposed to be good. From what I saw, I believe that in most instances a fair price is given, remembering that the very best land in America may be had for a dollar and a quarter an acre; and that those who are willing to build a house, may have so-called bit-land for about sixpence per acre. Since the Fijis have become a field for immigration the land has considerably risen, and I have seen, as already stated, £10 per acre refused. The greatest landed proprietor was perhaps the late Mr. Williams, United States Consul. Mr. Binner,

Wesleyan training-master, also owns large tracts and a great many small islands. The land is paid for in barter, cotton prints, cutlery, muskets, powder and shot. Parties desirous of establishing plantations will have no difficulty in obtaining any amount of good land near rivers or the sea. Labour can be had to some extent in Fiji, but Polynesians will work much better if they are not in their own islands; and hands might be had by running over to Rotuma, Fotuna, Were, Raratonga, and the New Hebrides; indeed some of the best working men and women I saw in Fiji were obtained from those sources.

On the 2nd of November we returned to Lado, from our voyage around Vanua Levu. We had left Nukubati on the 30th of October, and called at Solevu and Levuka. On the 7th of November the 'Staghound,' Captain Sustenance, arrived from Tahiti and Samoa, and, as I had seen as much of Fiji as was accessible and gathered all the information I had been directed to accumulate, I engaged a passage in her for Sydney. There were several passengers on board; two having come from Tonga, where they had established sheep-runs; and one had been over a great part of Fiji, to judge for himself about the capabilities of the group for colonization. From what I could gather from conversation, he had been sent out by a party of friends, all of whom were desirous of investing capital in the islands if his report should prove favourable. He spoke in high terms of the country, and its resources.

I left Levuka on the 16th of November, and two days after lost sight of Kadavu and the Fiji group. On the

22nd we were out of the tropics, on the 26th near Norfolk Island, and on the 3rd of December off Lord Howe Island. Here we encountered a series of the most awful electric storms it has ever been my misfortune to pass through. The wind and waves were very high, the peals of thunder truly terrific, and sheet and flash lightning without interruption from dusk till dawn. Our vessel was struck several times by the lightning, and two men were seriously injured. I was fully prepared for going down, as it seemed almost impossible to survive a storm, to which all I had previously witnessed in the tropics could not be compared in intensity and violence. The St. Elmo's fire on the masthead and rigging gave a peculiarly ghastly appearance to the vessel when the darkness of night was restored by the momentary cessation of the lightning. The men got terribly frightened, and the rope's-end had to be used freely to make them do their duty. Captain Sustenance, every inch a sailor, took the helm himself, and never quitted his post till all was safe. His powerful voice could be heard through the storm, and was almost the only thing that inspired confidence, when all the elements seemed to be bent upon our destruction.

Otherwise our passage was a very pleasant one. Captain Sustenance had been in the Royal Navy, and seen, heard, and read a good deal, so that we were never hard up for topics of conversation. When on the 10th of December we dropped anchor in Sydney Harbour, we had as much to talk about as when first stepping on board at Levuka. To ascertain a man's mental calibre, no place is better suited than on board a ship. The generality of men are very dull company after the first few days; they have exhausted their little store of conversation, and, having no newspapers and clubhouses to supply them with fresh matter, they have absolutely nothing to say, even their autobiographies refusing to yield any new or interesting matter.

The collections I had dispatched to Sydney had safely arrived and were well taken care of by Mr. Moore, the director of the Botanic Garden. As the 'Jeddo,' the next " Peninsular and Oriental" steamer for England, did not leave before the 22nd of December, I took advantage of my stay to arrange and repack my treasures, and Mr. Moore's library and commodious premises were of the greatest service to me for that purpose. I remained all the time Mr. Moore's guest, as I had been on a former occasion, and enjoyed very much the fine garden in which his house is situated. Mr. Moore delivers every season a series of lectures on botany, and during my stay the distribution of prizes took place in the presence of a numerous assembly. Dr. George Bennett having only recently given a graphic description of the Sydney garden in his 'Gatherings of a Naturalist in Australasia,' I shall not dwell on a subject to me so tempting, and one that confers great credit upon the zealous director of the institution.

Leaving Sydney on the 22nd of December, we made Melbourne on Christmas Eve, and King George's Sound on the 31st of December. Thence my voyage led to Point de Galle, Ceylon, Egypt, and Malta, whence I took the French steamer and paid a visit to Sicily and Italy, ascending Vesuvius in company of Mr. and Mrs. George

Macleay, and, returning again to Malta, reached Southampton on the 12th of March, 1861, with no other accident than the breaking of the main shaft of the engine, between Valetta and Gibraltar.

The war in New Zealand continuing, it soon became apparent that the British Government had no inclination to accept the cession of Fiji, but the fact was not officially known until May, 1862, when the Wesleyan body had intimation of it. They had written, it appears, a letter asking for information, and stating at the same time that if her Majesty's Government should accept the cession, they should feel very much pleased if Colonel Smythe was appointed Governor of the new colony. Since then the official correspondence relative to the Fijian islands has been laid before Parliament; and the public has now ample materials to form an opinion on the whole subject. I have simply written an unvarnished account of all I heard and saw, and refrained from discussing the rejection of so fine a country from a political point of view. I have no doubt as to the future of Fiji. The importance of the group once recognized, nothing will stop our race from taking possession of it, and replacing barbarism and strife by civilization and peaceful industry.

APPENDIX.

I. REPORT OF ADMIRAL WASHINGTON, R.N.
II. REPORT OF COLONEL SMYTHE, R.A.
III. SYSTEMATIC LIST OF ALL THE FIJIAN PLANTS AT PRESENT KNOWN.

APPENDIX.

I.—REPORT OF ADMIRAL WASHINGTON, R.N.

In accordance with the Board Minute, to report upon the Colonial Office letter of the 9th instant, I have to state that—

The Fiji, or more properly the Viti group, in the south-western Pacific, consists of some 200 islands, islets, and rocks, lying between latitude $15\frac{1}{2}°$ and $19\frac{1}{2}°$ south, at about 1900 miles, N.E. of Sydney, and 1200 north of Auckland, at the north end of New Zealand. The two largest islands may be some 300 miles in circumference, or each is about the size of Corsica; 65 of the islets are said to be inhabited, and the whole population of the group may be 200,000.

I propose to reply categorically to the queries contained in the Colonial Office letter:—

Q. 1. If the Fiji Isles be obtained, are all the available harbours obtained in that part of the Pacific?

A. 1. Certainly not *all*, but a great part of them. The Friendly or Tonga Islands, only 400 miles to the south-east, possesses good harbours, as Tonga-tabú and Vavau. The Samoa or Navigator Isles, the same distance to the north-east, have good harbours, as Sangopango and Apia. Some of the Society Islands also may be available, but lying 1800 miles to the eastward, they may not be considered as within the limits named: none of the harbours, however, are superior to those of the Fiji Islands.

Q. 2. Do the natural harbours now existing require much, if any, artificial development for naval purposes? Whether such harbours are few or many?

A. 2. There are several roadsteads and harbours in the Fiji group, the principal of which is the extensive harbour of Levuka, on the eastern side of Ovalau; this harbour has good holding-ground, is easy of access, and has every facility for the supply of fruit, vegetables, wood, and water. Gau, on its western side, has a sheltered roadstead of large extent. Totoga is surrounded by a coral reef, within which is a spacious sheltered anchorage, with good holding-ground and an entrance for ships. All the above harbours have been thoroughly surveyed by order of the Admiralty, and plans of them, on a large scale, are available when required. These natural harbours will not require any artificial development for naval purposes.

3. There is nothing unusual in the tides and currents around the Fiji group; they depend chiefly on the prevailing winds; nor are they of sufficient strength to render the entrance into or egress from the harbours dangerous. There is no present necessity for buoys, beacons, or lights, but should trade greatly increase, or should mail-steamers call by night, a light would become necessary.

4. The Fiji Islands lie nearly in the direct track from Panama to Sydney, as will be seen by the annexed chart of the Pacific Ocean, on which I have shown that track, as also one by calling at the Fijis, whence it appears that the steamer, if she touched at one of the Fiji isles for coal, would lengthen her voyage only about 320 miles, or one day's run out of 32 days, on a distance of 8000 miles. In like manner it appears, that on the voyage from Vancouver Island to Sydney, the touching at Fiji would lengthen the distance 420 miles in a voyage of 7000 miles. An intermediate station between Panama and Sydney will be most desirable; indeed, if the proposed mail route is to be carried out, it is indispensable. One of the Society Islands, as lying halfway, would be a more convenient coaling station; but as they are under French protection it seems doubtful if one could be obtained. The Consul at Fiji, in the enclosed papers, hints at the possibility of coal being found in one of the islands; if this

should prove to be the case, it would at once double their value as a station.

In the above statements I have confined myself to answering the questions in the Colonial Office letter, but on looking into the subject I have been much struck by the entire want by Great Britain of any advanced position in the Pacific Ocean. We have valuable possessions on either side, as at Vancouver and Sydney, but not an islet or a rock in the 7000 miles of ocean that separate them. The Panama and Sydney mail communication is likely to be established, yet we have no island on which to place a coaling station, and where we could insure fresh supplies. * * * * And it may hereafter be found very inconvenient that England should be shut out from any station in the Pacific, and that an enemy should have possession of Tongatabú, where there is a good harbour, within a few hundred miles of the track of our homeward-bound gold-ships from Sydney and Melbourne. Neither forts nor batteries would be necessary to hold the ground; a single cruizing ship should suffice for all the wants of the islands; coral reefs and the hearty goodwill of the natives would do the rest.

I have, etc.,

(Signed)   JOHN WASHINGTON,
Admiralty, March 12th, 1859.   *Hydrographer.*

II.—REPORT OF COLONEL SMYTHE, R.A., TO COLONIAL OFFICE.

The Fiji group of islands is situated in the Pacific Ocean, between the meridians of 176° east and 178° west longitude, and between the parellels of 15° and 20° south latitude. It is composed of about 200 islands and islets, of which less than one-half is inhabited. Two of the islands (Viti Levu and Vanua Levu) are of unusual size for the Pacific Ocean, having each a circumference of 250 miles. The islands rise in general abruptly from the sea, and present in their bold and irregular outlines the peculiar characters of the volcanic formation to which they belong. With the exception of some tracts on the two larger islands, but little level land is anywhere to be seen. Almost every island is surrounded by a coral reef, either fringing the shore, or separated from it by a channel more or less narrow.

The inhabitants belong to the darker of the two great Polynesian races, but living on the confines of the lighter-coloured race, have received from it some admixture. One language, with some varieties of dialect, prevails throughout the group. The population is estimated at 200,000, of whom 60,000 are numbered as Christian converts. [67,489 according to exact returns, B. S.] The men are generally above the middle height, robust, and well-built. Their principal occupation is the cultivation of their yam and taro plots, which affords periodical but easy employment, sailing in their canoes, fishing, and frequently fighting. The chief articles of food are yams, taros, fish, and coco-nuts, breadfruit, bananas, and other fruits, the spontaneous productions of the soil. Their clothing is extremely scanty, consisting of a narrow strip of cloth, or rather paper, prepared from the bark of the paper-mulberry. Their houses are constructed of reeds and grass on a framework of poles. The floor is the natural soil covered with fern leaves and mats; in the middle is a sunken hearth, the smoke from which escapes through the walls and roof. Apertures for light other than the doorways are very rare. The houses are never isolated, but are crowded together in towns or "koros," which are frequently surrounded by a ditch and an earthen mound. The natives have raised no permanent structures. Although the coral reefs present an inexhaustible supply of lime, and they have discovered the art of burning it, they make no use of it except as paint, and to plaster their hair with. There are no beasts of burden or draught, and consequently no roads. The usual mode of moving about and of carriage is by canoes. The only mechanics among them are the carpenters or canoe-builders, who form an hereditary caste. The women, in a few favourable localities, manufacture a rude kind of pottery. There are in the group probably not less than forty independent tribes, twelve of which, from their superior influence, may be considered as virtually to govern it. The names of these are Bau, Rewa, Navua, Nadroga, Vunda, Ba, Rakiraki, and Viwa; round the coast of the largest island (Viti Levu), Bua, Macuata, and Cakadrove, or the other large island (Vanua Levu), and Lakeba, among the windward islands. The rule of the chiefs is absolutely despotic (see p. 231); the lives and goods, and to some extent the lands of their people, are at their mercy. The number of chiefs is very great; almost every "koro" has one or more. They differ greatly in rank and influence. In many instances there are two great chiefs at the same place, as at Bau. Here one of these is called "Rokotuebau," or "Great Chief of Bau," and the other "Na Vu-ni-valu," or the "root of war." They are both consecrated to their office. At Bau, the "Vu-ni-valu" is the principal personage; but in other places, where similar titles exist, the "Vu-ni-valu," although charged with special duties in the conduct of war, has but little power.

South-eastward of Fiji, at a distance of 250 miles, lie the Friendly or Tonga Islands. The inhabitants belong to the lighter-coloured Polynesian race. They have long had intercourse with the nearer islands of Fiji, attracted by the fine timber for canoes which they afford. Canoes are built on the spot where the material is found; the construction of a large one occupies several years.

In 1822 the English Wesleyan Methodist Society commenced a mission in Tonga, which led at a later period to the introduction of Christianity into Fiji. This event took place in 1835, when two missionaries from Tonga landed at Lakeba, the principal of the eastern islands, and where many Tonguese were located. The success of these missionaries was so encouraging, that their Society gradually added to their number, and eventually formed the Fiji group into a separate missionary district.

The number of Tonguese in Fiji fluctuates considerably, but may be taken at an average at from 300 to 400. Of late years they have taken an active part in Fijian wars, sometimes helping one chief, sometimes another, and invariably with success. They are distinguished by daring, coupled with unity and discipline,—qualities in which the Fijians are most wretchedly deficient.

They possess strong feelings of nationality, and own ready obedience to their chief, Maafu, a near relative to the king of Tonga. Native agency is largely employed by the missionaries in Fiji, and many of the most efficient teachers are Tonguese. In cases where Tonguese teachers have been ill-treated by the heathen natives, Maafu has interfered as the protector of his countrymen. In this manner, while extending his own influence, he has rendered safer the position of the native teachers. [Compare Chapter XV.] The presence of the Tonguese in Fiji has been far from an unmixed benefit. Their conduct has often been in direct contradiction to their profession of Christianity, and the help which they have afforded to the chiefs has occasioned much oppression to the people in the contributions levied to recompense their services. The population of the Tonga group does not exceed a tenth of that of Fiji; yet from the mental and physical superiority of the Tonguese, their courage and discipline, and the dread of them established among the Fijians, there is little doubt that they could easily make themselves masters of Fiji,—an enterprise which George, King of Tonga, has been said to meditate.

The permanent white residents in Fiji amount to about 200, composed chiefly of men who have left or run away from vessels visiting the islands. They are principally British subjects, citizens of the United States, with a few French and Germans; the two former are the most numerous. They traffic with the natives for produce, which they dispose of to vessels. They do nothing to civilize or improve the natives; on the contrary, they have in many instances fallen to a lower level. Whenever they can obtain spirits, most of them drink to excess. From false information given in the colonial journals regarding the acceptance by Her Majesty of the sovereignty of the islands, and their advantages for settlers, a considerable number of people were induced to visit them during last year. Discovering on their arrival the true state of affairs, many of them hastened to return to the colonies, and the greater number of the remainder will probably follow. They were generally of a much superior class to the old white residents. [The latest intelligence received from Fiji states the number of respectable white residents to be increasing.—*B. S.*]

Besides the British Consul, there is a Consul for the United States of America residing in Fiji.

The principal articles of produce are cocoa-nut oil, tortoise-shell, pearl-shell, and arrowroot. Formerly considerable quantities of sandal-wood and *bêche-de-mer* were carried to China, but this trade has now nearly ceased. The staple article of produce is cocoa-nut oil, of which about 200 tons are annually exported.

The sugar-cane and coffee-tree both grow well, and may in time contribute to the exports from Fiji. [Dr. Brower and Mr. Whippy, Americans, have, according to recent intelligence, set up a sugar-cane crushing-machine and coppers.—*B. S.*]

The climate of Fiji is not unhealthy; fevers are almost unknown. The most fatal disease to Europeans is dysentery. The mean temperature of the whole year is probably about 80°. Much rain falls, especially during the summer months of January, February, and March. At this season thunder-storms are frequent. Hurricanes scarcely ever occur except in these months, and frequently several years in succession pass without any. During the remainder of the year easterly winds prevail. Of the meteorology of Fiji more precise information will soon be obtained, as I brought out with me from the Meteorological Department of the Board of Trade a complete set of instruments.

The three principal reasons stated in my instructions as having been urged for accepting the sovereignty of the Fiji islands are—

> 1st. That they may prove a useful station for any mail steamers running between Panama and Sydney.
>
> 2nd. That they may afford a supply of cotton.
>
> 3rd. And, in close connection with the first reason, that their possession is important to the national power and security in the Pacific.

On the first head I beg to refer to the accompanying chart of the Pacific Ocean, on which I have traced the great circle lines joining Sydney, Panama, and Fiji, or, in other words, the lines of shortest distance on the globe between these places. The line from Sydney to Panama, it will be seen, crosses the northern island of New Zealand almost in the latitude of Auckland, and passes to the south of the great field of the Pacific Islands. The distance by this line from Sydney to Panama is 7626 nautical miles. The distance from Sydney to Fiji is 1735 miles, and from Fiji to Panama 6250, making the distance from Sydney to Panama, by way of Fiji, 7985 miles, or 359 miles longer than by the direct line. The latter line would be augmented by about 100 miles by the necessity of having to round the northern extremity of New Zealand. There would still remain a difference of 260 miles in favour of the Auckland route. The route by Fiji, besides being the longer, traverses the Pacific Archipelagoes, the navigation among which is undoubtedly difficult and dangerous, from the reefs and shoals in which they abound, and the occurrence of hurricanes at certain seasons. [Compare Admiral Washington's more favourable view, as expressed in his official report above.—*B. S.*]

2ndly. Regarding the supply of cotton. The cotton plant is not indigenous in Fiji.* From the concurring evidence of the natives in all parts of the group, its first introduction may be fixed at twenty-five years ago. As six different varieties are now found, it is probable that since its first introduction fresh seeds have from time to time been brought by vessels visiting these islands. The natives do not cultivate it, and make scarcely any use of it. Dr. Seemann brought out with him last year some cotton seed, presented by the "Manchester Cotton Supply Association," for distribution in Fiji. It was of two kinds, "Sea Island," and "New Orleans." None of the former kind germinated, but the New Orleans proved very good. In an experiment made under Dr. Seemann's own direction, the seed was sown on the 9th of June, and when he visited the plot again on the 18th of October, the plants were from four to seven feet high, and had some very fine ripe pods upon them. Since Mr. Pritchard's return from England at the end of 1859, some of the

* Most of the newspapers took this fact to be a serious drawback to the successful cultivation of cotton, quite forgetting that cotton is not indigenous to the United States and many other countries in which it flourishes. I made exactly the same statement ("cotton is not indigenous in Fiji"), but added that notwithstanding it had become almost wild in some parts, so well is the country adapted for its growth.—*B. S.*

native chiefs have been induced to encourage the growth of cotton, and a few young plants are now to be seen in the native gardens in various places. Very little, however, can be expected for some time from the natives. They will only be induced to raise cotton by meeting with a ready sale for the small quantities which they will bring in at first. The cultivation of cotton by white settlers is principally a question of land and labour. In a general way it may be said that there is not an acre of land in Fiji which is not private property, the ownership resting either in families or in individuals. A small portion of the land only at any one time is under cultivation, as a narrow patch of ground supplies the wants of a Fijian household, and the custom is to break up frequently new ground and abandon the old. On the subject of the purchase of land by whites, I made particular inquiry of the chiefs at each of the public meetings; the general reply was, that an agreement made with the owners, if approved by the chief, would hold good. In the older purchases of land by whites, when the quantity exceeded what was required for a house, the native residents were not interfered with, as no cultivation of land was attempted. In a few recent cases, where purchases have been effected by the whites who came last year to the islands, and who, with the view of forming plantations, wished to remove the natives from the land, opposition from the latter has been met with. By a clearer understanding with the owners before the purchase was concluded, these difficulties would probably have been avoided. The only mode hitherto of obtaining labour has been through the instrumentality of the chiefs, who send a party of their people to perform the work agreed upon and receive the payment, which they distribute at their pleasure. This system would not meet the daily demand of labour required in a cotton plantation. The general habits and sentiments of the Fijians are opposed to the acquisition of property by individuals. The chief seizes anything belonging to his people that takes his fancy, and as readily gives it away, and the people are equally ready to beg and to give. As the influence of Christianity increases, the rule of the chiefs will become more mild, and private rights will be more respected. It is very doubtful, however, whether the people will become more industrious, their wants being so few, and being so easily supplied. Although capable of making a considerable exertion for a short period, the natives dislike regular and continuous employment. On the whole, I am of opinion that whether by natives or by white planters with native labourers, the supply of cotton from Fiji can never be otherwise than insignificant. [Compare Chapter III., where the cotton question is regarded in a more favourable light.—*B. S.*]

3rdly. Regarding the importance of the possession of the Fiji Islands to the national power and security in the Pacific. Influence of a great power in the Pacific is dependent entirely on its naval force. By the possession of Australia and New Zealand England completely commands the western portion of the Pacific. In these colonies naval armaments can be recruited and equipped, and perhaps in a few years may even be created. No group in the Pacific can ever offer these advantages, and the possession of one, in the western section more especially, is not only not required, but would be a source of embarrassment in the event of war. [Compare Admiral Washington's opinion.—*B. S.*] The Fiji Islands do not lie in the path of any great commercial route. The whole of the Pacific Archipelagoes lie to the north of the direct line from the Australian colonies to Panama and South America, and south of the line from Panama and North America to China and India. All that it seems necessary for England to possess in the Pacific is an island with a good harbour, midway between Auckland and Panama, in the steam-packet route. Pitcairn's island is nearly in the required position, but it has no harbour. If a suitable island in its neighbourhood could be found, it would become, in addition to a coaling station for steam-vessels, the entrepôt of the pearl-shell and other trade which now centres in Tahiti, and afford a very favourable place of rendezvous for a squadron to protect our shipping homeward-bound from Australia and the Pacific.

Of the native population of Fiji, less than one-third profess the Christian religion; among the remainder cannibalism, strangulation of widows, infanticide, and other enormities, prevail to a frightful extent. Should the sovereignty of the islands be accepted by Her Majesty, the suppression of these inhuman practices would be put into immediate execution. For this service, and for the general support of the Government, a force of not less than the wing of a regiment would be required, in addition to a ship of war, with a tender of light draught, both steamers. The expenses of a civil establishment, composed on a sufficient scale to act efficiently on the condition of the natives, would probably not fall short of £7000 a year. The only mode of raising a revenue would appear to be by a capitation tax; customs duties would be so small as not to cover the cost of collection, if the importation of ardent spirits were prohibited (see p. 81), as a regard for the welfare of the natives would imperatively demand. For many years the Government would be necessitated to accept the tax in kind, as the natives have no circulating medium of exchange; and a still longer period would elapse before the islands became self-supporting. Looking solely at the interests of civilization, the forcible and immediate suppression of the barbarous practices of the heathen portion of the population might appear a very desirable act; yet, in beneficial influence on the native character, it might prove less real and permanent than the more gradual operation of missionary teaching. The success which has attended the missionaries in Fiji has been very remarkable, and presents every prospect of continuance. The principal tribes at present without missionaries or native teachers are willing to receive them, and there appears nothing wanting but time and a sufficiency of instructors to render the whole of the inhabitants professing Christians. Judging from the present state of the Sandwich Islands, and the former condition of Tahiti, it would seem that the resources of the Pacific Islands can be best developed, and the welfare of their inhabitants secured, by a native government aided by the counsels of respectable Europeans.

On a review of the foregoing considerations, and the conclusions derived from a personal examination of the islands and the people, I am of opinion that it would not be expedient that Her Majesty's Government should accept the offer which has been made to cede to Her Majesty the sovereignty over the Fiji Islands.

Having thus stated the conclusion to which my inquiries have led me regarding the offer to Her Majesty of the sovereignty of the Fiji Islands, I would beg leave to add a few suggestions towards the improvement of our relations with them. The great hindrance to the progress of civilization and Christianity among the inhabitants of the Pacific Islands, is the conduct and example of the whites residing or roving among them. Of the general character of these men in Fiji I have already spoken. During the few months I have been in the group, a case of arson, one of theft, one of burglary, and one of aggravated assault, have occurred among them. The great difficulty in these cases is the want of legal authority to arrest suspected persons, and of a proper and safe place in which to keep them. The only British functionary is the Consul, and he is powerless in these respects. To remedy these evils, I would suggest that the Consul have conferred on him some of the powers of a magistrate; that two constables (married men, selected either from the police or the army) be sent out from England; and that a stone lock-up house be erected for the safe custody of offenders, until there is an opportunity of sending them to the colonies for trial, or they are otherwise disposed of. The place of residence of the Consul is a matter of considerable importance. The principal white settlement in Fiji at present is at Levuka, on the island of Ovalau. It owed its selection to political causes in disturbed times. Its harbour may be considered good, but the hills rise abruptly from the beach and shut it in, and it is dependent on other places for much of its supplies. The present British Consul has an office at Levuka, but he resides at a further part of the island of Ovalau.

The locality best adapted in Fiji for a white settlement is the country round the harbour of Suva in Viti Levu, the largest of the islands. It is rich, level, and well-watered. The harbour is, perhaps, the best in the group; it is easy of access, can be entered and quitted with all the prevailing winds, and has communication within the reef with a great extent of coast. If the British Consulate were permanently established in this locality, a white settlement would spring up near it, which, if the Consul were armed with the powers suggested above, would not be disgraced by the scenes of drunkenness and rioting so prevalent at

Levuka, and would be of eminent service in developing the natural resources of the Fiji Islands.

Fiji Islands, May 1st, 1861.

III.—SYSTEMATIC LIST OF ALL THE FIJIAN PLANTS AT PRESENT KNOWN.

The Vitian Islands were until 1840 a virgin soil, and still offer a tempting field for botanical explorations. Absolutely nothing was known of their Flora until Messrs. Hinds and Barclay, who accompanied Sir Edward Belcher in H.M.S. Sulphur, collected a few specimens in the neighbourhood of Rewa, Viti Levu, and Bua Bay in Vanua Levu, afterwards described by Mr. Bentham in the 'London Journal of Botany,' vol. ii., and the Botany of H.M.S. Sulphur. About the same time (1840) Viti was visited by the United States Exploring Expedition, Commander Wilkes, and considerable collections were made by Messrs. Brackenridge, Rich, and Pickering, furnishing the materials for Professor Asa Gray's celebrated 'Botany of the United States Exploring Expedition.' In 1856, H.M.S. Herald, Captain Denham, R.N., explored different parts of the group, and Mr. Milne, his botanical collector, was enabled to add a good number of species to our knowledge. Another visit was paid to the group by that indefatigable botanist Professor Harvey, of Trinity College, Dublin, productive of many new types. In 1860 I collected about 800 species and made a great many notes of the country explored. Whilst part of the latter, relating to the resources and vegetable productions, were embodied in an official report, addressed to his Grace the Duke of Newcastle, and presented to Parliament by command of her Majesty, a preliminary list of the former was published by me in the 'Bonplandia,' vol. ix. p. 253 (1861). Since then I have had time to examine the plants more closely and correct a few errors crept in. Other botanists have also been led to study the materials collected by me and publish the result. Prof. A. Gray has carefully collated my plants with those published by him in the 'Botany of the United States Exploring Expedition' and the 'Proceedings of the American

Academy,' the result of which has been given in the 'Bonplandia,' x. 34 (1862), and also in the Proceedings of the Academy named. As there are very few original specimens in Europe of the numerous new types described by that eminent *savant*, these papers are invaluable to the working botanist. Mr. Mitten has examined all my Mosses and Hepaticæ (Bonpl. ix. 365, and Bonpl. x. 19); amongst the 35 species collected there being 20 new ones. For the determination of the Ferns I am indebted to Mr. Smith, at Kew; for that of the Fungi, to the Rev. M. J. Berkeley; for that of the Palms, to Mr. Wendland; the Lichens to the Rev. Churchill Babington, and the Aroideæ to Mr. Schott, at Vienna, who has also described the new species (Bonplandia, ix. 367, seq.); for my own part, I have begun to describe the new genera and species in the 'Bonplandia,' ix. and x., and given coloured illustrations drawn by the skilful pencil of Mr. Fitch. In the following catalogue will be found embodied the result of all these labours, and also all the species enumerated by previous authors. The numbers which follow the different species refer to my distributed collections, and those remitted to me by Mr. J. Storck, who was my able assistant, and is now a permanent resident in Fiji.

Ranunculaceæ.
Clematis Pickeringii, A. Gray (1).

Dilleniaceæ.
Capellia biflora, A. Gray; vulgo 'Kulava' vel 'Kukulava' (2).
C. membranifolia, A. Gray.

Anonaceæ.
Anona squamosa, Linn. Cultivated (3).
Richella monosperma, A. Gray.
Uvaria amygdalina, A. Gray.
U. odorata, Lam.; vulgo 'Makosoi' (5).
Polyalthia Vitiensis, Seem. (4).

Myristicaceæ.
Myristica castaneæfolia, A. Gray; vulgo 'Male' (6).
M. macrophylla, A. Gray; vulgo 'Male' (7).
M. sp.; vulgo 'Male' (866).

Cruciferæ.
Cardamine sarmentosa, Forst. (8).
Sinapis nigra, Linn. Cultivated and naturalized (9).

Capparideæ.
Capparis Richii, A. Gray.

Flacourtianeæ.
Xylosma orbiculatum, Forst. (10).

Samydaceæ.
Casearia disticha, A. Gray (11).
C.? acuminatissima, A. Gray.
C. Richii, A. Gray.

Violaceæ.
Agathea violaris, A. Gray, et var. (12).
Alsodeia? sp.; vulgo 'Sesirakavono' (867).

Mollugineæ.
Mollugo striata, Linn. (230).

Portulaceæ.
Portulaca oleracea, Linn.; vulgo 'Taukuka ni vuaka' (13).
P. quadrifida, Linn.; vulgo 'Taukuku ni vuaka' (14).
Talinum patens, Willd. (15).
Sesuvium Portulacastrum, Linn.

APPENDIX.

Malvaceæ.

Sida linifolia, Cav.
S. rhombifolia, Linn. (16).
S. retusa, Linn.
Urena lobata, Linn. (17).
U. moriifolia, De Cand.
Abelmoschus moschatus, Mœnch; vulgo 'Wakiwaki' (19, 869).
A. canaranus, Miq. ? (20).
A. Manihot, Med.; vulgo 'Bele,' vel 'Vauvau ni Viti' (18).
A. esculentus, Wight et Arn. Cultivated, according to A. Gray.
Hibiscus Rosa-Sinensis, Linn.; vulgo 'Kauti,' 'Senitoa,' vel 'Seniciobia' (22).
H. Storckii, Seem.; vulgo 'Seqelu' (23).
H. diversifolius, Jacq.; vulgo 'Kalauaisoni,' vel ' Kalakalauaisoni' (21).
Paritium purpurascens, Seem.; vulgo 'Vau damudamu' (24).
P. tiliaceum, Juss.; vulgo 'Vau dina' (25).
P. tricuspis, Guill. vulgo 'Vau dra' (26).
Thespesia populnea, Corr.; vulgo 'Mulomulo' (7).
Gossypium religiosum, Linn.; vulgo ' Vauvau ni papalagi' (28).
G. Peruvianum, Cav.; vulgo 'Vauvau ni papalagi' (29).
G. Barbadense, Linn.; vulgo 'Vauvau ni papalagi' (30).
G. arboreum, Linn. et var.; vulgo ' Vauvau ni papalagi' (31, 32).

Sterculiaceæ.

Heritiera littoralis, Dryand.; vulgo ' Kena ivi na alewa Kalou' (33).
Firmiana diversifolia, Gray.

Buettneriaceæ.

Commersonia platyphylla, De Cand. (34).
Büttneriacearum gen. nov. aff. Commersoniæ (83).
Kleinhovia hospita, Linn.; vulgo ' Mamakara' (35).

Waltheria Americana, Linn. (36).
Melochia Vitiensis, A. Gray (37).

Tiliaceæ.

Triumfetta procumbens, Forst. (38).
Grewia persicæfolia, A. Gray (= G. Mallococca, var. ?); vulgo 'Siti' (39).
G. prunifolia, A. Gray; vulgo 'Siti' (40).
G. Mallococca, L. fil.
Trichospermum Richii, Seem. (= Diclidocarpus Richii, A. Gray); vulgo ' Maku' (41, 870).
Elæocarpus laurifolius, A. Gray.
E. cassinoides, A. Gray.
E. pyriformis, A. Gray.
E. Storckii, Seem. sp. nov. (E. aff. speciosi, Brongn. et Gris.); vulgo ' Gaigai' (874).

Ternstrœmiaceæ.

Draytonia rubicunda, A. Gray; vulgo ' Kau alewa' (42, 872).
Eurya Vitiensis, A. Gray (43).
E. acuminata, De Cand. (44).
Ternstrœmiacearum gen. nov. (45).

Guttiferæ.

Discostigma Vitiense, A. Gray.
Calysaccion obovale, Miq. (= Garcinia Mangostana, A. Gray in United St. Expl. Exped.); vulgo ' Vetao' vel ' Uvitai' (46).
Calophyllum Inophyllum, Linn.; vulgo ' Dilo' (48, 873).
C. Burmanni, Wight; vulgo 'Damanu' (49).
C. (polyanthum, Wall. ? v. lanceolatum, Bl. ? = C. spectabile, United St. Expl. Exped.; vulgo ' Damanu dilodilo') (47).
Garcinia sessilis, Seem. (Clusia sessilis, Forst. 51).
G. pedicellata, Seem. (Clusia pedicellata, Forst. 50).

Pittosporeæ.

Pittosporum arborescens, Rich.
P. Richii, A. Gray; vulgo 'Tadiri' (54).
P. Brackenridgei, A. Gray (55).

P. tobiroides, A. Gray (56).
P. Pickeringii, A. Gray (53).
P. rhytidocarpum, A. Gray (52).

Aurantiaceæ.

Micromelum minutum, Seem. (M. glabrescens, Bth.; Limonia minuta, Forst.); vulgo 'Qiqila' teste Williams (57).
Citrus vulgaris, Risso (C. torosa, Picker.); vulgo 'Moli kurikuri' (58).
C. Aurantium, Risso ; vulgo ' Moli ni Tahaiti.'—Cult.
C. Decumana, Linn.; vulgo 'Moli kana.' Cultivated and naturalized.
C. Limonum, Risso; vulgo 'Moli kara.'

Meliaceæ.

Aglaia edulis, A. Gray (Milnea edulis, Roxb.) ; vulgo 'Danidani loa.'
A. ? basiphylla, A. Gray.
Didimochyton Richii, A. Gray.
Xylocarpus Granatum, Kœn.; vulgo 'Dabi' (61).
X. obovatus, A. Juss. (var. præcedent.? 62).
Vavæa amicorum, Benth. (63).
Meliæ sp. nov. (64).

Sapindaceæ.

Cardiospermum microcarpum, H. B. et K.; vulgo 'Voniu' (65).
Sapindus Vitiensis, A. Gray (66).
Cupania falcata, A. Gray (70).
C. Vitiensis, Seem. (an var. præced.? 68).
C. rhoifolia, A. Gray ; vulgo 'Buka ni vuda' (74, 69).
C. apetala, Labill. (67).
C. Brackenridgei, A. Gray.
C. leptobotrys, A. Gray.
Nephelium pinnatum, Camb.; vulgo 'Dawa,' et var. plur. (71).
Dodonæa triquetra, Andr.; vulgo 'Wase' teste Williams (72).

Malpighiaceæ.

Hiptage Javanica, Bl. ?
H. myrtifolia, A. Gray.

Ampelideæ.

Vitis saponaria, Seem. (= Cissus geniculata, A. Gray, non Bl.); vulgo 'Wa Roturotu' (76).
V. Vitiensis, Seem. (Cissus Vitiensis, A. Gray).
V. acuminata, Seem. (Cissus acuminata, A. Gray) (77).
Leea sambucina, Linn. (78).

Rhamneæ.

Smythea pacifica, Seem. Bonpl. t. 9 (79).
Ventilago ? Vitiensis, A. Gray (an Smytheæ spec.? = cernua, Tul.).
Colubrina Asiatica, Brongn.; vulgo 'Vuso levu' (80).
C. Vitiensis, Seem. sp. nov. (85).
Alphitonia zizyphoides, A. Gray (= A. franguloides A. Gray); vulgo 'Doi' (81).
Gouania Richii, A. Gray (82).
G. denticulata, A. Gray.
Rhamnea dubia (84).

Chailletiaceæ.

Chailletia Vitiensis, Seem. sp. nov. (876).

Celastrineæ.

Catha Vitiensis, A. Gray (86).
Celastrus Richii, A. Gray.

Aquifoliaceæ.

Ilex Vitiensis, A. Gray (87).

Olacineæ.

Ximenia elliptica, Forst.; vulgo 'Somisomi,' 'Tumitomi,' vel 'Tomitomi' (88).
Stemonurus ? sp.; vulgo 'Duvu' (877).
Olacinea ? (878).

Oxalideæ.

Oxalis corniculata, Linn. ; vulgo 'Totowiwi' (89).

Rutaceæ.

Evodia hortensis, Forst.; vulgo ' Uci,' vel 'Salusalu' (91).
E. longifolia, A. Rich. (92).
E. drupacea, Labill. ? (90).
Acronychia petiolaris, A. Gray.

APPENDIX.

Zanthoxylon varians, Benth. (= Acronychia heterophylla, A. Gray (102, 879).
Z. Roxburghianum, Cham. et Schlecht. (103).
Z. sp. (n. 104).

Simarubeæ.
Soulamea amara, Lam.
Amaroria soulameoides, A. Gray (880).
Brucea ? sp. (105).

Ochnaceæ.
Brackenridgea nitida, A. Gray (93).

Anacardiaceæ.
Oncocarpus atra, Seem. (O. Vitiensis, A. Gray; Rhus atrum, Forst.); vulgo 'Kau Karo' (94, 881)
Buchanania florida, Schauer (882).
Rhus simarubæfolia, A. Gray (95).
Rh. Taitensis, Guill. ? (96).

Burseraceæ.
Canarium Vitiense, A. Gray (97).
Evia dulcis, Comm.; vulgo 'Wi' (98).
Dracontomelon sylvestre, Blum.; vulgo 'Tarawau' (99).
Dr. sp. ? (100).

Connaraceæ.
Rourea heterophylla, Planch.
Connarus Pickeringii, A. Gray (101).

Leguminosæ.
I. Papilionaceæ :—
Crotalaria quinquefolia, Linn.
Indigofera Anil, Linn. (106).
Tephrosia purpurea, Pers. (T. piscatoria, Pers. 107).
Ormocarpus sennoides, De Cand.
Uraria lagopodioides, De Cand. (108).
Desmodium umbellatum, W. et Arn. (109).
D. australe, Bth. (Hedysarum, Willd.)
D. polycarpum, De Cand. (111).
Abrus precatorius, Linn.; vulgo 'Qiri damu,' 'Lere damu,' vel 'Diri damu' (110).
Canavalia obtusifolia, De Cand. (122).

C. turgida, Grah. (112).
C. sericea, A. Gray.
Glycine Tabacina, Bth. (123).
Mucuna gigantea, De Cand. (119).
M. platyphylla, A. Gray (200).
Erythrina Indica, Linn.; vulgo, 'Drala dina,' (125) et var. fl. albis.
E. ovalifolia, Roxb.; vulgo 'Drala kaka' (124).
Strongylodon ruber, Vogel (113).
Phaseolus rostratus, Wall.
Ph. Mungo, Linn. ?
Ph. Truxillensis, H. B. et K. (116).
Vigna lutea, A. Gray (121).
Lablab vulgaris, Savi; vulgo 'Dralawa' (118).
Cajanus Indicus, Spr. Introd. (115).
Pongamia glabra, Vent.; vulgo 'Vesivesi, v. 'Vesi ni wai' (126, 884).
Derris uliginosa, Benth.; vulgo 'Duwa gaga' (127, 883)
Dalbergia monosperma, Dalz. (128).
D. torta, Grah.
Pterocarpus Indicus, Willd.; vulgo 'Cibicibi' (129).
Sophora tomentosa, Linn.; vulgo 'Kau ni alewa' (130, 886).
II. Cæsalpineæ :—
Guilandina Bonduc, Ait.; vulgo 'Soni' (132).
Poinciana pulcherrima, Linn.—Cult.
Storckiella Vitiensis, Seem. in Bonpl. t. 6; vulgo 'Marasa' (133).
Cassia occidentalis, Linn. vulgo 'Kau moce' (134).
C. obtusifolia, Linn.; vulgo 'Kau moce' (135).
C. lævigata, Willd.; vulgo 'Winivikau' (136).
C. glauca, Lam.
Afzelia bijuga, A. Gray; vulgo 'Vesi' (137).
Cynometra grandiflora, A. Gray (138).
C. falcata, A. Gray.
Inocarpus edulis, Forst.; vulgo 'Ivi' (371).
III. Mimoseæ :—
Entada scandens, Bth.; vulgo 'Wa lai,' v. 'Wa tagiri' (139).

Mimosa pudica, Linn. Naturalized (140).
Leucæna glauca, Bth. (141)
L. Forsteri, Benth. (142).
Acacia laurifolia, Willd.; vulgo 'Tatakia' (143).
A. Richii, A. Gray; vulgo 'Qumu' (144).
Serianthes myriadenia, Planch.
S. Vitiensis, A. Gray; vulgo 'Vaivai' (145, 887).

Chrysobalaneæ.

Parinarium laurinum, A. Gray (= P.? Margarata, A. Gray = P. insularum, A. Gray); vulgo 'Makita' (146).

Rosaceæ.

Rubus tiliaceus, Smith; vulgo 'Wa gadrogadro' (147).

Myrtaceæ.

Barringtonia speciosa, Linn.; vulgo 'Vutu rakaraka' (148).
B. Samoensis, A. Gray; vulgo 'Vutu ni wai' (149).
B. excelsa, Blume; vulgo 'Vutu kana' (150).
B. sp.
Eugenia (Jambosa) Malaccensis, Linn.; vulgo 'Kavika:' var. α, floribus albis, vulgo 'Kavika vulovulo;' var. β, floribus purpureis, vulgo 'Kavika damudamu' (161).
E. (Jambosa) Richii, A. Gray; vulgo 'Bokoi' (164).
E. (Jambosa) sp. (an Richii var.?); vulgo 'Sea' (165).
E. (Jambosa) quadrangulata, A. Gray.
E. (Jambosa) gracilipes, A. Gray; vulgo 'Lutulutu,' vel 'Bogibalewa' (158).
E. (Jambosa) neurocalyx, A. Gray; vulgo 'Leba' (159).
E. rariflora, Bth. (160).
E. Brackenridgei, A. Gray (155).
E. confertiflora, A. Gray.
E. sp. nov. confertiflor. proxima (156).
E. effusa, A. Gray (151).
E. amicorum, Benth. (152).
E. rubescens, A. Gray; vulgo 'Yasi dravu' (154).

E. corynocarpa, A. Gray (153).
E. rivularis, Seem.; vulgo 'Yasi ni wai' (162).
E. Grayi, Seem. sp. nov. fl. purpureis (163).
Nelitris fruticosa (A. Gray).
N. Vitiensis, A. Gray; vulgo 'Nuqanuqa' (166, 888).
Acicalyptus myrtoides, A. Gray.
A. Seemanni, A. Gray (168).
Metrosideros collina, A. Gray; vulgo 'Vuga' (169, 889).
M. sp. fl. luteis (170).
M. sp. fl. coccineis (171).

Melastomaceæ.

Memecylon Vitiense, A. Gray et var. (172).
Astronia Pickeringii, A. Gray.
A. confertiflora, A. Gray (174).
A. Storckii, Seem., sp. nov.; vulgo "Cavacava" (890).
Astronidium parviflorum, A. Gray (465).
Anplectrum? ovalifolium, A. Gray.
Medinilla heterophylla, A. Gray (175).
M. rhodochlæna, A. Gray; vulgo 'Cararaca ra i resiga' (177, 891).
M. sp. (182).
M. sp. (75).
M. sp. (175).
Melastoma Vitiense, Naud. (180).
M. polyanthum, Bl.? (179).
Melastomacea (181).

Alangieæ.

Rhytidandra Vitiensis, A. Gray.

Rhizophoreæ.

Haplopetalon Richii, A. Gray.
H. Seemanni, A. Gray (184).
Crossostylis biflora, Forst.
Rhizophora mucronata, Lam.; vulgo 'Dogo' (185).
Bruguiera Rhumphii, Bl. (186).

Combretaceæ.

Lumnitzera coccinea, Willd.; vulgo 'Sagali' (189).
Terminalia Catappa, Linn.; vulgo 'Tavola' (187).

APPENDIX.

P. Moluccana, Lam.; vulgo 'Tivi' (188).
T. glabrata, Forst.?

Passifloreæ.

Passiflora, sp. fl. viridibus (190).

Papayaceæ.

Carica Papaya, Linn.; vulgo 'Oleti,' Introd. (190).

Cucurbitaceæ.

Karivia Samoensis, A. Gray (192).
Luffa insularum, A. Gray (193).
Cucumis pubescens, Willd. (194).
Lagenaria vulgaris, Ser. (195).

Saxifrageæ.

Spiræanthemum Vitiense, A. Gray.
Sp. Katakata, Seem., sp. nov.; vulgo 'Katakata' (196).
Weinmannia affinis, A. Gray, (197,) et var. (199 et 200).
W. Richii, A. Gray.
W. spiræoides, A. Gray.
W. sp. (198).
Geissois ternata, A. Gray; vulgo 'Vuga' (201).

Umbelliferæ.

Hydrocotyle Asiatica, Linn.; vulgo 'Totono' (202).

Araliaceæ.

Aralia Vitiensis, A. Gray (203).
Panax fruticosum, Linn.; vulgo 'Danidani' (204).
Paratropia? multijuga, A. Gray; vulgo 'Danidani' (205).
Plerandra Pickeringii, A. Gray.
P. Grayi, Seem., sp. nov. (206 et 209).
P.? sp. nov. (208).
P. sp. (207).

Loranthaceæ.

Loranthus insularum, A. Gray; vulgo 'Saburo' (211).
L. Vitiensis, Seem. (210).
L. Forsterianus, Schult.
Viscum articulatum, Burm. (212).

Balanophoreæ.

Balanophora fungosa, Forst.

Rubiaceæ.

I. Coffeaceæ:—
Coprosma persicæfolia, A. Gray.
Geophila reniformis, Cham. et Schlecht. (239).
Chasalia amicorum, A. Gray? (241).
Psychotria Brackenridgei, A. Gray.
P. Forsteriana, A. Gray, var. Vitiensis, A. Gray (236).
P. turbinata, A. Gray.
P. tephrosantha, A. Gray.
P. parvula, A. Gray.
P. gracilis, A. Gray.
P. calycosa, A. Gray? (246).
P. macrocalyx, A. Gray (243).
P. filipes, A. Gray.
P. hypargyræa, A. Gray.
P. (Piptilema) cordata, A. Gray.
P. (Piptilema) Pickeringii, A. Gray (251).
P. (Piptilema) platycocca, A. Gray (249).
P. insularum, A. Gray? (250).
P. collina, Labill. (244 et 254).
P. sarmentosa, Blum. (245).
P. sp.; vulgo 'Wa kau:' ramis scandentibus sarmentosis (895).
P. sp. foliis bullatis (248).
P. sp. nov. aff. filipedis (253).
P. sp. nov. aff. Brackenridgei (255).
P. sp. aff. Brackenridgei (259).
Calycosia petiolata, A. Gray.
C. pubiflora, A. Gray (214).
C. Milnei, A. Gray; vulgo 'Kau wai' (213, 892).
Ixora Vitiensis, A. Gray (247); Pavetta triflora, De Cand.; Coffea triflora, Forst.; Cephaëlis? fragrans, Hook. et Arn.
I. sp. nov. (258).
I. sp.; vulgo 'Kau sulu' (893).
Canthium sessilifolium, A. Gray.
C. lucidum, Hook. et Arn.; Coffea odorata, Forst. (220 et 221).
Morinda umbellata, Linn. (222).
M. myrtifolia, A. Gray; foliis majoribus (an v. M. umbellatæ?) (223).
M. mollis, A. Gray (224).
M. phillyreoides, Labill. (226).

M. citrifolia, Linn.; vulgo 'Kura,' v. 'Kura kana' (225).
M. lucida, A. Gray.
M. bucidæfolia, A. Gray.
Hydnophytum longiflorum, A. Gray (= Myrmecodia Vitiensis, Seem.) (216).
Vangueria? sp. (257).
Guettarda speciosa, Linn.; vulgo 'Buabua' (237).
G. (Guettardella) Vitiensis, A. Gray (= 257?).
Timonius sapotæfolius, A. Gray.
T. affinis, A. Gray.
Coffeacea; vulgo 'Kau lobo' (893).
II. Cinchoneæ :—
Hedyotis tenuifolia, Sm. (231).
H. deltoidea, W. et Arn.? (232).
H. paniculata, Roxb. (233).
H. paniculata, Roxb. var. crassifolia, A. Gray (234).
H. bracteogonum, Spr. (235).
Ophiorrhiza laxa, A. Gray (227).
O. peploides, A. Gray (228).
O. leptantha, A. Gray (229).
Lindenia Vitiensis, Seem. Bonpl. t. 8 (217).
Lerchea calycina, A. Gray.
Dolicholobium oblongifolium, A. Gray.
D. latifolium, A. Gray.
D. longissimum, Seem. (215).
Stylocoryne Harveyi, A. Gray.
St. sambucina, A. Gray (S. pepericarpa, Bth.) (242).
Griffithiæ sp.? (260).
G.? sp. v. gen. nov. (240).
G. sp. fl. odoratis.
Gardenia Vitiensis, Seem. (218).
G.? (an gen. nov.?) (240).
Mussænda frondosa, Linn.; vulgo "Bovu."

Compositæ.

Monosis insularum, A. Gray.
Lagenophora Pickeringii, A. Gray.
Erigeron albidum, A. Gray; vulgo 'Wavuwavu,' v. 'Co ni papalagi' (261).
Adenostemma viscosum, Forst. (262).
Siegesbeckia orientalis, Linn. (263).
Dichrocephala latifolia, De Cand. (264).
Myriogyne minuta, Linn. (265).
Sonchus oleraceus, Linn. (n. 266).
Ageratum conyzoides, Linn.; vulgo 'Botebotekoro,' vel 'Matamocemoce' (267).
Wollastonia Forsteriana, De Cand.; vulgo 'Kovekove' (268).
Eclipta erecta, Linn.; vulgo 'Tumadu' (269).
Bidens pilosa, Linn.; vulgo 'Batimadramadra (270).
Glossogyne tenuifolia, Cass. (271).
Blumea virens, De Cand. (272).
B. Milnei, Seem. (sp. nov. aff. B. aromaticæ, De Cand. 273).

Goodeniaceæ.

Scævola floribunda, A. Gray (S. saligna, Forst.?); vulgo 'Totoirebibi' (274, 896).
S. Kœnigii, Vahl (275).

Cyrtandreæ.

Cyrtandra acutangula, Seem. (276).
C. Vitiensis, Seem.; vulgo 'Betabiabi' (277).
C. anthropophagorum, Seem. (278).
C. involucrata, Seem. (279).
C. coleoides, Seem. (280).
C. Milnei, Seem. (281).
C. ciliata, Seem. (282).
C. Pritchardii, Seem. (283).

Vaccineæ.

Epigynum? Vitiense, Seem. (284).

Epacrideæ.

Leucopogon Cymbula, Labill.; vulgo 'Tagatagalesa.'

Myrsineæ.

Mæsa Pickeringii, A. Gray.
M. persicæfolia, A. Gray (287?).
M. corylifolia, A. Gray (288).
M. nemoralis, A. Gray (286?).
Myrsine myricæfolia, A. Gray (290 ex parte).
M.? Brackenridgei, A. Gray.
M. capitellata, Wall.? (289).
Ardisia? capitata, A. Gray.

A. grandis, Seem. (293).
A. sp. (292, 897).
A. sp. (291).

Styraceæ.

Symplocos spicata, Roxb.; vulgo 'Ravu levu.'

Ebenaceæ.

Maba foliosa, Rich.
M. elliptica, Forst.; vulgo 'Kau loa' (295, 296, 297, 898).

Sapotæ.

Sapota? pyrulifera, A. Gray.
S. ? Vitiensis, A. Gray.
S. sp. (ex A. Gray).

Jasmineæ.

Jasminum tetraquetrum, A. Gray.
J. gracile, Forst.; vulgo 'Wa Vatu' (298).
J. didymum, Forst.; J. divaricatum, R. Brown (299).

Loganiaceæ.

Geniostoma rupestre, Forst. (301).
 var. puberulum, A. Gray (G. crassifolium, Bth.) (300).
G. microphyllum, Seem. (304).
Strychnos colubrina, Linn. (302).
Courthovia corynocarpa, A. Gray (= Gærtnera pyramidalis, Seem.); vulgo 'Boloa' (303).
C. Seemanni, A. Gray (Gærtnera barbata, Seem.) (305, 899).
Fagræa gracilipes, A. Gray (F. viridiflora, Seem.) (306).
F. Vitiensis, Seem. (307).
F. Berteriana, A. Gray; vulgo 'Bua' (308).

Apocyneæ.

Alyxia bracteolosa, Rich; vulgo 'Vono' (310, 900); var. α macrocarpa, A. Gray (A. macrocarpa, Rich.); var. β angustifolia, A. Gray (A. stellata, Seem.); var. γ parviflora, A. Gray.
A. stellata, Labill.
Cerbera lactaria, Ham.; vulgo 'Rewa' vel 'Vasa' (309).
Melodinus scandens, Forst. (311).

Tabernæmontana Vitiensis, Seem.; T. citrifolia, Forst. non L. = ? T. Cumingiana, A. De Cand.
T. sp.
Rejoua scandens, Seem. sp. nov.; vulgo 'Wa rerega' (901).
Ochrosia parviflora, Hensl. (O. elliptica, Labill. ?) (318).
Alstonia plumosa, Labill. (318).
A? sp. (317).
Echites scabra, Labill. ? (315).
Lyonsia lævis, A. Gray.

Asclepiadeæ.

Tylophora Brackenridgei, A. Gray.
Gymnema subnudum, A. Gray.
G. stenophyllum, A. Gray; vulgo 'Yauyau' (322).
Hoya bicarinata, A. Gray; Asclepias volubilis, Forst.; vulgo 'Wa bibi' vel 'Bulibuli sivaro' (319).
H. diptera, Seem. (320).
H. pilosa, Seem. (321).

Gentianeæ.

Erythræa australis, R. Brown.
Limnanthemum Kleinianum, Griseb.; vulgo 'Bekabekairaga' (323).

Convolvulaceæ.

Ipomœa campanulata, Linn.; vulgo 'Wa vula' (324).
I. peltata, Chois.; vulgo 'Wiliao' teste Seemann, 'Veliyana' teste Williams (325).
I. Pes capræ, Sw.; vulgo 'Lawere' (326).
I. Turpethum, R. Brown; vulgo 'Wa kai' (327).
I. sepiaria, Kœn. (328).
I. cymosa, Rœm. et Schult.; vulgo 'Sovivi' (334).
Aniseia uniflora, Chois. (329).
Batatas paniculata, Chois.; vulgo 'Wa Uvi' vel 'Dabici' teste Storck (330, 902).
B. edulis, Chois.; vulgo 'Kumara' vel 'Kawai ni papalagi.'—Cult.
Pharbitis insularis, Chois.; vulgo 'Wa Vuti' (331).
Calonyction speciosum, Chois. (332).
C. comosperma, Boj. (333).

Boragineæ.
Tournefortia argentea, Linn. (335).
Cordia Sprengelii, DeCand.; vulgo 'Tou' (336).
C. subcordata, Lam.; vulgo 'Nawanawa' (337).

Solaneæ.
Physalis Peruviana, Linn. (338).
P. angulata, Linn. (339).
Solanum viride, R. Brown? (340).
S. anthropophagorum, Seem. (sp. nov. Bonpl. t. 14); vulgo 'Borodina' (341).
S. repandum, Forst.; vulgo 'Sou,' 'Sousou,' vel 'Boro sou' (342).
S. inamœnum, Benth. Lond. Journ. ii., p. 228 (343).
S. oleraceum, Dun.; vulgo 'Boro ni yaloka ni gata' (344).
S. sp. (S. repand. var.? (345).
Capsicum frutescens, Linn.; vulgo 'Boro ni papalagi' (346).
Nicotiana Tabacum, Linn.—Cultivated (347).
Datura Stramonium, Linn.—Introd. (348).

Scrophularineæ.
Vandellia crustacea, Benth. (349).
Limnophila serrata, Gaud. (350).

Acanthaceæ.
Eranthemum laxiflorum, A. Gray (351, ex parte).
E. insularum, A. Gray (351, ex parte).
Adenosma triflora, Nees ab Esenb.; vulgo 'Tamola' (352).

Verbenaceæ.
Clerodendron inerme, R. Brown; vulgo 'Verevere' (353).
Vitex trifolia, Linn.; vulgo 'Vulokaka' (354).
Premna Tahitensis, Schauer (Scrophularioides arborea, Forst.); vulgo 'Yaro' (355).
P. Tahitensis, Schauer; var.? (356).
Gmelina Vitiensis, Seem. (sp. nov.).

Labiatæ.
Leucas decemdentata, Sm. (357).
Ocimum gratissimum, Linn. (358).

Plectranthus Forsteri, Benth.; vulgo 'Lata' (359).
Teucrium inflatum, Swartz (360).

Plumbagineæ.
Plumbago Zeylanica, Linn. (361).

Plantagineæ.
Plantago major, Linn.—Introd. (362).

Nyctagineæ.
Pisonia Brunoniana, Endl. (363).
P. viscosa, Seem. (sp. nov.) (364).
Boerhaavia diffusa, Linn., var. pubescens (365).

Amarantaceæ.
Amarantus melancholicus, Moq., var. tricolor; vulgo 'Driti damudamu' (366).
A. paniculatus, Moq., var. cruentus, Moq.; vulgo 'Driti.'—Introd. (367).
Euxolus viridis, Moq.; vulgo 'Driti' vel 'Gasau ni vuaka' (368).
Cyathula prostrata, Blum. (369).

Polygoneæ.
Polygonum imberbe, Sol. (370).

Laurineæ.
Hernandia Sonora, Linn.; vulgo 'Yevuyevu' vel 'Uviuvi' (372).
Cassytha filiformis, Linn.; vulgo 'Waluku mai lagi' teste Williams (373).
Cinnamomum sp.; vulgo 'Macou' (376).
Laurinea. Arbor 15–20 ped. (374).
Laurinea (375).
Laurinea (377).
Laurinea; vulgo 'Siqa' vel 'Siga' (378).
Laurinea; vulgo 'Lidi' (903).

Thymeleæ.
Drymispermum sp. (379).
D. montanum, Seem. (sp. nov.) (380).
D. subcordatum, Seem. (sp. nov.); vulgo 'Matiavi' (381).
D.? sp. (382).
Leucosmia Burnettiana, Benth. (= Dais disperma, Forst.); vulgo 'Sinu damu' vel 'Sinu dina' (383).
Wikstrœmia Indica, C. A. Mey.; vulgo 'Sinu mataiavi' (384).

APPENDIX. 441

Santalaceæ.
Santalum Yasi, Seem. (sp. nov.); vulgo 'Yasi' (385).

Ceratophylleæ.
Ceratophyllum demersum, Linn. (386).

Euphorbiaceæ.
Euphorbiacea?? (387).
Acalypha? (388).
Acalypha Indica, Linn.? (389).
A. sp. (390).
A. rivularis, Seem. (sp. nov.); vulgo 'Kadakada' (391).
A. virgata, Forst. (= A. circinata, A. Gray); vulgo 'Kalabuci damu' (392).
A. grandis, Benth.; vulgo 'Kalabuci' (393).
Claoxylon parviflorum, Juss. (394).
Mappa Molluccana, Sprengl.? (395).
M. macrophylla, A. Gray; vulgo 'Mavu' (396).
M. sp. (397).
M. sp. (419).
M. sp. (420).
Excœcaria Agallocha, Linn.; vulgo 'Sinu gaga' (398).
Manihot Aipi, Pohl.; vulgo 'Yabia ni papalagi' (399).
Curcas purgans, Juss.; vulgo 'Wiriwiri ni papalagi' (400).
Ricinus communis, Linn.; vulgo 'Bele ni papalagi' (401).
Omalanthus pedicellatus, Bth.; vulgo 'Tadauo' (402).
Aleurites triloba, Forst.; vulgo 'Lauci,' Tutui,' vel 'Sikeci' (403).
Euphorbia Norfolkica, Bois.; vulgo 'Soto' (404).
E. pilulifera, Linn.; vulgo 'De ni osi' (405).
E. Atoto, Forst. (E. oraria, F. Muell.) (406, 904).
Rottlera acuminata, Vahl. (407).
Croton metallicum, Seem. (sp. nov.) (408).
C. sp.; vulgo 'Sacasaca loa' (409).
C. sp. (an. var. n. 409?) (410).
C. Storckii, Seem. sp. nov. aff. C. Hillii, F. Müll.; vulgo 'Danidani' (905).

Codiæum variegatum, A. Juss.; vulgo 'Sacaca' vel 'Vasa damu' (411).
Melanthesa sp. (aff. M. Vit. Idææ) (412).
M. sp.; vulgo 'Molau.' Arbor (413).
Glochidion sp. (414).
G. ramiflorum, Forst.; vulgo 'Molau' (415).
G. cordatum, Seem. (sp. nov.); aff. G. mollis (416).
Bischoffia sp.; vulgo 'Koka.' Arbor (417).
Phyllanthus fruticosa, Wall. (418).

Urticeæ.
Elatostemma? nemorosa, Seem. (sp. nov.) (422).
Gironniera celtidifolia, Gaud.; vulgo 'Nunu' (423).
Missiessya corymbulosa, Wedd.; vulgo 'Matadra' (424).
Maotia Tahitensis, Wedd.; vulgo 'Waluwalu' (425).
Laportea Harveyi, Seem. (sp. nov.); vulgo 'Salato.' Arbor 30–40 ped. (426).
L. Vitiensis, Seem. (sp. nov.); aff. L. photinifol.; vulgo 'Salato' (427).
Fleurya spicata, var. interrupta, Wedd.; vulgo 'Salato ni koro' vel 'Salata wutivali' (428).
Pellionia elatostemoides, Gaud. (429).
Procris integrifolia, Don, Hook., Arn (430).
Bœhmeria Harveyi, Seem. (sp. nov.) vulgo 'Rere' (431).
B. platyphylla, Don (432).
B. platyphylla, Don, var. virgata, Wedd. (433).
Malaisia? sp.; Arbor (434 a).

Moreæ.
Morus Indica, Linn.—Introd. (434 b).
Trophis anthropophagorum, Seem. (sp. nov.); vulgo 'Malawaci' (435).
Ficus obliqua, Forst.; vulgo 'Baka' (436).
F. tinctoria, Forst. (437).
F. sp.; vulgo 'Loselose.' Frutex fruct. edul. (438).
F. sp.; vulgo 'Loselose ni wai.' Frutex rivularis (439).

F. sp. (440).
F. sp. Frutex 16 ped., caule subsimpl. (441).
F. sp. (442).
F. sp. (443).
F. sp. (444).
F. scabra, Forst.; vulgo 'Ai Masi' (445).
F. aspera, Forst. (446).
F. sp. (447).
F. sp. (448).

Artocarpeæ.

Antiaris Bennettii, Seem. Bonpl. t. 7. (sp. nov.); vulgo 'Mavu ni Toga' (449).
Artocarpus incisa, Linn., var. integrifolia, Seem. (aff. A. Chaplashæ, Roxb.); vulgo 'Uto lolo' v. 'Uto coko coko' (450).
A. incisa, Linn. var. pinnatifida, Seem.; forma vulgo 'Uto dina' dicitur (551).
A. incisa, forma vulgo 'Uto Varaqa' (452).
A. „ „ „ 'Uto Koqo' (453).
A. „ „ „ 'Balekana' (454).
A. „ „ „ 'Uto buco' (455).
A. „ „ „ 'Uto assalea' (456).
A. „ „ „ 'Uto waisea' (457).
A. „ „ „ 'Uto Bokasi' (458).
A. „ „ „ 'Uto Votovoto' (459).
A. incisa, Linn. var. bipinnatifida, Seem.; vulgo 'Uto Sawesawe' vel 'Kalasai' ((560).

Gyrocarpeæ.

Gyrocarpus Asiaticus, Willd.; vulgo 'Wiriwiri' (561).

Celtideæ.

Sponia orientalis, Linn. (562).
Sp. velutina, Planch. (563).

Chloranthaceæ.

Ascarina lanceolata, Hook. fil. (564).

Piperaceæ.

Peperomia sp. (565).
Macropiper latifolium, Miq. (566).
M. puberulum, Benth.; vulgo 'Yaqoyaqona' (567).
M. methysticum, Miq.; vulgo 'Yaqona' (568).
Piper Siriboa, Forst.; vulgo 'Wa Gawa.' Frutex scandens (569).

Casuarineæ.

Casuarina equisetifolia, Forst.; vulgo 'Nokonoko' (570).
C. nodiflora, Forst.; vulgo 'Velao' (571).

Cycadeæ.

Cycas circinalis, Linn.; vulgo 'Roro' (572).

Coniferæ.

Dacrydium elatum, Wall.; vulgo 'Leweninini' vel 'Dakua salusalu' (573, 906).
Podocarpus (elatus, R. Br.?); vulgo 'Kuasi' (574).
P. (polystachya, R. Br.?); vulgo 'Gagali (575).
P. cupressina, R. Brown; vulgo 'Kau tabua.'
P.? v. gen. nov.; vulgo 'Kau solo' (576).
Dammara Vitiensis, Seem.; vulgo 'Dakua' (577).

Orchideæ.

Dendrobium Mohlianum, Reichb. fil. (sp. nov.) (578).
D. crispatum, Swartz (579).
D. (580).
D. Millingani, F. Muell. (581).
D. biflorum, Sw. (582).
D. sp. (an var. præced.?) (583).
D. Tokai, Reichb. fil. (sp. nov.); vulgo 'Tokai' teste Williams (584).
D. sp. (591).
Limodorum unguiculatum, Labill. (585).
Bletia Tankervilliæ, R. Brown (586).
Oberonia (587).

O. brevifolia, Lindl. (Epidendrum equitans, Forst. (588).
O. Myosurus, Lindl. (589).
Microstylis Rheedii, Lindl. (Pterochilus plantagineus, Hook. et Arn.) (590).
Appendicula (592).
Tæniophyllum Fasciola, Seem. (Limodorum Fasciola, Swartz); vulgo 'De ni caucau' (593, 907).
Saccolabium sp. (594).
S. sp. (595).
Eulophia macrostachya, Lindl.? (596).
Eria sp., aff. E. baccatæ, Lindl.? (597).
Cirrhopetalum Thouarsii, Lindl. (598).
Rhomboda (599).
Sarcochilus (600).
Dorsinia marmorata, Lindl. (601).
Monochilus sp. (602).
Corymbis disticha, Lindl. (603).
Pogonia biflora, Wight (604).
Calanthe (605).
C. sp. florib. pallide aurantiacis (606).
C. veratrifolia, R. Brown (607).
Habenaria (608).
Orchidea (609).
O. (610).
O. (611).
O. (612).
O. (613).
O. (614).
O. (615).
O. (616).
O. (617).
O. (618).

Scitamineæ.

Musa Troglodytarum, Linn.; vulgo 'Soqo' (619).
Gen. nov.; vulgo 'Boia' (620).
Alpinia sp. (621).
Curcuma longa, Linn.; vulgo 'Cago' (622).
Zingiber Zerumbet, Linn.; vulgo 'Beta' (623).
Amomum sp.; vulgo 'Cevuga' (624).
Canna Indica, Linn.; vulgo 'Gasau ni ga' (625).

Dioscoreæ.

Helmia bulbifera, Kth.; vulgo 'Kaile' (626).

Dioscorea alata, Linn.; vulgo 'Uvi' (627).
D. nummularia, Lam.; vulgo 'Tivoli' (628).
D. aculeata, Linn.; vulgo 'Kawai' (629).
D. pentaphylla, Linn.; vulgo 'Tokulu' (630).

Smilaceæ.

Smilax sp.; vulgo 'Kadragi' vel 'Wa rusi' (631).

Taccaceæ.

Tacca sativa, Rumph.; vulgo 'Yabia' (632, 909).
T. pinnatifida, Forst.; vulgo 'Yabia dina' (633, 908).

Liliaceæ.

Cordyline (634).
C. sp.; vulgo 'Ti kula.'—Colitur (635).
C. sp.; vulgo 'Qai' v. 'Masawe.'—Colitur (636).
Allium Ascalonicum, Linn.; vulgo 'Varasa.'—Colitur (637).
Geitonoplesium cymosum, Cunn.; vulgo 'Wa Dakua' (638).
Dianella ensifolia, Red. (639).

Amaryllideæ.

Crinum Asiaticum, Linn.; vulgo 'Viavia' (640).

Astelieæ.

Astelia montana, Seem. (sp. nov. bacca trilocul.); vulgo 'Misi' (641).

Commelyneæ.

Commelyna communis, Linn. (= C. pacifica, Vahl?); vulgo 'ai Rorogi' vel 'Rogomatailevu' (642).
Aneilema Vitiense, Seem. (sp. nov.; florib. pallide cœruleis) (643).
Flagellaria Indica, Linn.; vulgo 'Sili Turuka' vel 'Vico' (644, 910).
Joinvillea elegans, Gaud. (= Flagellaria plicata, Hook. fil., 645).

Typhaceæ.

Typha angustifolia, Linn.; vulgo 'De ni ruve' (646).

Bromeliaceæ.
Ananassa sativa, Lindl.; vulgo 'Balawa ni papalagi.'
A. sativa, var. prolifera.

Pandaneæ.
Freycinetia Vitiensis, Seem. (sp. nov.) (647).
F. Milnei, Seem. (sp. nov.) (648).
F. Storckii, Seem. (sp. nov.) (695).
F. sp. (696).
Pandanus odoratissimus, Linn.; vulgo 'Balawa' vel 'Vadra' (649).
P. caricosus, Rumph.; vulgo 'Kiekie' vel 'Voivoi' (650).

Aroideæ.
Alocasia Indica, Schott; vulgo 'Via mila,' 'Via gaga,' 'Via sori,' v. 'Via dranu' (651).
Amorphophallus? (sp. nov.); vulgo 'Daiga' (652).
Cyrtosperma edulis, Schott (sp. nov.); vulgo 'Via kana' (653).
Raphidophora Vitiensis, Schott. (sp. nov.); vulgo 'Wa lu' (654).
Cuscuaria spuria, Schott (sp. nov.) (655).
Colocasia antiquorum, Schott, var. esculenta, Schott; vulgo 'Dalo' (655 b).
Aroidea (911).

Lemnaceæ.
Lemna gibba, Linn.; vulgo 'Kala' (656).
L. minor, Linn.; vulgo 'Kala' (657).

Palmæ.
Cocos nucifera, Linn.; vulgo 'Niu dina.'
Sagus Vitiensis, Herm. Wendl. (Cœlococcus Vitiensis, Herm. Wendl.); vulgo 'Niu soria' vel 'Sogo' (658).
Pritchardia pacifica, Seem. et Herm. Wendl. (gen. nov.); vulgo 'Sakiki,' 'Niu Masei,' vel 'Viu' (659).
Kentia? exorrhiza, Herm. Wendl. (sp. nov.); vulgo 'Niu sawa' (660).
Ptychosperma Vitiensis, Herm. Wendl. (sp. nov.) (662).
P. filiferum, Herm. Wendl. (sp. nov.); vulgo 'Cagecake' (661, 663).

P. Seemanni, Herm. Wendl. (sp. nov.); vulgo 'Balaka' (664).
P. perbreve, Wendl.
P. pauciflorum, Wendl.
P. Pickeringii, Wendl.

Cyperaceæ.
Baumia sp. (665).
Hypolytrum giganteum, Roxb. (666).
Lepironia mucronata, Rich. (667).
Cyperus sp. (668).
C. sp. (912).
Mariscus lævigatus, Rœm. et Schult. (669).
Kyllingia intermedia, R. Brown (670).
K. sp. (671).
Lamprocarya affinis, A. Brongn. (672).
Gahnia Javanica, Zoll. (673).
Fimbrystylis marginata, Labill. (674).
F. stricta, Labill. (675).
Scleria sp. (676).
S. sp. (677).
Elæocharis articulata, Nees ab Esenb.; vulgo 'Kuta' (678).

Gramineæ.
Zea Mays, Linn.; vulgo 'Sila ni papalagi.'—Cult.
Oplismenus sp. foliis purpurascentib.; vulgo 'Co damudamu' (679).
O. sp. foliis albo-maculatis.—Cum præcedente colitur (680).
O. compositus, Rœm. et Schult. (681).
Paspalum scrobiculatum, Linn.; vulgo 'Co dina' (682).
Eleusine Indica, Gærtn. (683).
Centotheca lappacea, Desv. (684)
Andropogon refractum, R. Brown (= A. Tahitense, Hook. et Arn.) (685).
A. acicularis, Retz. (686).
A. Schœnanthus, Linn.; vulgo 'Co boi' (687).
Cenchrus anomoplexis, Labill. (688).
Sorghum vulgare, Pers.—Colitur (689).
Digitaria sanguinalis, Linn. (690).
Saccharum floridum, Labill. (691).
Coix Lacryma, Linn.; vulgo 'Sila' (692).
Panicum pilipes, Nees ab Esenb. (693).
Bambusa sp.; vulgo 'Bitu' (694).

APPENDIX. 445

Equisetaceæ.

Equisetum sp.; vulgo 'Masi ni tabua' (697).

Lycopodiaceæ.

Psilotum complanatum, Sw. (698).
P. triquetrum, Sw. (699).
Lycopodium cernuum, Linn.; vulgo 'Ya Lewaninini' (700).
L. flagellare, A. Rich. (701).
L. Phlegmaria, Linn. (702).
L. varium, R. Br. (703).
L. verticillatum, Linn. (704).
L. sp. (705).
L. sp. (706).
L. sp. (707).
L. sp. (708).

Filices.

Acrostichum aureum, Linn.; vulgo 'Boreti,' vel, teste Williams, 'Caca' (709).
Stenochlæna scandens, J. Smith. (710).
Lomariopsis leptocarpa, Fee (711).
L. cuspidata, Fee (712).
Lomogramme polyphylla, Brack. (713, 421).
Goniophlebium subauriculatum, Blum. (714).
Hemionitis lanceolata, Hook. (716).
H. elongata, Brack. (715).
Antrophyum plantagineum, Kaulf (717).
Diclidopteris angustissima, Brack.; vulgo 'Mokomoko ni Ivi' (718, 914).
Vittaria revoluta, Willd. (719).
V. elongata, Sw. (720).
Arthropteris albopunctata, J. Smith (721).
Prosaptia contigua, Presl (722).
Phymatodes stenophylla, J. Smith (723).
Niphobolus adnascens, Sprengel, Sw., J. Sm. (724).
Loxogramme lanceolata, Presl (725).
Hymenolepis spicata, J. Smith (726).
Pleuridium cuspidiflorum, J. Smith (727).
P. vulcanicum, J. Smith (729).
Phymatodes Billardieri, Presl (730).
P. alata, J. Sm. = Drynaria alata, Brack.) (731.)

P. longipes, J. Smith; vulgo 'Caca,' teste Williams (732).
Drynaria musæfolia, J. Smith (728).
D. diversifolia, J. Smith; vulgo 'Bevula,' 'Teva,' vel 'Vuvu' (733).
Dipteris Horsfieldii, J. Smith; vulgo 'Koukou tagane' (734).
Meniscium sp. (735).
Nephrodium simplicifolium, J. Smith (736).
N. sp. (737).
N.; vulgo 'Watuvulo' (738).
N. sp. (739, 740).
Lastrea sp. (741).
Polystichum aristatum, Presl (742).
Nephrolepis ensifolia, Presl (743).
N. hirsutula, Presl (744).
N. repens, Brack. (745).
N. obliterata, J. Smith (831).
Dictyopteris macrodonta, Presl (746).
Aspidium latifolium, J. Smith; vulgo 'Sasaloa' (v. Saloa?) (747).
A. decurrens, J. Smith (748).
A. repandum, Willd (749).
Oleandra neriiformis, Cav. (750).
Didymochlæna truncatula, Desv. (751).
Microlepia polypodioides, Presl (751 b).
M. sp. (752).
M. papillosa, Brack. (753).
M. Luzonica, Hook. (gracilis, Blum.) (754).
M. flagellifera, J. Smith (Wall.) (755).
M. (fructif.) (An var. n. 751 b? B. Seem.) (756.)
Humata heterophylla, Cav. (759).
Davallia elegans, Sw. (757).
D. Fijiensis, Hook. (758).
D. fœniculacea, Hook. (760, 762).
D. gibberosa, Sw. (761).
D. Moorei, Hook. (830).
Schizoloma ensifolia, Gaud. (763).
Synaphlebium davallioides, J. Smith (764).
S. Pickeringii, Brack. (765).
S. repens, J. Smith (766).
Sitolobium stramineum, J. Smith (767).
Cyathea medullaris, Sw. (768).
Trichomanes javanicum, Blum. (769).
T. rigidum, Sw. (780, 829).
T. meifolium, Bory (781).

T. bilingue, Blum. (= n. 780 ?) (782).
T. angustatum, Carm. = T. caudatum, Brack. (783).
T. erectum, Brack. (784 ex parte).
Hymenophyllum (784).
H. formosum, Brack. (785).
H. parvu-lum, Poir. (786).
Todea Wilkesiana, Brack. (787).
Marattia sorbifolia, Sw.; vulgo ' Dibi ' (788).
Angiopteris evecta, Hoffm. (789).
Lygodictyon Forsteri, J. Smith; vulgo ' Wa Kalou ' (790).
Gleichenia dichotoma, Hook. (791).
Schizæa dichotoma, Sw.; vulgo ' Sagato ni tauwa ' (792).
Actinostachys digitata, Wall. (793).
Ophioglossum pendulum, Linn. (794).
Blechnum orientale, Linn. (795).
Lomaria attenuata, Willd. (796).
L. elongata, Blume (797).
Pellæa geraniifolia, Fee (798).
Cheilanthes tenuifolia, Sw. (799, 800).
Adiantum lunulatum, Sw.; vulgo 'Kau ni vi vatu' (801, 915).
A. hispidulum, Sw. (802).
A. aff. A. setulonervi, J. Smith (803).
Pteris quadriaurita, teste Hook. Sp. Fil. (804).
P. sp. (Litobrochia divaricata, Brack. ?) (805).
P. tripartita, Sw. (806, 913).
P. esculenta, Forst. (809).
P. crenata, Sw.; vulgo ' Qato,' teste Williams (811).
Litobrochia sinuata, Brack.; vulgo 'Wa Rabo' (807).
L. sinuata var. (808).
L. comans, Presl (810).
Neottopteris australasica, J. Smith (812).
Asplenium vittæforme, J. Smith (813).
A. falcatum, Lam. (814).
A. sp. (815).
A. brevisorum, Wall. (827).
A. obtusilobum, Hook. (828).
A. induratum, Hook. (816).
A. lucidum, Forst. (817).
A. sp. (820).
A. resectum, Sm. (821).

A. laserpitiifolium, Lam. (822).
A. (Darea) sp. (784 ex parte).
Callipteris ferox, Blum. (= C. prolifera, Hook. var.) (818).
C. (sine fructif.) (819).
Cryptosorus Seemanni, J. Smith (= Polypodium contiguum, Brack. non Sw. (823).
Diplazium melanocaulon, Brack. (824).
D. bulbiferum, Brack. (825).
D. polypodioides, Blume. (826).
Tænitis blechnoides, Sw. (? abnormal.) (832).

Musci.

Leptotrichum flaccidulum, Mitt. sp. nov. (841).
L. trichophyllum, Mitt. sp. nov. (inter 862).
Leucobryum laminatum, Mitt. sp. nov. (844).
Leucophanes densifolius, Mitt. sp. nov. (inter 862).
L. smaragdinum, Mitt. sp. nov. (inter 863).
Syrrhopodon tristichus, Nees (inter 846).
S. scolopendrius, Mitt. sp. nov. (843).
Meteorium longissimum, Dzy. et Molk (inter 863).
M. (Esenbeckia) setigerum, Mitt. (Pilotrichum, Sullivant) (846).
Trachyloma Junghuhnii, Mitt. (Hypnum C. Mueller) (842).
T. arborescens, Mitt. (845).
Neckera flaccida, C. Muell. (836).
N. Lepineana, Montagn. (863).
N. dendroides, Hook. (838).
Spiridens Reinwardti, Nees. (840).
Trachypus helicophyllus, Mont. (838).
Leskea glaucina, Mitt. (inter 847).
L. ramentosa, Mitt. sp. nov. (inter 863).
Racopilum spectabile, Hsch. (inter 863).
Sphagnum cuspidatum, Ehrh. (839).

Hepaticæ.

Cheiloscyphus argutus, Nees (inter 862).
Plagiochila arbuscula, L. et L. (inter 862).

P. Vitiensis, Mitt. sp. nov. (862).
P. Seemanni, Mitt. sp. nov. (864).
Trichocolea tomentella, Nees (inter 862).
Radula amentulosa, Mitt. sp. nov. (inter 837).
R. scariosa, Mitt. sp. nov. (inter 837).
R. spicata, Mitt. sp. nov. (inter 837).
Lejeunia (Bryopteris) Sinclairii, Mitt. sp. nov. (inter 843).
L. eulopha (Phragmicoma, Tay.) (inter 846).
Frullania deflexa, Mitt. sp. nov. (inter 834).
F. meteoroides, Mitt. sp. nov. (inter 834).
F. cordistipula, Nees (inter 846).
F. trichodes, Mitt. sp. nov. (inter 846).
Sarcomitrium plumosum, Mitt. (847).
Marchantia pileata, Mitt. (838).

Lichenes.

Sticta damæcornis, var. caperata, Nyl. (848).
S. (Stictina) filicinella, Nyl. (849).
Ramalina calicaris, Nyl.; vulgo 'Lumi' (ni Vanua) (851).
Coccocarpia molybdæa, Pers. (852).
Leptogium tremelloides, Fries (853).
Sticta (Stictina) quercizans, Ach. (854).
Sticta Freycinetii, Del. (861).
Verrucaria aurantiaca, Nyl. (865).
Parmelia peltata, Ach. var.

Fungi.

Rhizomorpha sp.; vulgo 'Wa loa' (855).
Lentinus sp. (856).
Polyporus sanguineus, Fries (857).
P. affinis, Fries (858).
P. hirsutus, Fries (859).
Hoomospora transversalis, Brebisson (860).
Agaricus (Pleuropus) pacificus, Berk.
Schizophyllum commune, Fries.
Xylaria Feejeensis, Berk.

Algæ.

Hoomonema fluitans, Berk. (gen. nov.) (860).

THE END.